How to Diagnose and Fix Everything Electronic

Also by Michael Jay Geier

How to Get the Most from Your Home Entertainment Electronics:
Set It Up, Use It, Solve Problems

How to Diagnose and Fix Everything Electronic

Third Edition

Michael Jay Geier

Sponsoring Editor
Lara Zoble

Production Supervisor
Richard Ruzycka

Acquisitions Coordinator
Olivia Higgins

Project Manager
Patricia Wallenburg, TypeWriting

Copy Editor
James Madru

Proofreader
Alison Shurtz

Indexer
WordCo Indexing Services, LLC

Art Director, Cover
Anthony Landi

Composition
TypeWriting

This book is dedicated to all the wonderful people who have encouraged my love of electronics throughout my lifetime. Family, friends, employers, colleagues and amateur radio operators, all of you have made it a joy for me to explore, learn about and pass along my fascination with the wacky world of electrons and their sometimes complex and downright baffling behaviors. My adventures in technology have led me from the micro universe of circuit boards and chips to travels all over the world, and dinners and meetings with some of the biggest innovators in the consumer electronics business. I've enjoyed every minute of it, from fixing a wide variety of products in service shops to covering conferences in the tech press and writing magazine articles and these books. It's been a heck of a great ride all these years.

It's my sincere hope that you'll savor the enjoyment and catch the electronics bug as you learn to navigate circuits and think like a technician. Repair work is all about logic and sleuthing, and it's a ton of fun!

About the Author

Michael Jay Geier has been an electronics technician, designer and inventor since age six. He took apart everything he could get his hands on and soon discovered that learning to put it back together was even more fascinating. By age eight, he operated a neighborhood electronics repair service that was profiled in the *Miami News*. He went on to work in numerous service centers in Miami, Boston and Seattle, frequently serving as the "tough dog" tech who solved the cases other techs couldn't. At the same time, Michael was a pioneer in the field of augmentative communications systems, helping a noted Boston clinic develop computer speech systems for children with cerebral palsy. He also invented and sold an amateur radio device while writing and marketing software in the early years of personal computing.

Michael holds an FCC Extra-class amateur radio license. His involvement in ham radio led to his writing career, first with articles for ham radio magazines and then with general technology features in *Electronic Engineering Times, Desktop Engineering, IEEE Spectrum* and *The Envisioneering Newsletter*. His work on digital rights management has been cited in several patents. Michael earned a Boston Conservatory of Music degree in composition, was trained as a conductor and is an accomplished classical, jazz and pop pianist and a published songwriter. Along with building and repairing electronic circuitry, he enjoys table tennis, restoring antique mopeds, camping, bicycling and banging out a jazz tune on his harpsichord.

Contents

Foreword

I've known Michael since we were in middle school together. We hit it off immediately, with shared interests in shortwave radio listening, walkie-talkies, tape recorders, playing in the school band, electronic kit making, pizza (hey, we were hungry teens!) and early video recording.

It was great having a friend like Michael, who was more than eager to grab his wire cutters, hop on his bike and ride with me around the neighborhood as we hunted for old TV and stereo consoles that had been dumped on the curb, awaiting trash pickup. To us, they were a treasure trove of electronic parts. We'd snip off resistors, capacitors, chokes, transistors and switches, and stash 'em in our parts trays.

Even so, we didn't always find what we needed to attempt those cool projects in the electronics magazines. Michael would analyze the circuit diagram and alter it to accommodate the parts we did have—and make it work!

Creativity has always been one of Michael's best qualities. When we wanted our little, low-power kids' walkie-talkies to have greater range, he came up with a way to increase the effectiveness of the audio section, giving them greater talk power without actually increasing the transmitter output, which wouldn't have been practical or legal. After the modification, we got two miles out of those things, far more than they were designed for. We both went on to become licensed amateur radio operators and have continued to practice the art of electronics to great fulfillment, both professionally and personally, over many decades.

Plenty of electronics experts were educated "by the book" and can quote chapter and verse on theory and its application. But few possess the knowledge and experience *and* the ability to convey this information to others in a way that's understandable and lots of fun. Michael's writing reads like he's actually speaking to me—exactly as he has all these decades.

Michael is the MacGyver of creative electronic repair, the Mister Rogers' Neighborhood of patient teaching and communicating and the Carl Sagan of explaining practical electronic theory, which may help save you "billions and billions!" Or at least a few bucks, anyway.

Enjoy reading the latest edition of *How to Diagnose and Fix Everything Electronic*. If you're like me, you'll keep it on your repair bench, right next to your variable power supply and your soldering iron.

Gregory T. Nesper, N4PSA
Senior Technician and Technical Trainer
Purified Water Industry

Foreword

Michael lives just up the road from me. We met a few years ago when he saw my ham radio antenna and stopped in to say hi. It wasn't long before we started nightly conversations over our ham radio walkie-talkies. I quickly realized that Michael is one of those guys everyone needs to know because of his vast knowledge set in the field of electronics. And when I say "vast," I don't mean just the ability to hook up a Blu-Ray player to an audio amplifier. This guy knows the inner workings of both and can diagnose and repair things such as these with ease. Over time, and via many conversations with him, some of that wide range of expertise has leached into my brain. Thanks to Michael, I now know how to work with many of the various diagnostic tools required for just about any repair and actually succeed at some repairs on my own. Michael convinced me to buy an oscilloscope and learn to use it! And it wasn't hard after all.

Sometimes, though, having this kind of knowledge can be a curse. Having put myself through school as a mechanic, and working currently as an information technology networking engineer, I can relate to the fact that it's easy to become the go-to guy for tech support. Family, friends and even *their* friends can consume significant amounts of time, thought and effort when their stuff stops working. In his brilliance, Michael has found a solution: Write a book! He currently has several books published, and they've been of tremendous help in my quest to understand circuitry and become adept at fixing it.

If you're struggling with a complicated schematic, going around in circles chasing a diagnosis, or you can't figure out what the device's manual is actually telling you to do, pick up one of Michael's books. There you'll find the knowledge you need, explained in a way that anyone can comprehend. His books help me on a regular basis, and I'm sure they will help you too.

Marcus Florido, K4VBB
Network Engineer

Acknowledgments

To Lara Zoble, Patty Wallenburg and the other wonderful folks at McGraw Hill, many thanks for recognizing the value of this material and shepherding it into existence. It takes a team to raise an idea. You've been a great team, and I'm honored to have been part of it. Thanks also to James Baker for some great LCD TV tips.

Introduction

Wow, so much has changed since the last edition of this book! Darned near everything is digital now, and moving parts in consumer products are rare. Many specialized devices have been replaced by phone apps, and the whole world, from cars to pocket gadgets, runs on lithium batteries. Our products are all made with high-density, incredibly small parts on tiny circuit boards, making changing components an exacting task, and sometimes an impossible one. The days of simplicity in consumer electronics are long gone.

So, is there a point to trying to repair things anymore, beyond changing out a phone screen or replacing a tablet battery? I think there is. For one thing, some of what's inside today's products is still accessible, and it tends to be the stuff that needs fixing. Rarely does a microprocessor chip die, but power supply components fizzle out all the time. And you can get to those! Computer peripherals, too, such as printers, gaming controllers and flight simulator pedals can often be repaired without too much trouble. While peripherals are not covered specifically in this book, the principles and techniques you'll learn here apply quite well to them.

Some application-specific products are still being made. While most of us stream music from our phones and videos from online sources, there's enough interest in CDs, DVDs and Blu-Rays that manufacturers still make the players, and discs can be rented and purchased or checked out at the library. Bluetooth speakers are popular, and TVs and projectors dominate home theater electronics, along with stereo and multichannel audio setups.

Plus, lots of products no longer available are out there, and their owners covet them and want to keep them going, especially because replacements are so difficult to come by. Even seriously retro formats like vinyl records have made a comeback and are popular not just with old fogeys but with their grandkids. Go take a look at what a stereo reel-to-reel tape recorder from the 1970s goes for on eBay! The better ones are fetching far more than they cost new. Somebody wants these things.

Finally, if you're into hobbies like 3D printing or amateur radio, or you're hanging onto your old VCR or camcorder so you can dub your home movies to digital, you're going to want to keep that equipment alive. Just because it's not the latest thing doesn't mean we don't care about it!

We're not going to delve into very old technologies like vacuum tubes or items using them, though, such as guitar amplifiers, high-end stereos and CRT (picture-tube) TVs. Tube gear incorporates dangerous voltages and requires specific techniques and precautions for safe, effective servicing. There are plenty of books dedicated to them.

The aim of this book is to teach you to think like a professional service technician so you can approach the repair of just about any product you might find. While it covers taking things apart in general terms, it is *not* a tear-down manual for specific items. If you need to replace a tablet screen, there are plenty of online videos and tutorials that will step you through the process for just about every model. You don't need a book for that.

What you won't learn from those is *why*. Why do various parts fail, why does a bad transistor in a stereo or multichannel receiver cause blown fuses, why does a skipping disc player have trouble reading the disc? Why can it play DVDs but no longer recognizes Blu-Rays? And what do you do about these things so you can enjoy your equipment again?

The art of electronic service—and it is an art as much as a science—is in understanding how things are supposed to work and what could be causing them not to. It's a diagnostic process very much like a doctor's, and it can lead to many hours of engrossing intellectual stimulation as you solve the puzzles and experience the satisfaction of seeing your prized possessions come back to life. It has done this for me for a lifetime, and my house is filled with the results of my enjoyment. I hope this book will help do the same for you.

Chapter 1

Prepare for Blastoff: Fixing Is Fun!

Electronics is a lifelong love affair. Once its mysteries and thrills get in your blood, they never leave you. I became fascinated with circuits and gadgets when I was about five years old, not long after I started playing the piano. There may have been something of a connection between the two interests—both involved inanimate objects springing to life by the guidance of my mind and hands. Building and repairing radios, amplifiers and record players always felt a little like playing God, or perhaps Dr. Frankenstein: "Live, I command thee!" A yank on the switch, just like in the movies, and, if I had figured out the puzzle correctly (which was far from certain at that age), live it would! Pilot lights would glow, speakers would crackle with music and faraway voices, and motors would turn, spinning records that filled my room with Haydn, Berlioz and The Beatles. It was quite a power trip (okay, a little pun intended) for a kid and kept me hankering for more such adventures.

By age eight, I was running my own neighborhood fix-it business, documented in an article in the *Miami News* titled, "Little Engineer Keeps Plugging Toward Goal." Repairs usually ran about 25 cents, and I had customers! Neighborhood pals, their families and my dad's insurance business clients kept me busy with malfunctioning radios and tape recorders. I even fixed my pediatrician's hearing tester for 50 cents. If only I'd known what *he* was charging!

My progression from such intuitive tinkering to the understanding required for serious technician work at the employable level involved many years of hands-on learning, poking around and deducing which components did what, and tracing signals through radio stages by touching solder joints with a screwdriver while listening for the crackling it caused in the speaker. Later came meters, signal tracers and, finally, the eye-opening magic window of the oscilloscope. I got my first scope when I was a teen and have never been without one since.

Ah, how I treasure all the hours spent building useful devices like intercoms and fanciful ones like the electroquadrostatic litholator (don't ask), fixing every broken gadget I could get my hands on, and devouring *Popular Electronics, Electronics Illustrated* and

Radio-Electronics—great magazines crammed with construction articles and repair advice columns. Only one issue a month? What were they waiting for? C'mon, guys, I just *have* to see the last part of that series on building your own color TV camera, even though I'll never attempt it. But now I know how a vidicon tube works! And, thanks to my parents' wise and strict rule that I experiment only on battery-powered items, I survived my early years to share my enthusiastically earned expertise with you, the budding tech.

After graduating from the Boston Conservatory of Music, I did what any highly trained, newly certified composer/conductor does: I completely abandoned my field of study and started working in electronics! I was a tech in repair shops, I programmed computers, and I developed circuitry and software for several companies around Boston and New York while building my own inventions and running a little mail-order company to sell them. All of those experiences integrated into the approach I will present in this book, which includes inductive and deductive reasoning, concepts of signal flow and device organization, taking measurements, practical skills and tips for successful repair, a little bit of art and even a touch of whimsy here and there.

No book can make you an expert at anything; that takes years of experience and squirreling away countless nuggets of wisdom gleaned from what did and didn't work for you. My hope is that this distillation of my own hard-won understanding will infect you with the love of circuits and their sometimes odd behaviors, and start you on the very enjoyable path of developing your skills at the wonderful, wacky world of electronics repair.

So, warm up your soldering iron, wrap your fingers around the knobs of that oscilloscope and crank up the sweep rate, 'cause here we go!

Repair: Why Do It?

When I was a kid, there were radio and TV servicers in many neighborhoods. If something broke, you dropped it off at your local electronics repair shop, which was as much a part of ordinary life as the corner automotive service garage. These days, those shops have all but disappeared as rising labor costs and device complexity have driven consumer electronics into the age of the disposable machine. When it stops working, you toss it out and get a new one. So, why fix something yourself? Isn't it cheaper and easier just to go out to your local discount store and plunk down the ol' credit card?

It might be easier, but it's usually not cheaper! Sans the cost of labor, repair can be quite cost effective. There are lots of other good reasons to become a proficient technician, too:

- *It's fun.* You'll get a strong sense of satisfaction when your efforts yield a properly working gadget. It feels a bit like you're a detective solving a murder case, and it's more fun to use your noodle than your wallet.
- *It's absorbing.* Learning to repair things is a great hobby to which you can devote many fruitful hours. It's good for your brain, and it beats watching TV any day (unless you fixed that TV yourself!).

- *It's economical.* Why pay retail for new electronics when you can get great stuff cheap or even for free? Especially if you live in or near a city, resources like craigslist.org will provide all the tech toys you want, often for nothing. Lots of broken gadgets are given away because bringing them in for repair costs so much. They're yours for the taking. All you have to do is fix 'em!
- *It can be profitable.* Some of the broken items people nonchalantly discard are surprisingly valuable. When your tech skills become well developed, you'll be able to repair a wide variety of devices and sell what you don't want for yourself.
- *It can preserve rare or obsolete technology.* Obsolete isn't always a negative term! Some older technologies were quite nice and have not been replaced by newer devices offering the same features, utility or quality. The continued zeal of analog audio devotees painstakingly tweaking their turntables offers a prime example of the enduring value of a technology no longer considered up-to-date.
- *It's green.* Every product kept out of the landfill is worth two in ecological terms: the one that doesn't get thrown away and the one that isn't purchased to replace it. The wastefulness of tossing out, say, a video projector with a single bad capacitor is staggering. To rip off an old song, "Nothing saves the green'ry like repairing the machin'ry in the morning."
- *Your friends and family will drive you crazy.* Being a good tech is like being a doctor: Everyone will come to you for advice and help. Okay, maybe this one isn't such an incentive, but it feels great to be able to help your friends and loved ones, doesn't it? Being admired as an expert isn't such a terrible thing either.

Is It Always Worth It?

Often, it's sensible to repair malfunctioning machines, but sometimes the endeavor can be a big waste of time and effort, either because the device is so damaged that any repair attempt will be futile or the cost or time required is overwhelming. Part of a technician's expertise, like a doctor's, lies in recognizing when the patient can be saved and when it's time for last rites and pulling the plug—in this case, literally! Luckily, in our silicon and copper realm, those destined for the hereafter can be recycled as parts. A stack of old circuit boards loaded with capacitors, transistors, connectors and other components is as essential as your soldering iron, and you'll amass a collection before you know it.

Chapter 2

Setting Up Shop:
Tools of the Trade

To repair anything, you will need some basic test gear and a suitable place to use it. Because electrons and their energy flows are invisible, test equipment has been around almost as long as human awareness of electricity itself. The right test instruments and hand tools enable you to get inside a product without damaging it, find the trouble, change the bad parts and reassemble the case correctly and safely.

Must-Haves

Electronics work can involve a seemingly unending array of instruments, but you don't need them all. Some of them are insanely expensive and only rarely useful. Others cost a lot less and find application in almost every circumstance. Some items are absolutely essential, so let's start by looking at the things you can't live without, and how and where to set them up for the most effective, efficient service environment.

A Good Place to Work

Like surgery, tech work is exacting; there's little room for error. A slip of the test probe can cause a momentary short that does damage worse than the problem you were trying to solve. One of the most important elements of effective, conscientious repairing is an appropriate workspace set up to make the task as easy and comfortable as possible, minimizing the likelihood of catastrophic error.

First, consider your location. If you have young children, it's imperative that your workbench is set up in a room that can be locked. Opened electronic products and the equipment used to service them are not child-safe, and the last thing you or your kids need is an accident that could injure them. Dens and basements can be suitable locations, but garages are probably best avoided if the kids are still at that "poke in a finger and see what happens" age. Pets, too, can wreak all kinds of havoc on disassembled machinery. Cats love to climb on and play with things, particularly if those things are warm. The effects can range from lost screws and broken parts to a pet funeral in the back yard! Let's face it, cats are not big readers, and a "Danger! High Voltage!" sticker looks about the same to them as "Cat Toy Inside." Keep kitty away from your repair work, even when you're in the room. You just never know when the little angel sitting there so placidly will make a sudden leap at your project and turn it, or Fluffy herself, into op-art.

Many of us have our workshops in the basement. This location is a mixed bag. It keeps the somewhat messy business of repair out of your living space, but it has some drawbacks. If you live in a cooler clime, it can get mighty chilly down there in the wintertime! Worse, basements tend to be damp, which is bad for your test gear. In damp environments, oscilloscopes and meters have a way of not working if you haven't used them for awhile because moisture gets into connectors and redirects normal current paths in unpredictable ways. Still, the basement may be your best bet. Just be sure to fire up your gear now and then to dry it out, and run a dehumidifier if the humidity climbs above 70 percent or so. Use an electric heater in the winter; kerosene and propane heaters designed for indoor operation still emit quite a bit of carbon dioxide that will build up in the unventilated spaces of most basements. They also use up oxygen. And should such a heater malfunction and put out a little carbon monoxide . . . well, you can imagine.

The workbench itself should be as large as you can manage, with plenty of space for your test equipment, soldering iron, power supplies and other ancillary gear along the edges; you'll need to keep the center clear for the item to be repaired. Wonderful prefab test benches can be mail-ordered, but they're fairly expensive and are most often found in professional shops. If you have the means, go for it. Get one with shelves and lots of power strips. If, like most of us, you'd rather not spend hundreds of dollars on a bench, there are plenty of alternatives. You can make your own in the time-honored way from an old solid door (hollow doors aren't strong enough) and some homemade wooden legs and braces. If you're not the woodworking type, a big desk will do the trick, and you can always add a hutch for test gear.

Sturdy desks and tables suitable as workbenches can often be had for very little from thrift stores or for free from online trading boards because of one factor in your favor: They don't have to be pretty. In fact, avoid spotless, fancy furniture because you'll feel bad when you nick, scrape, singe and accidentally drill holes in it. An Ikea-style desk works great as long as it's well braced and sturdy. A light, solid-color surface is nice too, because you can see dropped screws and such much more easily than with a darker, textured covering (see Figure 2-1). Don't even consider covering the bench with carpeting; you'll lose so many parts in it that you could eventually shake it out and build a fusion reactor from what you find! Also, carpeting can build up static electric charges lethal to circuitry, so you don't want to do your service work directly on it. Besides, soldering irons melt it, making a mess and possibly releasing toxic fumes.

Carpeting on the floor around the bench has its pros and cons. It's easy to lose small parts in it, but it also helps prevent them from bouncing away into oblivion when they

FIGURE 2-1   The well-appointed workbench.

fall. If you do choose to have a carpeted floor, pick a light color and as shallow and tight a pile as possible. This is no place for a thick carpet with loose fibers. Also, keep in mind the static problem. Especially if you live in a dry climate, be extra careful to discharge yourself by touching a ground point after sitting down, before doing anything else. Those little zaps we get when touching a doorknob in a carpeted room are up in the thousands of volts! Even charges too small to feel can destroy transistors and chips.

You'll need modern three-wire (grounded) electrical service at your bench. This is critical for safety! A grounding adapter plugged into a 1920s two-wire outlet will not do, even if you screw the adapter's ground lug to the wall plate. Most of those plates are not properly grounded, and a bad ground can get you *killed* in certain circumstances.

The current (amperage) requirement is not high for most service work. Your scope and other instruments won't eat a lot of power, so most benches can be run quite safely using a single, modern 15-amp plug fanned out by a couple of hardware store–variety power strips. Also, this arrangement has the advantage that all ground points are at exactly the same voltage level, which helps prevent ground loops (unwanted current between ground points), which can cause false readings on your test equipment and even damage what you're fixing. Again, be sure the strips are three-wire, grounded types.

Lighting is another very important factor that shouldn't be ignored. While it might seem obvious that the entire room should be brightly lit, that is not the most productive approach, as it can actually make it harder to see small details that need to be scrutinized and, therefore, brighter than their surroundings. Average lighting in the room is adequate. What you need most is spot lighting, and the best solution is a fluorescent light on a swing arm, as shown in Figure 2-1. If it has a magnifier, all the better, but you'll be wearing one anyway, so it's not necessary.

Forget about using an incandescent bulb up close; the heat it produces will cook your hands, your face and the gadget you're servicing. An inexpensive way to obtain the necessary lighting is to get a swing-arm desk lamp and replace its incandescent bulb with an LED bulb. Avoid using spiral "eco bulb" fluorescents. They have a rather yellowish tint and also put out a fair amount of ultraviolet light, so using one close to your eyes may not be comfortable or safe. Plus, they operate at a high frequency and can emit significant short-range radio-frequency energy capable of interfering with some kinds of measurements or even the circuit under test. An LED bulb or a good old circular bluish-white fluorescent lamp is still your best bet.

Digital Multimeter

A *multimeter* (pronounced "mull-TIH-mih-ter" meaning multi-meter) is a device that can test several electrical parameters. The most common and important items you'll need to measure are the three basic electrical quantities: voltage (volts), resistance (ohms) and current (amps or, more typically, milliamps, which are thousandths of an amp). The analog incarnation of this test device, recognizable by its big meter needle and multiple-stop selector knob, used to be called a VOM (volt-ohm-milliammeter), as shown in Figure 2-2.

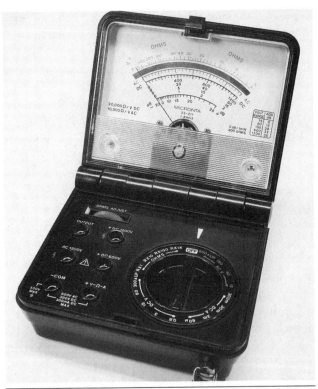

FIGURE 2-2   Analog VOM.

Now that the meters are digital, they're called DMMs (digital multimeters). They do the same thing, except that the readout is numerical instead of something interpreted from the position of a meter needle. Some of them add extra capabilities, too, like testing transistors and other components (see Figure 2-3).

DMMs began as very expensive, high-end laboratory instruments, but they're cheap now and much more widely available than analog meters. The market positions have reversed: VOMs have become the exotic technology, with a good one selling for considerably more than a DMM. VOMs can be useful, but a DMM is essential. Hardware and electronics supply stores offer DMMs for around $20 to $50, and they're on sale on occasion for as little as $5. Some, however, can still be in the range of $200 or more. The expensive ones may have the ability to test various components, and some have ultra-low resistance scales that can read a fraction of an ohm. That's a surprisingly useful attribute, but you can do the same thing with an inexpensive ESR (equivalent series resistance) meter, as we'll discuss a little later. Mostly what the fancy, pricey DMMs offer are much higher precision and accuracy.

Say Whatcha Mean, Mean Whatcha Say

Precision and accuracy are two different things. Precision is the fineness to which a measurement is specified, and accuracy is how truthful the measurement is. For instance, if I say, "It's between 60 and 80 degrees outside," and the actual temperature is 72 degrees, my

FIGURE 2-3   Low-cost DMM.

statement is not very precise, but it's quite accurate. If, however, I say, "It's 78.69 degrees outside," and it's really 82 degrees, my statement is very precise but not at all accurate. In this case, the accuracy does not support the precision.

So, for a DMM to specify that it measures voltages to three digits to the right of the decimal point, it has to have a basic accuracy of somewhere around a thousandth of a volt. Otherwise, those pretty digits won't mean much! Who on earth would build an instrument that displayed meaningless numbers?

Makers of low-cost DMMs do it all the time. The digits make one manufacturer's unit look more desirable than another's, but the basic accuracy doesn't support them. Does it matter? Not really, as long as you are aware of the limitations of the instrument's basic accuracy so you know what to ignore toward the right side of the display. In any event, even inexpensive DMMs are both more precise and more accurate than any VOM ever was.

Just how much precision do we need? For general service work, not a lot. When things break, they don't do so in subtle ways. For example, if you're checking the output of a 5-volt power supply and your DMM reads 5.126 volts, that's not cause for concern. If it reads 3.5 or 7 volts, then perhaps you've found a problem! Bottom line: You don't need a $200 DMM, but avoid the five- or ten-buck specials. I've seen their readings be wildly off, which could confuse your diagnostic process, leading you down frustrating dead ends. Some of 'em also skimp on insulation and might be dangerous if you probe something that turns out to have a few hundred or more volts on it. The $20 to $50 instruments will do fine.

ESR Meter

Capacitors, also called *caps* in tech slang, are components that store electricity, somewhat like a rechargeable battery but without nearly as much storage capability. They abound in every electronic device and are a frequent source of problems. The *electrolytic* types used in power supplies to smooth the output power are some of the most trouble-prone components of all, mostly because they contain a liquid that helps increase the storage capacity, and that liquid dries out or leaks over time. In addition to suffering complete failures like opens (no connection inside) and shorts (one lead directly connected inside to the other), these capacitors can gradually lose their ability to store energy. Worse, their internal *equivalent series resistance* (ESR) can rise, in which case the capacitor will measure just peachy keen on a capacitance meter but will act like it has a partial blockage between it and the circuit when it's in use. The resistance slows down its rate of charge and discharge like a kink in a hose, rendering it ineffective at smoothing out fast changes in the incoming power because the capacitor can't take in current and put it back out quickly enough. Excessive capacitor ESR is one of the most common causes of oddball circuit behaviors, and it rears its ugly head more often than any other fault, thanks to the fast-changing energy used in today's products. All that rapid charging and discharging simply wears out the capacitors after a while.

Measuring ESR takes more than a standard ohmmeter because as the capacitor charges from the test voltage, its resistance naturally goes up anyway, making reading the "true" resistance impossible. An ESR meter gets around this by testing the capacitor with a low-voltage alternating current (one in which the direction flip-flops back and forth), slightly charging and then discharging the cap so fast that it never builds up enough charge to affect a reading. During each charge cycle, a resistance measurement is taken as current is first applied, before the cap can charge. That finds the resistance inherent in the capacitor, not the result of any charging.

Until recently, ESR meters were pricey luxuries. With older products that didn't stress their capacitors so hard, the meters weren't needed often anyway, so few of us had them. Now that ESR is the bane of every service tech's life, you can find some nice ESR meters online for very little. Get one! You will use it often, I promise (see Figure 2-4).

Oscilloscope

Many hobbyists feel intimidated by the oscilloscope, but it is the best buddy any tech can have. Repeat after me: "My scope is my friend." Come on, say it like you mean it! Once you get the hang of using one, you will love it, I assure you. It's the only instrument that actually lets you see what is going on in an electronic circuit.

The basic function of an oscilloscope is to generate a graph of voltage versus time, as in Figure 2-5. As a spot sweeps from left to right across the screen at a constant rate, it moves up and down in relation to the incoming signal voltage, drawing a *waveform*, or representation of the signal that shows you how the voltage is changing. How fast the

FIGURE 2-4   ESR meter.

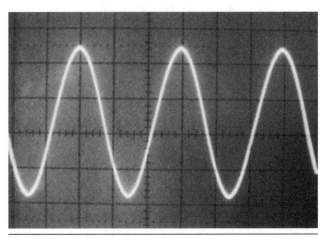

FIGURE 2-5 Waveform of a signal.

scope's circuits can respond determines the *bandwidth*, or how rapidly changing a signal the scope can display. Most scopes in our range of interest have bandwidths of around 100 megahertz (MHz, or 100 million cycles per second). They also have two *vertical input* channels, meaning they can display two waveforms at the same time. Now and then, that can be remarkably useful! In Chapter 6 we'll explore how to use a scope—it really isn't hard—but first you have to get one. There are several types.

Analog

This is the classic scope with a green CRT (cathode-ray tube, a.k.a. picture tube). It's old technology but still very nice. An analog scope displays signals as they occur and has no memory functions to store waveforms. It just lets you see 'em as they happen, plain and simple. A classic analog scope is shown in Figure 2-6.

Analog scopes have been available since around the 1940s, and they really got good in the 1970s. Some are still being made today, though digital scopes have been at the forefront of the marketplace for a decade or more. Newer is better, right? Not always. The oscilloscope is a good example of an older analog technology that was superior in some ways to its replacement. For most general service work, an analog scope is the simplest to use, and its display is the easiest to interpret. Further, it shows details of the signal that digital scopes may miss.

The lowest-end analog scopes have just one channel of input, and they lack features like *delayed sweep*, a very handy function that lets you zoom in on any part of a waveform you want and expand it for detailed viewing. Avoid them. There are tons of great analog scopes with all the nice features on the used market at ridiculously cheap prices, so there's no need to skimp on the goodies. Make sure any analog scope you buy has two channels (some have even more, but two are standard) and delayed sweep. Look for two input connectors marked "A" and "B" or sometimes "Channel 1" and "Channel 2," indicating

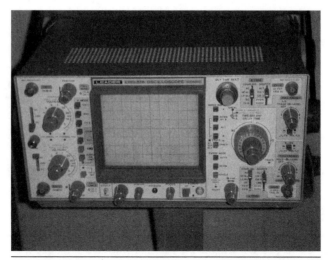

FIGURE 2-6   Leader LBO-518 100-MHz analog
oscilloscope.

two vertical input channels. If you see a knob marked "Delay Time Mult" that can be
turned multiple times and slowly advances a number imprinted on it, or you see "A"
and "B" buttons or similar labels on the big knob marked "Sec/Div" or "Horiz Sweep,"
or buttons labeled "B after A," "B ends A," or "B delayed," or any other reference to "A"
and "B" in the horizontal control area, then the unit has delayed sweep. Another tipoff
is a knob marked "Trace Sep." If you're still not sure, look up the model on the internet,
and you can probably find its specs or even download a free PDF of the manual. Typical
delayed-sweep controls are shown in Figure 6-4.

Digital

Like everything else, oscilloscopes have gone digital. Most digital scopes offer delayed
sweep and two channels, except for a few handheld models. Though some early examples
used CRTs, modern digital scopes can be recognized by their shallow cases and LCD
(liquid-crystal display) screens (see Figure 2-7). With a digital instrument, you can grab a
waveform and examine it in detail long after it has ceased. Thus digital scopes are ideal for
working on devices with fleeting signals you need to be able to snag that may zip by only
once.

Such is rarely the case in the kind of service work you'll be doing, though. The vast
majority of the time, you will be looking at repetitive signals that don't have to be stored,
and the limitations of a digital scope can get in the way sometimes.

One significant shortcoming arises from the basic nature of digital *sampling*, or
digitizing, in which a voltage is sampled, or measured, millions of times per second, and
the measured value of each sample is then plotted as a point on the screen. Alas, real-life
signals don't freeze between samples, so digital scopes miss some signal details, which can

FIGURE 2-7   Tektronix TDS-220 100-MHz digital oscilloscope.

result in a phenomenon called *aliasing*, in which a signal may be seriously misrepresented. (The same effect causes wagon wheels in old westerns to appear to rotate backward—the movie camera is missing some of the wheels' motion between frames.) If the sampling rate is considerably faster than the rate of change of the signal being sampled, aliasing won't occur. The sampling rate goes down, though, as you slow the *sweep rate* (speed of horizontal motion on the screen) down to compress the graph and squeeze more of the signal on the screen. As a result, when using a digital scope, you must always keep in mind that what you're seeing might be a lie, and you find yourself turning the sweep rate up and then back down, looking for details in the waveform that got missed at the slower rate. It takes some experience to be certain that what's on the screen is a true representation of the signal. Even so, sometimes aliasing is unavoidable at lower sweep rates, limiting how much of the signal you can view at once—a conundrum that never occurs with analog scopes.

Another big limitation arises from the screen itself, and it also limits how much you can see at one time. Unlike the continuously moving beam of the analog scope, the digital scope's display is a grid of dots, so it has a fixed *resolution*, and nothing can be shown between those dots. (This is why the sampling rate goes down at lower sweep rates; there's no point in taking samples between dots because there's no place to plot them anyway.) When examining complex waveforms like analog video signals, the result is a blurry mess unless you turn the sweep rate way up and look at only a small part of the signal. While an analog scope can show a useful, clear representation of an entire field of video, a digital instrument simply can't; all you see is an unrecognizable blob. Of course, it's unlikely you'll be playing around with analog video these days, unless you're into restoring old TVs, VCRs or camcorders. However, I just ran into some while troubleshooting a rear-view camera system for my car that used AHD, or analog high-definition, to send video from the camera to the mirror's display.

What You Can and Cannot See

Probably the most profound difference between an analog and a digital scope is that an analog instrument actually writes the screen at the sweep speed you select, while a digital unit does not. A digital scope collects the data at that speed, but it updates the screen much more slowly because LCDs don't respond very fast. For many signals, that's fine, and it can even help you see some signal features that might be blurred by repetitive overwrites on an analog screen.

Sometimes, however, those overwrites are exactly what you want to examine. When viewing the signals coming from video and laser heads or troubleshooting radio gear, for instance, you need to evaluate the *envelope*, or changes in the overall shape of the waveform over many cycles, rather than the individual waves. These changes represent *modulation*, which is information impressed on a signal by varying it more slowly than the rate at which its waves occur. So, each change encompasses many cycles of the original waveform, and the individual waves are too crammed together to see when your scope is sweeping slowly enough to let you view the changes. The left side of Figure 2-8 shows the envelope of a fraction of a second of speech modulating the strength of a radio transmitter's 14-megahertz (millions of cycles per second) *carrier* waves.

The right side of the figure shows the envelope of the signal from a VCR's two video heads while it plays a tape. This is FM (frequency modulation), in which the frequency of the waves wiggles back and forth slightly to represent the video information. Those wiggles are too tiny to see here, but there's still something important that's plainly visible. The smaller section in the middle is because one head is putting out a weaker signal than the other—a possible fault, and precisely the sort of useful information you can glean from looking at envelopes. You'd never see it by looking at the individual waves.

The overwriting and true-to-life writing speeds inherent in an analog scope make envelopes stand out clearly. Some envelopes can't be viewed *at all* with a digital scope because it misses too much between screen updates. You'll see individual cycles of the waveform but not their outer contour, unless you slow down the sweep rate so low that all you get is a featureless blur.

On the plus side, digital scopes are naturals at measuring waveforms, not just displaying them. They can provide a numerical readout of voltage at any spot in the signal,

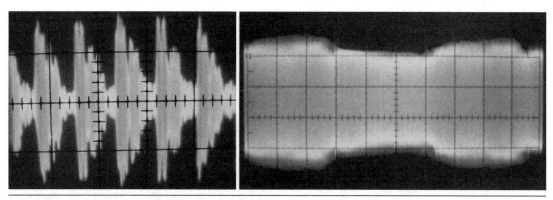

FIGURE 2-8   **Signal envelopes on an analog scope.**

frequency and other parameters. While all of these measurements can be done with an analog scope, the computer in that case is your brain; you have to do the math, based on what you're seeing on the screen and the settings of the controls. With a digital scope, you position cursors on the displayed waveform, and the scope does the work for you. Having quick and easy measurement of signal characteristics can greatly speed up troubleshooting. Even better, some digital units offer an "Auto" button that sets everything to show you the signal clearly, saving more time and effort. Just press the button, and there it is!

When choosing a digital scope, look for the sample rate and compare it to the vertical bandwidth. The sample rate should always be higher than the bandwidth so the scope can perform *real-time sampling*. The Tektronix TDS-220, for example (Figure 2-7), samples at 1 gigasample (billion samples) per second, with a bandwidth of 100 MHz. Thus, one cycle of the fastest waveform it can display will be broken into 10 segments, which is pretty good. At a minimum, the sample rate should be four times the bandwidth or signals near the maximum displayable rate will look highly distorted, in the same way a TV picture would if it were made up of only a few pixels (dots).

It is possible to sample repetitive waveforms that are faster than the sample rate using a technique called *equivalent-time sampling*, in which the repeating waveform is sampled at successive points, advancing the sampling position in the wave each time it repeats until the full digital representation is assembled. Equivalent-time sampling was developed when the analog-to-digital converters that process the incoming signals were too pokey for real-time sampling of fast waveforms. It is an inferior technique because developing an accurate representation requires the incoming signal to remain unchanged from cycle to cycle for as many cycles as it takes to assemble the digitized version. To provide, say, 500 points on the graph of a waveform requires 500 repetitions of it to collect all the data. So, what you see is never a true picture of any one particular cycle. Avoid scopes depending on equivalent-time sampling to reach their bandwidth specs. Real-time sampling is the only way to fly. Few new digital scopes use equivalent-time sampling, but you might see it in an older model or an especially low-cost instrument.

Lots of inexpensive digital scopes from Chinese manufacturers are being sold on eBay and through other online sources. Many of them are pretty good and are well worth the cost. You'll find desktop, handheld and pocket models with lots of features and decent bandwidth at low prices. Be sure to check the reviews before buying one, though. Some of the handheld and pocket units, especially, are more novelty items than useful, accurate instruments. They can be handy for checking whether signals are there, but they're not good enough to be your only scope.

Analog with Cursor Measurement

This is the best of both worlds: an analog scope capable of performing many of the measurements available in a digital instrument. This style of scope doesn't digitize signals, thus avoiding all of the limitations associated with that process. Like a digital unit, though, it uses movable cursors to mark spots on the displayed waveform and calculate measurements. This is my favorite type of scope, but it's fading into history along with the rest of the analog scopes.

Analog with Storage

Before the advent of digital scopes, some analog units were made with special CRTs that could freeze the displayed waveform, enabling a crude form of signal storage. These scopes were expensive and always considered somewhat exotic, and their bandwidth was on the low side. Digital storage has completely supplanted them. You may run across one on the used market. Don't buy it.

PC-Based

The PC-based scope uses your general-purpose computer as a display and control system for a digitizing scope. It seems like a great idea because you get a nice, big, high-resolution screen, and you can use your computer's keyboard and mouse to control the features. Plus, PC scopes are cheaper because you're not paying for knobs and an LCD. In practice, PC scopes are the worst option for service work. They're awkward to use and usually offer the lowest performance in terms of sampling rate. In fact, you're most likely to run into equivalent-time sampling in one of these. I recommend you avoid them.

Buying a Digital Oscilloscope   Prices have come down so much on modern digital scopes that this is probably what you'll get. There are many brand names for these, but Rigol, Siglent and Owon are pretty popular among the less expensive units. Check reviews on the model you're looking at before buying!

Desktop digital scopes with LCD screens are much smaller than the old analog units, saving valuable workbench space. They're also more menu-driven, with far fewer knobs, but a well-designed one is not hard to use. Most of the screens are in color, making it easy to keep track of all the settings and measurements. Color might seem like a frill, but it adds a lot!

Handheld and pocket digital scopes have come a long way, and inexpensive ones are abundant. You can find very simple, low-performance versions for $20! Don't buy one of those, though. Your scope is your most important piece of test gear, so get a real instrument, not a toy. Expect to pay at least a couple hundred bucks for a good one.

Among handheld units is the *scopemeter*, which is a DMM that also offers scope facilities (see Figure 2-9). This type is primarily a meter, and the scope part is more of an add-on than a primary function. If it features only the usual two test leads with banana plugs for signal input, like a standard DMM, rather than a *BNC* connector for a scope probe, it's not really a full-fledged scope, and you'll get frustrated with it when working on anything but the simplest low-speed circuitry (see Figure 2-10). Scopemeters can be handy for quick verification of the presence of signals, but think of them as DMMs with some casual signal-viewing ability, not true oscilloscopes. A pair of unshielded wires cannot provide accurate signal transfer to the scope for signals above a few hundred kilohertz (kHz, or thousands of cycles per second). So, even if the meter's stated bandwidth is in the megahertz range, what you see on the screen won't have much meaning at those frequencies. Scopemeters are not proper oscilloscopes unless they have BNC connectors for input and a calibrator terminal to adjust a scope probe for accurate signal transfer. See Chapter 6 for more about this.

FIGURE 2-9 Scopemeter with banana plug input.

FIGURE 2-10 BNC and banana connectors.

Buying an Analog Oscilloscope   If you do choose to go old-school, you might be surprised that new analog scopes are fairly expensive. Expect to pay around $1,000 for a 100-MHz instrument from a well-established, reputable maker. But why spend a lot when there are so many nice scopes on the used market for next to nothing? It's quite possible to get a good used analog scope for about $100 to $200.

There are plenty of scope manufacturers, but the gold standard in the oscilloscope world is Tektronix. Tek has dominated the scope market since the 1970s, and for good reason. Many of its models from that era are still going strong, 50 years later! If you find one of those gems in good working order for less than $150, it's worthy of consideration. Models from more recent decades are also available, including the 2200 and 2400 series, and they're pretty cheap too. Other good classic scopes were made by Hitachi, Hewlett-Packard, B&K Precision, Kikusui, Sencore and Leader.

Where do you find a used oscilloscope? Good old eBay is loaded with them, and they show up now and then on craigslist. Try going to your area's "hamfest," a periodic swap meet put on by ham radio operators and electronics aficionados, and you'll see plenty of scopes. Just be sure you can check that the instrument works properly before you plunk down your cash, because troubleshooting a broken oscilloscope is tough. And, of course, you'd need another scope to work on it! Read Chapter 6 first so you'll know how to test a potential purchase. Look for a nice, sharp trace and no lines burned into the display tube.

Probes

Along with the scope, you'll need a pair of probes, or a single probe if your scope has only one input channel. Most new scopes come with probes, but not all. Used scopes usually won't include probes. Scope probes are more than just pieces of wire; they are specialized devices designed to enable accurate signal measurement. Most divide the incoming voltage by 10 (you'll see why later) and are called *10X* probes. Some have switches to remove the division and are known as *switchable 10X/1X* probes (Figure 2-11). Like scopes themselves, probes are rated by bandwidth, and the high-end ones can cost a lot. Luckily, 100-MHz probes can be found on eBay brand new for around $15 each. New probes will include slip-on covers with handy hooks on the ends for grabbing onto the leads of components in the devices you're servicing. If you buy used probes, try to get the hooks with them because every brand has different sizes, so they don't fit each other, and you're unlikely to find any separately.

Soldering Tools

Soldering Iron

Soldering irons come in various shapes and sizes, and you'll probably wind up with more than one. The smallest, with heating elements in the 15-watt range, are great for getting into very tight spots and working on tiny surface-mount parts (the kind with no leads), at

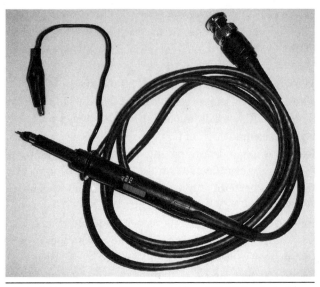

FIGURE 2-11   Oscilloscope probe, 10X/1X switchable.

least on boards assembled with low-temperature solder. These mini-irons don't generate enough heat to solder a power transistor, though; it's just too big and absorbs too much of the heat. The largest irons, usually pistol-shaped guns with elements of 100 watts or more, put out lots of heat and have sizable tips to transfer it to the part you're soldering. Those big guns can be real life savers, but you sure don't want to try soldering tiny components with them. Even if you could fit the tip where you needed it, the excessive heat would destroy the part and probably the circuit board as well.

The best choice for general soldering work on printed circuit boards is an iron with a medium-sized tip and a heating element in the range of 40 to 70 watts. Melting leaded solder requires a tip temperature of about 375 to 400°F. The newer, lead-free solder needs a much hotter tip, in the area of 675 to 700°F. Some inexpensive irons in the 20-watt range are about the same size, but steer clear of those. Supplying inadequate heat can cause lots of harm; you may easily pull up copper traces and severely damage the printed circuit board if things aren't hot enough, especially when removing components. Plus, not using enough heat can result in "cold" solder joints that don't transfer electrical energy properly, causing your repair to fail.

Many inexpensive irons of medium size plug directly into the wall. This is not the best way to go, as it may expose the circuitry you're soldering to small leakage currents from the AC (alternating-current) power line, and having the cord go off to a power strip can be awkward as you move the iron around. Finally, should you accidentally lay the iron on the cord and melt through the insulation, you'll cause a short directly across the AC line, which is likely to be spectacular and unpleasant—and possibly dangerous. Don't laugh, it happens!

A far better solution is an iron that plugs into a base unit with a step-down transformer. This kind of setup runs the heating element at low voltage and isolates the tip from the AC line's voltage. The base unit gives you a nice stand to hold the iron and a sponge for wiping the tip, too. Some bases have variable heat controls, and some even have digital temperature readouts. Before the age of lead-free solder, I never found such things to be useful, since the heat pretty much always needed to be turned all the way up on smaller irons anyway. These days, a variable-heat soldering station that can hit the temperatures required for lead-free soldering is well worthy of consideration because you can turn it down for leaded solder and crank it up for the newer stuff. And with this comes the advantage of a display showing you the temperature.

Numerous companies make soldering irons, but two make the nicest, most durable irons, the ones found in service shops for decades: Weller and Ungar. Another popular brand getting good reviews is Hakko. These irons can cost from $50 to more than $200, but they are worth every penny and will last for many years. Your soldering iron is usually the first thing you turn on and the last you turn off, so it will run for thousands of hours and needs to be well made. Don't be tempted by those $20 base-unit irons flooding the hobbyist market. They just don't hold up, and you'll be needing a new one before you know it. I tried one and got a year out of it before it fizzled out.

As with scopes, good used irons often show up at great prices at hamfests. Wherever you get your iron, plan on buying a spare tip or two. Tips wear out and become pitted and tarnished to the point that they no longer transfer heat well, so they must be replaced every few years. The heating elements can go bad too, but it's rare. I've seen them last for decades on the good irons.

The big guns are cheap, typically under $30, so buy one. There will be situations in which you will be very glad you did.

Plastic-Melting Iron

Sooner or later, you'll want to melt some plastic to repair a crack or a broken post. It's unhealthy to breathe in molten plastic fumes, but we all melt the stuff now and then, being as careful as we can with ventilation. If you're going to melt plastic, don't do it with the same iron you use for soldering! The plastic will contaminate and pit the tip, making it very hard to coat it with solder, or *tin* it, for subsequent soldering work. Instead, pick up a cheap iron in the 20- to 30-watt range and dedicate it for plastic use. For this one, you don't need a base unit or any other fancy accoutrements. You should be able to get a basic iron and a stand to keep it from burning its surroundings for around $10. Just watch out for that cord!

Solder

Traditional solder is an alloy of tin and lead with a rosin core that facilitates the molecular bond required for a proper solder joint. In the past, the alloy was 60 percent tin and 40 percent lead. More recently, it has shifted to 65 percent tin and 35 percent lead. This newer type of solder is better suited to the lower temperatures associated with tiny surface-mount parts, and it's getting hard to find the old 60/40 stuff anymore. The old proportions were better for the higher-heat environment of power transistors and voltage regulators. If you can find some 60/40 solder, it's worth getting. If not, you can live with the newer variety.

Lead is a toxic metal, and lead-free solder has become available and is widely used in the manufacture of new electronics, to comply with the legal requirements of some countries and U.S. states. The European "Restriction of Hazardous Substances," or RoHS, standard has been adopted around the world. All products displaying the RoHS mark are made with lead-free solder. You can buy the stuff for your repair work, but I recommend against doing so because it's hard to make good joints with it. It doesn't flow easily, and poor joints that don't conduct well often result. Plus, the higher heat required to melt it invites damage to the components you're installing.

Lead vaporizes at a much higher temperature than that used for soldering. The smoke coming off solder is from the rosin and does not contain lead you could inhale. Handling solder, however, does rub some lead onto your hands. So, never snack or touch food while soldering, and always wash your hands thoroughly after your repair session ends.

There is a variety of solder, found in hardware stores and intended for plumbing applications, with an acid core instead of rosin. *Never* use acid-core solder for electronics work! The acid will corrode and destroy your device. By the same token, the rosin flux paste used with acid-core solder is not needed for normal electronics solder because rosin is already in its core.

Solder comes in various diameters. A good choice for normal work is around 0.03 inches. Very small-diameter solder, in the 0.01-inch range, can be useful now and then when working with tiny parts, but not often. For most jobs, it's so undersized that you have to feed it into the work very fast to get enough on the joint, making it impractical to use. My own roll of the skinny stuff has been sitting there for a decade, and most of it is still on the roll.

Solder is like ketchup: You'll use a lot of it. Buy a 1-pound roll because it's a much better bargain per foot than those little pocket packs of a few ounces. A pound of solder should last you a good few years.

Desoldering Tools

Removing solder to test or replace parts is as vital to repair as is soldering new parts to the board. Desoldering ranges from easy to tricky, and it's a prime opportunity for doing damage to components and the copper traces to which they're attached. Fancy desoldering stations with vacuum pumps can cost considerably more than even top-end soldering irons. For most service work, though, you don't need anything exotic. There are some low-cost desoldering options that usually do the trick.

Solder Wick

One of the best desoldering tools is *desoldering braid*, commonly called *solder wick*. It's made of very fine copper wire strands woven into a flat braid. Usually it is coated with rosin to help solder flow into it. (You may run across some cheap wick with no rosin. Don't buy it; it doesn't work.) Wick can be purchased in short lengths on small spools from various hobbyist-oriented mail-order companies. Electronics supply houses offer it in much longer lengths on bigger spools. As with solder, the bigger spools are the far better deal. Always be sure to keep some wick around; it's some of the most useful stuff in your workshop.

Bulbs

Another approach to solder removal is to suck it up with a rubber solder bulb or solder sucker. Bulbs come in two forms: stand-alone and integrated with a soldering iron (see Figure 2-12). Both have their uses, but with the integrated type you're limited by the heating power of the built-in iron, which is usually not especially strong. Stand-alone bulbs are cheap, so get one even if you also get an integrated type.

Spring-Loaded Solder Suckers

One of the handiest solder removal tools for larger components, the spring-loaded solder sucker is another inexpensive option (see Figure 2-13). These bad boys have a fast, almost violent snap action and are a bit harder to control than bulbs. They suck up a lot of solder in one motion, though. Get one.

FIGURE 2-12   Bulb-type desoldering iron.

FIGURE 2-13   Spring-loaded solder sucker.

Hand Tools

The range of available hand tools seems practically infinite. Most likely you'll build up a significant collection of them as the years go by. My own assortment fills several drawers. While nobody needs six pairs of needlenose pliers, there is a core set of tools necessary for disassembling and reassembling the items to be repaired.

Screwdrivers

Today's gadgetry uses a wide range of types and sizes of screws. Some of the screws are incredibly tiny. A set of jeweler's screwdrivers is a necessity. Both Phillips and flat-blade screws are used, though Phillips types dominate. More and more, hex and Torx heads are showing up too (see Figure 2-14). The latter shapes started out as a way to prevent consumers from opening their gadgets, but the drivers have gradually become available, defeating that objective. In response, newer types have come along. One of the most recent is the Trigram, which looks like a center point with three lines radiating out toward the perimeter of the screw head. Those drivers are getting easier to find as well.

Get a good selection of small screwdrivers in all these form factors. At the very least, get Phillips, flat-blade and Torx. Pick up a few medium-sized Phillips and flat-blades too. You'll use the smaller ones much more often than the larger ones, but it pays to have as many sizes as you can find. Really big ones are rarely needed, though.

Cutters

Diagonal cutters, or *dikes*, are used to clip the excess lead lengths from newly installed components that aren't surface-mounted. Most techs also use them to strip insulation off wire and trim bare wire ends to the needed size. Again, smaller beats bigger. Get a couple of pairs of these things because they tend to get bad nicks in their cutting edges and gradually

FIGURE 2-14   Hex and Torx screwdriver tips.

become useless. A pair of dikes shouldn't cost more than about $7. Oh, and be sure the handles are insulated. They usually are.

Needlenose Pliers

A pair of needlenose pliers is essential for grabbing things, reaching into cramped spots and holding parts steady while you solder them. A length of 2 or 3 inches from the fulcrum to the tips is about right. Any shorter and they may not reach where you need them. Any longer and they'll probably be a bit too flexible, reducing their usefulness when twisting is required. Unlike cutters, needlenose pliers rarely wear out or need replacement. Still, get two pairs so you can hold one in each hand and use them at the same time. You'll need to do that now and then.

Hemostats

They're not just for surgeons anymore! Hemostats are much like needlenose pliers except that they lock, providing a firm grip without your having to keep squeezing the handle. Some have corrugated gripping ends, while others are smooth. Get one pair of each style. These things are indispensable for pulling a component lead from a board while heating the solder joint on the other side. They're great for installing new parts, too. You can find hemostats at most electronics and medical supply houses.

Magnifier

With the size of today's electronics, human eyes have hit their resolution limit for comfortable close-up work. It's essential that you have some magnification. Even if your spot lamp has a magnifying lens, you'll still need a head-worn magnifier because the lamp will get in your way when it's placed between your face and a small gadget. Glass is better than plastic, which gets scratched and can even melt when you try to solder under it. Be sure the magnifier you choose can be flipped up out of the way, because sometimes you need to step back a bit from the work and take in a longer view. If you wear glasses, look for a magnifier with adjustable focus to keep it compatible with your eyewear.

Chemicals

Contact Cleaner Spray

There are many brands of spray, each claiming superiority, but they all do pretty much the same thing: remove oxidation and dirt from electrical contacts. One of the more popular brands is DeoxIT. Whichever brand you choose, get a can or two. It's handy stuff, and you'll be using it, especially if you work on older gear.

Alcohol

Alcohol is good for cleaning excess rosin off circuit boards after you've replaced a component. It also comes in handy for removing sticky stuff and other cleaning tasks. Use isopropyl alcohol, and look for the highest percentage you can find. The 70 percent solution sold in drugstores is 30 percent water, which is bad for electronics. Some stores sell 91 percent, which is much better, and I've run across 99 percent on occasion. Don't use ethyl or any other type of alcohol. Be aware that all alcohols can damage some types of plastic rather badly. When working with alcohol, keep it away from plastic casings, LCD screens and control panels.

Naphtha

Sold in little yellow bottles as "cigarette lighter fuel" at grocery stores, and in bigger containers as "VM&P Naphtha" at hardware stores, naphtha is an amazing solvent that will effortlessly remove grime, sticker adhesive, solder rosin, tobacco tar and other general filth from just about any surface. I've never seen it harm plastic, either, not even LCD screens. It's used by dripping a very small amount on a tissue, paper napkin or swab and then gently rubbing the surface to be cleaned. Naphtha is flammable, so *never* use it on anything to which power is applied or near a flame or a hot soldering iron. It's best to use gloves, too, to keep it off your skin. Make sure of proper ventilation because it evaporates quickly and should not be inhaled. Even if you buy a big can of it, also buy one of the little yellow bottles so you can squirt out tiny doses as needed. It takes very little naphtha to do the job, and you can always refill the bottle from the big can *outside your house* later on. Keep naphtha containers tightly closed so evaporated fumes won't build up and become hazardous.

Computer

Not that long ago, a significant stack of reference books was required for looking up transistor types, cross-referencing replacement components and finding disassembly hints and diagrams for various products. Now we can do all that and more on the internet. If you want to see a transistor's characteristics or the pinout (arrangement of connections) on a chip, go to any of the major parts retailers' sites, such as mouser.com or digikey.com, and you'll find data sheets for pretty much every part there is. Or, just do a web search on the part number followed by "data sheet." Some sites offer schematics, typically charging for them, but you can find some free ones. Even if you have to pay a little bit for the diagram, it may be well worth it. There's plenty of free info out there about how to take apart certain products without breaking them, too. Having a computer nearby with internet access is essential these days.

Also, most of the service manuals and schematics you'll obtain will be downloaded PDFs you'll view on the screen while you work. You could print them before you dig into your project, but there can be many pages with complex schematics, making printing impractical. Working with electronic versions is cumbersome but unavoidable these days. Consider setting up two screens so you can view a pictorial layout and a schematic (see Chapter 8) at the same time. It's handy enough to border on being necessary. Notice mine on the right in Figure 2-1.

Accessories

These are the little items you don't think about until you need them! But you do need them, and you'll use them often.

Clip Leads

Frequently, testing involves making temporary connections. For this, nothing beats a batch of clip leads, which are wires about a foot long with *alligator clips* at both ends (see Figure 2-15). Get at least 10, making sure that the clips are small and have rubber insulating covers. You can buy the clips and make them yourself, but assembled ones are readily available and inexpensive. Just be aware that the premade leads are usually not soldered; the wires are merely crimped to the clips. After some use, they break inside the insulation, which leads to some head scratching when a clip-leaded connection doesn't produce the expected results. You can never trust the integrity of a clip lead completely. The quick test

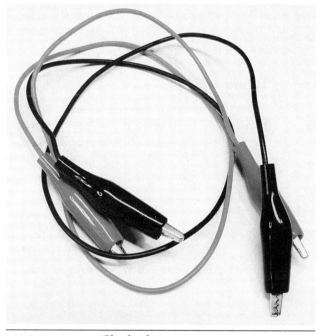

FIGURE 2-15   Clip leads.

is to pull the lead taut while holding onto the clips. If broken, one end will fall apart, after which you can solder it back on. In time, you'll wind up soldering all of them.

Swabs

Cotton swabs are very useful in the shop. Get the kind with paper sticks, not plastic ones. The paper type can be bent into shapes that will let you poke them into odd corners. If you can find them, also get some chamois swabs. Unlike the cotton type, chamois swabs don't leave little fibers behind. For some uses, especially cleaning video heads on VCRs and camcorders, the fibers can be problematic, and it's even possible to break a video head if the fibers get snagged on it while you clean. As devices with video heads have become obsolete, chamois swabs have gotten hard to find, though. Don't worry if you can't find any.

Heatsink Grease

This silicone-based grease is used between transistors, voltage regulators and other heat-producing parts and their metal heatsinks to help transfer the heat to the sinks. It fills in the tiny gaps between imperfect surfaces, providing more contact area through which the heat can travel. Even when mica insulators are used, as they are with power transistors, heatsink grease is still required for most parts. (The exception is an installation using a special rubber heat-transfer gasket; most of those do not require grease.) When you replace a heatsinked part, you'll need the grease; omitting it will result in an overheated component that will fail quickly. A small tube of heatsink grease lasts a very long time because only a thin film is required, and too much grease can actually reduce heat transfer.

Silicone grease is inappropriate for use with microprocessors and graphics chips. These especially hot-running parts require special silver-bearing grease, which you can find at computer stores and mail-order suppliers.

Heat-Shrink Tubing

This stuff looks like ordinary plastic tubing, but it has a wonderful trick up its synthetic rubber sleeve: It shrinks in diameter when you heat it up, forming itself around joints and damaged insulation spots in wires. It's much more permanent than electrical tape, which tends to get gooey and let go after a while. Get some lengths of heat-shrink tubing in various small diameters.

Electrical Tape

Despite its impermanence, electrical tape still has uses in situations where tubing won't fit or can't be slipped over what needs to be insulated. Plus, for extra insulative peace of mind, you can wrap a connection in tape and then put tubing over it.

Small Cups

If you eat yogurt or pudding, start saving the little plastic cups. However you obtain them, these cups are incredibly useful for temporary storage of screws and other small parts as you disassemble machines. Make sure the cups fit into each other. Most will.

Nice-to-Haves

Here are some items that can help you get repair work done more easily. You can live without 'em, but you might want to add some to your arsenal as time goes on.

Digital Camera

How's your memory? If it's imperfect like mine, a digital camera can save your rear end when you look up from the bench and realize you've removed 35 screws from four layers of a laptop and you're not sure where they all go. And what's that funny-looking piece over there? The one you took off three days ago, just before you answered the phone and took the dog for his emergency walk? Take pictures as you disassemble your devices. Use the macro lens as necessary to get clear close-up shots; a blurry photo of a circuit board does you little good. Make sure you can see which plug went in which connector and what everything looked like before you removed it. Experiment a bit with the flash, too, and the angles required to get decent shots without too much glare.

   If your phone has a decent macro lens, that'll do fine. If not, you can find a slightly obsolete digital camera on craigslist for next to nothing. You don't need 12 megapixels for this. Any 5-megapixel or more camera is more than adequate, and the better brands all have pretty good macro functions.

Power Supply

Unless it has its own AC power supply built in, your repair item runs either on batteries or from an AC adapter. AC adapters themselves fail often enough that you can't assume the adapter isn't the problem. So, running the device under test from a variable-voltage power supply can really help. The most important issue when choosing a supply is how much current it can provide. While a little pocket radio might eat around 50 mA (milliamps), a power amplifier or radio transmitter may require hundreds of times as much current. For most work, if you have 5 amps available, you're covered.

   There are some fancy laboratory-grade power supplies with digital metering, ultra-precise regulation and price tags to match. You don't need one. Any decent hobby-grade supply will do. Most of the newer, inexpensive types include digital metering for voltage and current anyway, just like the high-end lab versions. Some of these cheaper units can supply 20 or 30 amps, too, which is probably more than you'll ever need.

Many of the items that demand high current are for use in the automotive environment, so a 12-volt (more typically 13.8 volts, the actual voltage of a car with its engine running) supply with 10 amps or more is great to have as well, if your main bench supply doesn't offer that kind of current. Because all auto gear runs on the same voltage, this one doesn't need to be variable.

Testers, Signal Generators and Meters

Transistor Tester

Although it's possible to test many characteristics of transistors with a DMM, some failure modes, like excessive reverse leakage (when a small current can flow backward through a defective transistor's junctions), aren't easy to find that way. Dedicated *dynamic* transistor testers use the transistor under test as part of an operating circuit, measuring how the part behaves in actual use. Many testers can check various transistor types, including MOSFETs (metal-oxide-semiconductor field-effect transistors), JFETs (junction field-effect transistors) and standard bipolar (NPN and PNP) parts. In subsequent chapters, you'll learn what these parts do and what to look for.

Basic transistor testers of this type are inexpensive and a great addition to your bench setup, if your DMM doesn't already include these functions.

Capacitance Meter

Some DMMs offer built-in capacitance measurement. If yours doesn't, consider a capacitance meter. It'll let you check for reduced capacitance (ability to store energy) in worn-out power supply capacitors. It'll also help you identify unmarked capacitors by showing you their capacitance values. This can really help with those tiny surface-mounted parts too small to be marked. Most capacitance meters will *not* test ESR, which is why it's nice to have one but not an absolute must. Where electrolytic capacitors are concerned, ESR is the most useful test, and most of the capacitors you'll test are of that type.

Signal Generator

More useful for servicing analog equipment than digital, a signal generator lets you inject a test signal into a device's circuits to see whether doing so causes the expected effect. It's very handy when you're working on audio and radio equipment. If you're servicing mostly digital products, this is not something you'll use very often, but it has some application in simulating the clock oscillators that make microprocessors step through their programs. Many entertainment products, like disc players and TVs, have both analog and digital sections, and a signal generator can come in handy with those if the audio circuitry malfunctions.

The generators are called *function generators* when they have the ability to create different kinds of waveforms, such as sine, triangle and square waves, so named because

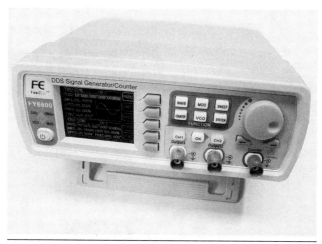

FIGURE 2-16   60-MHz function generator.

that's what their waveforms look like on an oscilloscope. While sine-wave-only generators are usually segregated by frequency band, either audio or radio, function generators may have a wide range encompassing both, though they don't offer high-frequency ranges anywhere near those of radio-only generators. Many function generators operate from a couple of Hz (hertz, or cycles per second) to around 2 MHz (megahertz, or millions of cycles per second), while radio-frequency (RF) generators might start at 100 kHz (kilohertz, or 100,000 cycles per second) and reach hundreds of megahertz. Some newer function generators can go from zero to around 60 MHz (see Figure 2-16).

Frequency Counter

A *frequency counter* does just that: count the frequency of a signal (see Figure 2-17). It does so by opening a gate for a precise period of time and counting how many cycles of the signal get through before the gate closes again. A counter is most useful when the frequency of a circuit's oscillator needs to be adjusted accurately. This is rarely the case with digital devices like cameras and computers, but it can be critical when calibrating the master oscillators that control the tuning of radio receivers and transmitters.

If you're considering getting a counter, look more at its low-frequency capabilities than at the high end, unless you plan to work on UHF (ultra-high-frequency) or microwave systems. Many of the inexpensive counters that can hit 1 gigahertz (billions of cycles per second) are optimized for radio work and have gate times too short to count audio frequencies accurately. (The slower the signal frequency to be counted, the longer the gate has to stay open to let enough cycles through for a proper count.) Oh, and counters are another product category, like DMMs, that may display lots of meaningless digits. Especially with the cheapies, there can be a long string of numbers to be taken with a significant grain of sodium chloride because the counters' master oscillators aren't accurate or stable enough to support the displayed resolution.

FIGURE 2-17 Frequency counter.

Frequency counters are capable of counting regular, continuous signals only; they're useless with complex, changing ones. To extract frequency information from those, you need . . . yup, a scope. See, I told you that darned scope was your friend!

Speaking of scopes, you'll need an extra scope probe for your counter unless you want to share one with the scope. Get one with a 10X/1X switch, as that is especially handy for counter use.

GPS-Disciplined Oscillator

Once an exotic laboratory instrument costing thousands, a *GPS-disciplined oscillator*, or *GPSDO*, can be had for around $100 these days (see Figure 2-18). It's a specialized receiver that picks up GPS signals and outputs a 10-MHz oscillator signal time-locked to the ultra-precise frequency references normally used to calculate your position for navigation. Having a very precise and accurate frequency reference lets you calibrate your signal generator and frequency counter accurately and easily. Some generators and counters feature 10-MHz reference inputs so they can be driven by a GPSDO directly, with no calibration required.

Multi-component Tester

This is a fairly new category of test equipment. Small testers that can determine and measure various types of components have become inexpensive and popular (see Figure 2-19). You

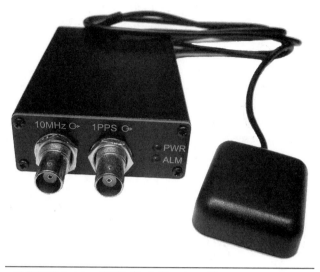

FIGURE 2-18   GPS-disciplined oscillator.

FIGURE 2-19   Multi-component tester.

hook up your part, and the tester tells you what it is, its characteristics and whether it is functional. Seems too good to be true, doesn't it?

Well, you know the old adage! My experience with these testers has been mixed. Simple items like resistors, capacitors and inductors (coils) are easy meat for these things, but transistors can get misidentified, leading to confusion and replacement of parts that don't need it. And, I had one tester that heated up certain kinds of diodes (two-lead semiconductors that let current flow in only one direction) so badly that I burned my fingers on them and even fried a few (diodes, that is, not fingers)! Gee, some tester, huh? "This part used to work, but now it's blown!" Uh, thanks for telling me.

Analog Meter

The moving meter needle of an old-fashioned analog VOM offers some info to the trained eye that a modern DMM can't (see Figure 2-2). Slowly fluctuating voltages, which you might encounter with, perhaps, a bad voltage regulator or a circuit pulling too much current, are easier to see with a VOM than with an oscilloscope. Little voltage dips or spikes cause a characteristic bounce of the needle that's very informative, too. You can even get a rough estimation of an electrolytic capacitor's condition with an analog *ohmmeter* (resistance checker) by watching the needle quickly rise and then slowly drop as the capacitor charges. With a DMM, such changes are just rapidly flashing numbers impossible to interpret.

A special type of VOM is known as a VTVM, for *vacuum-tube voltmeter*. A VTVM works like a VOM, but it contains an amplifier, making it considerably more sensitive to small signals and much less likely to steal meaningful amounts of current from the circuit under test, or *load it down*. VTVMs go back a long way, from before there were transistors, and early ones really did use vacuum tubes. Later models substituted the tubes with a very sensitive type of transistor called a *field-effect transistor* (FET). Those units were known as FET-VOMs, but most people continue to call any amplified analog meter a VTVM, whether it has a tube or not. If you can find a FET-VOM, you might want to snap it up because they're getting rare. True VTVMs are very, very old, and replacements for those small tubes are hard to find, but some working ones are still out there.

Decent VOMs, FET-VOMs and VTVMs occasionally turn up at hamfests. As long as it works, an old VOM is as good as a new one; there's not much in it to go wrong. If the meter needle moves without getting stuck and the selector switch works, you should be good to go. True tube-type VTVMs may run on AC power or on batteries, but the battery-operated units require cells nobody makes anymore, so avoid those meters. VOMs and FET-VOMs use batteries that may sit in them for years, so check inside the battery compartment to make sure an old cell hasn't leaked and corroded the contacts. Amplified meters (VTVMs and FET-VOMs) have a lot more in them than do VOMs, so it's best to test their functions before buying.

Many companies have made VOMs, but the best oldies were made by Simpson, which also made the best VTVMs. Some of those ancient Simpsons still go for real money, and they're worth it to those who still have a use for them, mostly for restoring antique products or working on guitar or other vacuum-tube amplifiers. You pretty much have to shoot those old meters when you don't want them anymore. Triplett was another company that made great meters.

Isolation Transformer

It used to be that most AC-powered gear had a *linear power supply*, which shifted the incoming voltage down to a lower one through a *transformer* operating at the 60-Hz line frequency. The transformer, an assembly of two or more coils of wire on an iron core, had no electrical connection between its input and output; the energy was transferred via a magnetic field. This arrangement helped with safety because it meant that the circuitry you might touch was not directly connected to the house wiring and thus couldn't find a path to complete a circuit through that most delicate of all resistors, you. Unfortunately, to provide substantial power required a big, heavy transformer.

Today's *switching power supplies*, or *switchers* (see Chapters 8 and 15), chop the incoming power into fast bursts to push lots of energy through a small, light transformer. This lets them provide lots of power efficiently in a small package. Everything from TVs to your phone's charger uses the switching technique now; the only places you'll still find linear power supplies are in especially low-noise applications like audiophile-grade stereo equipment and some radio receivers. Switchers are much more dangerous to work on because some of their circuitry is connected directly to the AC line, and it may have several hundred volts on it—and often it's the section that needs repair.

An isolation transformer is just a big, old-style AC line transformer into which you can plug your device (see Figure 2-20). The transformer has a 1:1 voltage ratio, so it doesn't change the power in any way, but it isolates it from the AC line, making service of switchers a lot safer. If you're going to work on switching power supplies while they're connected to the AC line, you *must* have an isolation transformer. Many times you can fix switchers while they're unpowered, so having an isolation transformer is optional. Just don't *ever* consider working on a live switcher without one. Seriously! You don't want the power supply to wind up being the only thing in the room that's live, if ya know what I'm saying.

A seemingly similar item called a *variac*, for variable AC, lets you turn a knob to adjust the AC output voltage going to the item you're repairing. Variacs were once common for servicing vacuum-tube gear, but they have little to no use for working on modern equipment. The most important thing to know about variacs is that they are *not* isolation transformers! The output of a variac is not isolated from the AC power line and provides no safety advantage over plugging directly into the wall socket. *Never, ever make the mistake of using a variac in place of an isolation transformer.* I'm putting that warning in italics for a reason!

There are isolation transformers that also provide adjustable or switch-selectable voltage, like the one on the left in Figure 2-20, but they will be clearly marked regarding the isolation. If it doesn't say it's isolated, assume it isn't!

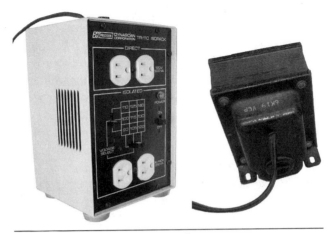

FIGURE 2-20 Isolation transformers.

Microscope

With electronics getting smaller and smaller, even a head-worn magnifier may not be enough for a comfortable view. More and more, techs are using microscopes to get a good, close look at solder pads on grain-of-salt-sized components. The pros have used them for decades. Stereo optical microscopes were the norm, but it was hard to solder under one because you had to hold your head still to keep your eyes properly placed over the eyepieces. Plus, decent microscopes were expensive.

These days, video microscopes are available for under $100 (see Figure 2-21). For electronics work, they're a huge boon. They come in two flavors: self-contained and USB for use with a computer, tablet or phone. Either will do the job, but I strongly recommend the self-contained style because it's much easier to solder under it when you're looking in the same direction you're working.

Their objective lenses are far enough from the circuit board that you can solder easily, with your face nowhere near the hot iron and the rosin smoke. The picture is sharp and can be zoomed for varying degrees of magnification. While soldering, you'll want it zoomed out a bit, but then you can get up very close to inspect the solder joint and make sure there aren't tiny solder bridges to other contact points that'll create short circuits, which can be disastrous when you apply power.

Many video microscopes can take still images and videos, saving them to an SD card. I've never found much use for the video feature, but being able to document my work with pictures is great, especially if I might repair the same model of product again sometime. Or, perhaps, if I get called away and need to remember what I was doing when I return to the project three days later.

FIGURE 2-21   Video microscope.

LCD Separator

The hardest part about servicing phones and tablets is that the screens of many of them are glued on, and you have to remove the screen to get at the guts. It's very easy to destroy the LCD when trying to get inside one of these things to replace the battery or swap out a shattered digitizer (the touch-sensitive layer on top of the LCD). LCD separators do away with the risky heat gun approach to LCD removal. They offer a uniform heating pad surface and carefully controlled temperatures. Best of all, they're inexpensive, typically around $50. If you plan to work on phones and tablets, an LCD separator is a wise investment.

Bench Vise

Wouldn't it be wonderful to be gifted with three hands? If you have only the standard two, you may find it difficult to hold a circuit board while pulling a component lead from one side and heating the solder pad on the other. Check out the PanaVise and similar small vises designed for electronics work. Some offer attachable arms perfect for gripping the edges of circuit boards, and they let you swivel the board to whatever angle you need (see Figure 2-22).

Vacuum Pump Desoldering Iron

These are like integrated bulb desoldering units, except that the suction is provided by a vacuum pump instead of a bulb (see Figure 2-23). The pros use these, and they're fast and powerful, but they're expensive. If you can snag a good used one at a hamfest, go for it. Just be sure the heating and vacuum systems work properly. The vacuum portions are prone to air leakage and worn parts because molten solder flows through them.

FIGURE 2-22 Circuit board vise.

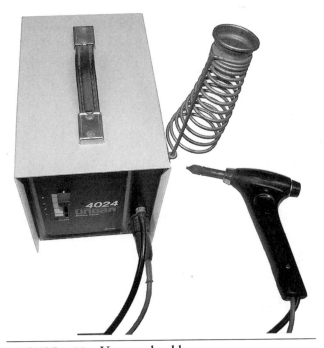

FIGURE 2-23 Vacuum desolderer.

Hot-Air Rework Station

You wouldn't think blowing hot air on solder could heat it enough to melt it, but it does, and fast! These stations can be used for both soldering and desoldering, getting the heat into tight spots hard to reach with the tip of an iron. The stations usually come with a few metal tips that fit around various sizes of integrated circuits (chips) so you can heat all their leads at once (see Figure 2-24). They get the components rather hot, so there is some risk of ruining the parts you're removing. Hot-air rework stations are great for removing surface-mount chips with lots of legs and for installing new parts with *solder paste*.

Soldering Tweezers

As shown in Figure 2-25, soldering tweezers are two soldering irons arranged as a set of tweezers for use in soldering and desoldering surface-mount devices, also called *SMD* components. These are the tiny, leadless parts that make up most of today's electronics. We'll get into them in Chapter 7.

There are two varieties of soldering tweezers: as an attachment to an existing soldering station, in place of the normal handle, and as a stand-alone soldering implement. Either way, the tweezers allow you to heat both sides of a part simultaneously, which is especially handy when removing one from a circuit board. As you'll find out in Chapter 7, though, SMD components keep getting smaller and smaller, and many are too tiny now to be

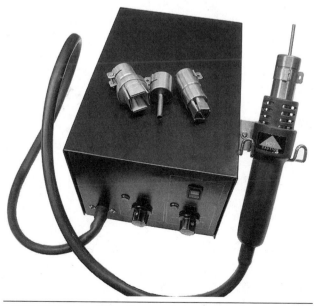

FIGURE 2-24 Hot-air rework station with tips.

FIGURE 2-25 Soldering tweezers.

manipulated with soldering tweezers. Still, this variation on the classic soldering iron can be quite useful with larger SMD parts.

Solder Paste

Once upon a not-so-distant time, solder was a wire wound on a spool. It still is, but solder paste has become the newer, easier way to install tiny SMD components. It melts at low temperature and is ideal for use with hot air. One of the paste's best features is how it clings only to the desired connection points, greatly reducing the chance of those potentially ruinous solder bridges on chips with leads too close together for manual, one-by-one soldering. We'll take a closer look at using paste in Chapter 12.

ChipQuik

This is a special low-temperature indium-silver alloy solder and flux kit used for desoldering surface-mount parts, especially with hot-air rework stations. When melted into the existing solder, the alloy keeps it molten at low temperatures, allowing you to get lots of pins hot enough simultaneously to remove even high-density chips with dozens of leads. After use, it must be removed, and solder paste or regular solder must be used to attach the new part.

Hot-Melt Glue Gun

A small glue gun can help you repair broken cabinet parts, and a dab of hot-melt glue is also great for holding wires down. Manufacturers sometimes use it for that, and you may have to remove the glue globs to do your work. Afterward, you'll want to replace the missing globs.

Magnet on a Stick

The first time you drop a tiny screw deep into a repair project and it won't shake out, you'll be glad you bought this tool. It's useful for pulling loosened screws out of recessed holes, too. Get one that telescopes open like a rod antenna. Just keep the darned tip away from hard drives, tape heads, video head drums and anything else that could be affected by a strong magnetic field. Also, keep in mind that the metal rod could contact voltage, so never use the tool with power applied, not even if the device is turned off. Charged electrolytic capacitors can impart a jolt after power is disconnected, so keep away from their terminals too. See Chapters 3 and 7 for more about capacitors.

Glue

Also known by the trademarks Super Glue and Krazy Glue, cyanoacrylate adhesive can be useful on some plastic and metal parts. It sticks things together pretty well, but it has poor shear strength and shouldn't be used for repair of mechanical parts that bear stress, such as plastic gears, because the repair will not last long. Some brands are a lot better than others. The Gorilla brand is thick and much stronger than the watery stuff you'll find at the dollar store. Just be aware that cyanoacrylate outgases a white film as it hardens that is tough to remove, so keep it away from lenses, display screens and other surfaces where this might be a problem. The outgassing is really rough on your eyes and lungs, too, so keep your face at a reasonable distance while working with this kind of glue.

Other glues worth keeping around are ordinary household cement, which can be stronger on plastic than cyanoacrylate, and good ol' epoxy. No glue is truly permanent, but epoxy comes closest. I've seen it last for decades, especially the type that takes overnight to set. The five-minute variety is considerably less strong and more likely to flake off after a while.

Component Cooler Spray

This stuff is colder than a witch's, um, iced tea, and it's used to put the deep freeze on suspected intermittent components, especially semiconductors (diodes, transistors and integrated-circuit chips). While it might seem primitive to blast parts instead of scoping their signals, doing so can save you hours of fruitless poking around when circuits wig out only after they warm up. One good spritz will drop the component's temperature by 50 degrees or more and can reveal a thermal intermittent instantly, returning the circuit to proper operation for a few moments until it heats up again.

Parts Assortment

In Chapter 7, we'll explore common components in depth. Having a supply of such parts is quite handy. You can strip old boards for parts, but it's time-consuming, and you wind up with very short leads that may be difficult to solder to another board. Plus, your stash will be hit or miss, with big gaps in parts values. Consider buying prepackaged assortments of small resistors and capacitors or going to a hamfest and stocking up for much less money. There, you're likely to find big bags of capacitors, transistors, chips, resistors, and so on for pennies on the dollar. Avoid buying transistors and chips with oddball part numbers you don't recognize, because they may be *house numbers*, which are internal numbering schemes used by equipment manufacturers. Those numbers are proprietary, and there's no way to determine what the original type number was. Thus, you can't look up the parts' characteristics, making them useless for repair work. Resistors and capacitors, luckily, almost always have standardized markings, and you can easily measure them if you have the appropriate meters.

Don't bother buying ceramic disc capacitors, because they almost never go bad, so you aren't likely to need any. If you do run into a suspicious one, you can pull its replacement off a scrap circuit board. Instead, focus on resistors, transistors, voltage regulators, fuses and, especially, electrolytic capacitors. You will be replacing lots of electrolytics. If you do too much repair work, you might start having nightmares in which you're chased around by hordes of swollen 'lytics screaming, "Change me! Change me!"

Electrolytic capacitors are the cylindrical ones with plastic sleeves around them and markings like "10μF 25VDC" (see Figure 7-7). There are several varieties of these things, but some are pretty common, and substitution of similar but not identical parts is feasible in many instances. (We'll explore how to do it in Chapter 12.) Get an assortment of caps in the range of 1 to 1,000 microfarads, with voltage ratings of 35 volts or more. The higher the ratings, both capacitive and voltage, the larger the cap will be. While it's fine in most cases to replace a cap with one of a higher voltage rating, get some rated at lower voltages too, in case there isn't room on the board for the bigger part.

Stocking up on transistors is tricky because there are thousands of types. Small-signal transistors, which don't handle a lot of current, are not hard to substitute, but power transistors, used in output stages of audio amplifiers, power supplies and other high-current circuits, present many challenges, and they're usually the ones that need replacement. Still, small transistors are very cheap—in the range of 5 to 25 cents each—and it's worth having some around. Get some 2N2222A or equivalent, along with some 2N3906. You may find hamfest bags of parts with similar numbers that start with "MPS" or other headers. If the number portion is 3906 or 2222, it's pretty much the same part and will do fine.

Diodes and rectifiers are frequent repair issues, so it pays to have some on hand. These components have two leads and permit current to pass only in one direction. They're used for many purposes, but the ones that handle a lot of current and are most likely to fail are found in power supplies, where they convert the incoming AC power to direct current (DC). The only difference between a diode and a rectifier is how much power it handles. Small-signal parts are called *diodes*, and larger ones made for use in power supply applications are dubbed *rectifiers*. Look for 1N4148 and 1N914 for the small fry and 1N4001 through 1N4007 for the big guns. Special high-speed rectifiers are used in switching power supplies, but there are too many types to preorder them. If you find a bad one, it's best to buy it then, replacing the dead part with an exact duplicate or verified substitute.

The *bridge rectifier*, which integrates four rectifiers connected in a diamond configuration into one plastic block with four leads, is commonly used in both linear and switching power supplies to save cost and space over placing four rectifiers on the board. Like other rectifiers, it's a power-handling part that fails fairly often. Get a few with current ratings in the 1- to 5-amp range and voltage ratings of 150 to 450 volts PIV (peak inverse voltage, which is how much voltage it can stand in the blocked direction before it breaks down). Bridges are all low-speed types.

To house components, most of us use those metal cabinets with the little plastic drawers sold at hardware stores. Sort resistors and capacitors by value. If you have too many values for the number of compartments in the drawers, arrange the parts into ranges. For example, one compartment can hold resistors from 0 to 1 kΩ, or kilohm (thousand ohms), while the next might contain those from just above 1 kΩ to 10 kΩ. Once you learn to read the color code (see Chapter 7), plucking the part you need from its drawermates is easy.

Scrap Boards for Parts

Despite what I said about stripping old boards, you *do* want to collect carcasses for parts. No matter how large your components inventory is, the one you need is always the one you ain't got! An old VCR or radio can provide a wealth of goodies, some of which are not easily obtained at parts houses, especially at 11:30 p.m. on a Sunday, when your hours of devoted sleuthing have finally unearthed the problem—at least you think so—and you would sell parts of your anatomy for that one darned transistor, just to see if it really brings your patient back to life.

If you have room, it's easy to pile up dead gadgetry until your spouse, conscience or neighbors intervene. Because they won't fit anything beyond the models for which they were made, you're highly unlikely ever to need cabinet parts, so saving entire machines is somewhat pointless and inefficient. The better approach is to remove circuit boards, knobs and anything else that looks useful, and scrap the rest. Don't bother stripping the boards; just desolder and pull off a part when you need it. If the leads are too short, solder on longer ones.

Wish List

For most service work, you can easily live without the following items, but they make for good drooling. Some advanced servicing requires them, but most of your jobs won't.

Inductance Meter

This meter reads the inductance value of coils, which is a measure of how they oppose AC currents because of the magnetic fields they create (see Chapter 7). Having a meter for this seems like something quite useful, right? It's really rare, though, for a coil to change its inductance without failing altogether. Usually, the coil will open (cease being connected from end to end) from a melted spot in the wire as a result of too much current overheating the windings. On occasion, though, insulation between coil windings can break down and arc over, causing a short circuit between a few windings but leaving most of them intact. That will reduce the coil's inductance, possibly causing a circuit malfunction and making an inductance meter useful. To get any benefit from it, however, you need to know what the correct inductance should be. Unless the coil is marked, which many aren't, or you have a schematic of the product, there's no way to ascertain the part's original value unless you have a known good coil with which to compare the suspect one. This, and the fact that coils don't wear out and show gradually declining performance the way electrolytic capacitors do, accounts for the inductance meter's place on the wish list, while the capacitance meter is a little higher up the chain of desire, and the ESR meter is a must-have.

LCR Meter

A popular option these days is the *LCR meter*, which stands for "inductance, capacitance and resistance" (L is the symbol for inductance). Your DMM already reads resistance on its ohms setting, but the L and C functions of an LCR meter can save you the cost and space of having two separate instruments (see Figure 2-26).

FIGURE 2-26   LCR meter.

Logic Analyzer

An offshoot of the oscilloscope, a logic analyzer has lots of input channels but shows only whether signals are on or off. It is used to observe the timing relationships among multiple digital signal lines. Getting benefit from it requires knowledge of what those relationships should be, information rarely provided in the service manuals of consumer electronics devices. It is unlikely you'll ever need one of these unless you're restoring antique computers and know a great deal about coding.

Spectrum Analyzer

This is a special type of oscilloscope. Instead of plotting voltage versus time, a spec-an plots voltage versus frequency, letting you see how a signal occupies various parts of the frequency spectrum. There are two types: audio and radio-frequency (RF). The audio variety is nice for checking the performance of stereo gear, especially if you work on old stuff like tape recorders, which require careful adjustment to preserve their ability to reproduce the higher frequencies found in cymbals and vocal sibilants like esses. For hobbyist use, though, there's little point in buying an audio spectrum analyzer because phone and laptop apps can perform this function easily using the microphone input. Freeware audio spec-an apps abound. The results from an app on a phone are not as precise as what you'd get from an expensive service instrument, but they're adequate for typical repair work, such as seeing if an amplifier has significant high-frequency loss that might indicate a need for some new capacitors.

Used extensively in design and testing of radio transmitters, RF spec-ans are overkill for most service work unless RF gear is your thing. Ham radio operators covet these instruments, but you won't need one to fix normal consumer electronics devices. Besides, it is illegal to service transmitters in any way that could modify their spectral output unless you're a licensed amateur radio operator working on ham gear or you hold a radiotelephone license authorizing you to work on transmitters.

Until recently, an RF spectrum analyzer was a very expensive device, typically costing thousands of dollars. Now, though, there are pocket-sized units for less than $100 that work surprisingly well. They're not as fast or versatile as the fancy stuff, but they offer tremendous value for those working on radio equipment (see Figure 2-27).

BGA Rework Station

A BGA, or *ball-grid array*, is a very small, dense grid of ball-shaped solder contacts underneath high-density chips like computer CPUs and graphics processors. Figure 2-28 provides side and underneath views of a BGA chip. The solder balls attach to mating pads on the circuit board, as shown in Figure 2-29. Over time, expansion and contraction from the heating and cooling cycles of these hot-running parts breaks a connection or two, and

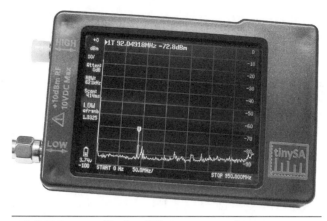

FIGURE 2-27 Pocket RF spectrum analyzer.

FIGURE 2-28 BGA chip side and underneath.

FIGURE 2-29 BGA board.

the product crashes. Their reliability has improved in recent years, but BGA problems have been a major cause of laptop, tablet, phone and video game machine failures.

Not long ago, BGA rework, called *reballing*, required around $50,000 in equipment and a fair amount of expertise. Now, there are kiosks in malls in some parts of the world performing while-you-wait BGA service on phones and tablets, and the technology has gradually filtered down to the hobbyist level. Simplified, home-level BGA rework stations can be had for around $500, with more professional versions costing $6,000 or so. These stations use infrared light to heat just the area required without desoldering nearby components.

If you work on high-density products like phones and tablets on a regular basis, it might be worth getting a BGA rework station. Most of us don't have them, though. The procedures for BGA rework are too involved to be covered in this book, but many of the stations come with tutorials. If you think you'd like to explore BGA repairs, go to youtube.com, search on "BGA rework" and see how it's done.

Chapter 3

Danger, Danger! Staying Safe

Before you start repairing electronics, get clear on one important fact: As soon as you crack open a product's case, you have left the government-regulated, "I'll sue you if this thing hurts me," coddled, protected world of consumer electronics behind. Once the cover comes off, you are on your own, and *you can get hurt or killed if you're not careful!* You've probably heard many times how dangerous CRT (picture-tube TV) sets were to service, but don't fool yourself into thinking that today's gear is all that much safer. True, modern consumer products don't have picture tubes with 40,000 volts on them (yes, color TVs really did!), but even some battery-operated devices step up the voltage enough to zap the living crud out of you.

That said, you can learn to navigate all kinds of repairs safely. Let's look at a few ways you could get injured and how to avoid it, followed by the inverse: how you might damage the product you're trying to service.

Electric Shock

This is the most obvious hazard and the easiest to let happen. It might seem simple to avoid touching live connection points, but such contact happens all the time because the insides of products are not designed for safety. Remember, you're not supposed to be in there! You may find completely bare, unprotected spots harboring dangerous voltage, and a slip of the tool can be serious.

Remove your wristwatch and jewelry before slipping your hand into a live electronic product. Yes, even a battery-operated one. Take off the wedding ring, too. They don't call metal contact points *terminals* for nothing!

In most devices, the electrical reference point called *circuit ground* is its metal chassis and/or metal shields. This is where old electrons go to die after having done their work, wending their way through the various components to get there. The trick is not to let them take you along for the ride! If you are in contact with the circuit ground point and also a point at a voltage higher than about 40 or 50 volts, you will get shocked. If your hands are moist, even lower voltages can zap you. The bodily harm from a shock arises more from the current (number of electrons) passing through you than from the voltage (their kick) itself. The higher the kick, though, the more electrons it forces through your body's resistance, which is why voltage matters. The path through your body is important as well, with the most dangerous being from hand to hand because the current will flow across your chest and through your heart. That, of course, is one electrically regulated muscle whose rhythm you don't want to interrupt. So, it's prudent to keep your hand away from the circuit ground when taking measurements, just in case your other hand touches some significant voltage. In the old CRT TV service days, techs lived by the "one hand rule," keeping one hand behind their backs while probing for signals in a powered set. If you're working on anything with even moderately high voltages, like a switching power supply, an audio receiver or an older LCD TV or monitor that has a fluorescent backlight, this is a procedure still worth following. Also, don't service electronics while you're barefoot or wearing socks; you're more likely to be grounded, offering a path through your body for wayward electrons. Always wear shoes.

Switching power supplies (see Chapter 15) have part of their circuitry directly connected to the AC house current line. As I mentioned in Chapter 2 in the section about isolation transformers, that's a very dangerous thing because lots of items around you in the room represent lovely ground points to which those unisolated electrons are just dying to go, and they don't mind going through you to get there. Once again, *never* work on circuitry while it is directly connected to the AC line. If there's no transformer between the AC line and the part of the circuit you wish to investigate, it's directly connected. Unplug it from the line even before connecting your scope's ground clip, because where you clip it may be at 120 volts or more, which will flow through the scope's chassis on its way to the instrument's ground connection, blowing fuses and possibly wrecking your scope.

Lots of AC-operated products have exposed power supplies, with no protection at all over the fuse and other items directly connected to the AC line. Touching one of those parts is no different than sticking a screwdriver in a wall socket. It's all too easy for the back of your hand to grant you a nasty surprise while your fingers and attention are aimed elsewhere. Even if the shock isn't serious (which it could be), you'll instinctively jerk your hand away and probably cut yourself on the machine's chassis. When probing a device with an exposed power supply, place something nonconductive over the board when you're not working on the supply itself. I like to use a piece of soft vinyl cut from the cover of a school notebook.

Capacitors, especially the large electrolytic types we discussed in Chapter 2, can store a serious amount of energy long after power has been removed. I've seen some that were still fully charged weeks later, though many circuits will bleed off their energy within seconds or minutes. The only way to be sure a cap is discharged is to discharge it yourself. *Never* do this by directly shorting its terminals! The current can be in the hundreds of amps, generating a huge spark and sometimes even welding your tool or wire to the terminals. Worse, that fast, furious flow can induce a gigantic current spike into the device's circuitry, silently destroying transistors and chips. Instead, connect a 10-ohm resistor (see Chapter 7) rated at a watt or two between a couple of clip leads, and clip the other ends across the capacitor's terminals to discharge the cap a little more slowly. Keep the clips connected for 20 seconds, and then remove one and measure across the cap with your DMM (digital multimeter) set to read DC voltage. It should read zero or close to it. If not, apply the resistor again until it does.

Before discharging a capacitor, look at its voltage rating. The voltage it could be charged to will always be less than the rating. If a cap is rated at 16 volts, it isn't going to be dangerous because it can't be storing anything higher than that. If it's rated at 150 volts, watch out. Even with the low-voltage part, you may want to discharge it before soldering or desoldering other components, to avoid causing momentary shorts that permit the cap's stored energy to flow into places it doesn't belong and damage something. Most of the time, low-voltage caps are in parts of circuits that cause the capacitors to discharge pretty rapidly once power is turned off, but not always.

The capacitance value tells you how much energy the capacitor can store. A 0.1-µF (microfarad, or millionth of a farad) cap can't store enough to cause you harm unless it's charged to a high voltage, but when you have tens, hundreds or thousands of microfarads of storage capacity, the potential for an electrifying experience is considerable at the lower voltages you're more likely to encounter.

CRTs (picture tubes) have gone the way of the dodo, so you probably won't work on them. Just be very, very careful if you get the urge to fix up an old TV set for, say, a video gaming system. Those tubes act like capacitors and have low enough leakage to store the high voltage applied to their anodes (the hole in the side with the rubber cap and the thick wire coming from it) for months. There isn't much capacitance—thus not a lot of current—but the voltage is so high that the stored charge can go through you fast and hard enough to cause a large, sad family gathering about a week later. The terminals at the back of the tube carry some pretty high voltages as well, enough to do you in.

The backlighting circuits of older, fluorescent-lit LCD monitors and TVs, along with much of the circuitry of plasma TVs, operate at high enough voltages to be treated with respect. It's unwise to try to measure the output of a running backlighting circuit at the point where it connects to the fluorescent lamp tube unless you have a high-voltage probe made for that kind of work. Without one, you may get shocked from the voltage exceeding the breakdown rating of the plastic handles of your probes, you're likely to damage your DMM or scope, and the added load caused by the unintended circuit path through you and your test gear might blow the backlighting circuit's output transistors.

Speaking of lamps, the high-pressure mercury vapor arc lamps used in many video projectors are "struck," or started, by putting around 1 kilovolt (1,000 volts) on them until they arc over, after which the voltage is reduced to about 100 volts. Keep clear of their connections during the striking period, and don't try to measure that startup voltage.

Physical Injury

The outsides of products are carefully designed to be user-safe. Not so the insides! It's easy to get sliced by component leads sticking up from solder joints, by the edges of metal shields, and even by plastic parts. Move deliberately and carefully; quickly shoving a hand into nooks and crannies leads to cuts, bleeding and cursing. That said, it still happens often enough that my years of tech work led me to coin the phrase, "No job is complete without a minor injury."

CD and DVD players and recorders (especially recorders) put out enough laser energy to harm your eyes, should you look into the beam. Video projectors use lamps so bright that you *will* seriously damage your vision by looking directly at them. The lamps and their housings get more than hot enough to burn you, and hot projection lamps are very fragile, so don't bounce the unit or hit anything against it while it's operating. An exploding lamp goes off like a little firecracker, oh-so-expensively showering you with fine glass particles and a little mercury, just for extra entertainment.

Speaking of eyes, yours will often be at rather close range to the work. Much of the time, you'll be wearing magnifying lenses offering some protection from flying bits of wire or splattered molten solder. When the magnifiers aren't in use, it's a good idea to wear goggles, especially if you don't wear glasses. Excess component leads clipped with diagonal cutters have an odd, almost magnetic tendency to head straight for your corneas at high speed. Plus, solder smoke gets drawn toward your face when you inhale, and it can be pretty irritating on the ol' peepers.

You can hurt your ears, too, particularly when working on audio amplifiers with speakers connected. Touching the wrong spot may produce a burst of hum or a squeal loud enough to do damage when your ears are close to the speakers. This sort of thing happens mostly with musical instrument amplifiers because their speakers are right in your face when you work on them, and those amps pack quite a wallop. Even a 15-watt guitar amp can get painfully loud up close. Don't think turning the volume knob down will protect you; there are plenty of places you can touch that will produce full power output regardless of the volume control's position.

Other opportunities for hearing damage involve using headphones to test malfunctioning audio gear. Even a phone or a little MP3 player with just a few milliwatts (thousandths of a watt) of output power can pump punishingly loud noises into your ears, particularly when ear buds that fit into the ear canals are used. If you must wear headphones to test a device, always use the over-the-ear type, and pull them back so they rest on the backs of your earlobes. That way, you can hear what's going on, but unexpected loud noises won't blast your eardrums.

Breathing in solder smoke, contact cleaner spray and other service chemicals isn't the healthiest activity. Keep your face away when spraying. When you must get close while soldering, holding your breath before the smoke rises can help you avoid inhalation.

Your Turn

Sure, electronics can hurt you, but you can hurt the equipment, too. Today's devices are generally more delicate than those of past decades. It was pretty hard to damage a vacuum-tube circuit with anything short of shattering the glass with a dropped wrench. Older solid-state gear was built at a large enough size scale to be serviceable comfortably, without too much risk of accidental short circuits or physical damage. Today's ultraminiaturized circuitry is an entirely different slab of silicon. Here are some ways you can make a mess of your intended repair.

Electrical Damage

Working with powered circuits is essential in many repair jobs. You can't look at signals on your scope when they're not there! Poking around in devices with power applied, though, presents a great opportunity to cause a short circuit, sending voltages to the wrong places and blowing semiconductors, many of which cannot withstand out-of-range voltages or currents for more than a fraction of a second. One of the easiest ways to trash a circuit is to press a probe against a solder pad on the board, only to have it slip off the curved surface and wind up touching two pads at the same time when you look up at your test instrument. Any time you stick a probe on a solder pad, be aware of this potential slip. At some point, it'll happen anyway, I promise you. Luckily, many times it causes no harm. Alas, sometimes the results are disastrous. If you experience this oops-atronic event and the circuit's behavior suddenly changes, and toggling the power doesn't restore it to its previous state, assume you did some damage.

Another common probing problem occurs when a scope probe is too large for where you're trying to poke it, and its ground ring, which is only a few millimeters from the tip, touches a pad on the board, shorting it to ground. Again, sometimes you get away with it, sometimes you don't. In small-signal circuit stages, it's more likely to be harmless. In a power supply, well, you don't want to do it, okay? It can be helpful in tight circumstances to cut a small square of electrical tape and poke the end of the probe through it, thus insulating the ground ring.

It's possible to cause electrical damage when taking measurements, even if you don't slip with the probe. Leaving your DMM set to read current when you mean to take a voltage measurement effectively places a short circuit across its leads, due to how current is measured (see Chapter 6). You connect one lead to the broken device's chassis, touch the other to, say, its power supply's output, and blam! Now you have new damage to chase down.

Now and then, you may want to connect a voltage to a point to see if it restores operation. That can be useful, but it requires consideration of the correct voltage and polarity, the amount of current required and exactly where that energy will go. Get one of those things wrong, and you could let some of that magic smoke out of the unit's components, with predictable consequences.

When your oscilloscope is set to AC coupling (see Chapter 6), it inserts a capacitor between the probe and the rest of the scope. After you probe a point with a DC voltage on it, that voltage remains on the capacitor and will be discharged back through the probe into the next point you touch. The amount of current is very, very small, but if you touch a connection to an especially sensitive IC (integrated circuit) chip or transistor that can't handle the stored voltage, you could destroy the part in the short time it takes to discharge the scope's capacitor. When using AC coupling, touch the probe to circuit ground between measurements to discharge the cap and prevent damage to delicate components. If the scope's trace (horizontal line) on the screen bounces, keep the probe touched to ground until the line settles back to where it started, at which point the capacitor will be fully discharged. It won't take more than a few seconds. You'll get a more accurate reading that way, as well as protecting sensitive circuits.

When operating a unit with your bench power supply, there are several things you can get wrong that will wreck the product. First and foremost, don't connect positive and negative backward! Nothing pops IC chips faster than reversed polarity. Products subjected to it are often damaged beyond repair.

Some devices, especially those intended for automotive use, have reverse-polarity protection diodes across their DC power inputs. The diode, deliberately connected backward with its anode to negative (−) and its cathode to positive (+) (see Chapter 7), doesn't conduct as long as the power is applied correctly. When polarity is reversed, the diode conducts, effectively shorting out the power input and usually blowing the power supply's fuse, protecting the product's sensitive transistors and chips from backward current. If the power supply has a lot of current available, the diode may rapidly overheat and short, requiring its replacement, but the rest of the unit should remain unharmed.

Few battery-operated products have protection diodes. Reverse-polarity protection is usually accomplished mechanically in the battery compartment by a recessed terminal design that prevents the battery's flat negative terminal from touching the positive contact. The AC adapter jack probably isn't protected either because it's assumed that you will use the adapter that came with the product.

So, be very careful to connect your power supply the right way around, and never hook it up while power is turned on, lest you even momentarily touch the terminals with your clips reversed. Be sure you've set your power supply's voltage correctly, too. Undervoltage rarely causes damage, but overvoltage is likely to do so if it's applied for more than a few seconds. We're not talking millivolts here; if you're within a quarter of a volt or so, that's usually good enough. A decade ago, most items ran on linear-type adapters that were nothing more than transformers and rectifiers. The voltage regulation was done inside the products, so they were fairly tolerant of having excessive voltage coming in, and you were fine if you were within 2 or 3 volts. These days, most products use regulated switching-type AC adapters with very steady voltage outputs, so the gadgets expect a pretty accurate voltage.

Automotive products are a special category. In many of them, the reverse-polarity protection diode is a *zener diode*. These special diodes act like regular diodes, except that they also begin to conduct in their reverse direction when a specific voltage, called the *zener voltage*, is reached (see Chapter 7). Just like all reverse-polarity protection diodes, they're connected backward, with cathode to positive. So, they actually start conducting in the *circuit's* forward direction above the preset voltage. The idea is to keep alternator spikes and overvoltage in a car's electrical system from damaging the device's circuitry by making them pass through the diode instead. Because the zener diode also conducts in the circuit's reverse direction, just like a regular protection diode, it guards against reversed polarity as well.

If you turn your power supply's voltage up too high, the zener diode will start conducting, resulting in excessive current through it that will overtax your power supply. It may heat up and destroy the diode, causing a short circuit across the supply. In that case, you'll have to replace or at least disconnect the diode before you can resume repairing the device. (If you disconnect the ruined diode, be sure to replace it before completing the repair!)

Most zener diodes used in this manner start conducting at 15 or 16 volts. When servicing products made for car use, keep your power supply as close to 13.8 volts as possible, as it's the correct voltage of a car's electrical system when the engine is running. Virtually all modern automotive electronic products are designed for this voltage.

A static charge from rubbing your shoes across the carpet, or just the dry air of winter, can put hundreds to thousands of volts on your fingertips and any tools you're holding. If you think you could be charged, and especially any time you handle CMOS (complementary metal-oxide-semiconductor) chips, MOSFETs (metal-oxide-semiconductor field-effect transistors), memory cards or other sensitive semiconductors, touch a grounded object first. The metal case of your bench power supply should do the trick as long as it's plugged into a three-wire outlet. I don't recommend using a scope as a discharge point, because it has plenty of sensitive chips inside, and you sure don't want to damage those!

Physical Damage

There are lots of ways you can break things when you're inside a machine. To replace bundles of wiring that would be way too big to fit in the case, modern products use ribbon cables to connect various circuit boards and control panels together (see Figure 3-1). These are thin, flat, flexible items with printed conductors lined up next to each other. They are specific to the products they're in, and they're delicate!

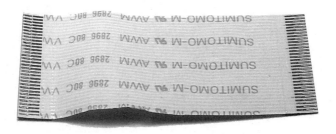

FIGURE 3-1   Ribbon cable.

One of the easiest ways to ruin a device is to tear a ribbon cable while disassembling the unit. Some products pop apart easily but may have hidden risks. I once serviced a video projector cleverly designed to snap open without a single screw, but a tiny ribbon cable linked the top and bottom, and I was lucky that it popped out of its connector without being torn in half when I removed the top case. Had I pulled a little harder, I'd probably have wrecked the projector. As careful as I was trying to be, I still didn't see that darned ribbon until it was too late.

Small connectors of the sorts used on laptop motherboards, phones, tablets and digital cameras can be torn from the circuit board. Today's products are soldered by machine, and the soldering to connectors isn't always the greatest because they are a bit larger than the components, so they don't get quite as warm during soldering. A little too much pressure when you disconnect the cable, and the connector can come right off the board. Depending on the size scale of its solder pads, it may be rather difficult to reattach it. Most ribbon connectors have a release latch you must flip up or pull out before removing the ribbon. Always look for it before pulling on the cable (see Figures 3-2 and 3-3).

Your soldering iron, that magic instrument of thermo-healing, can also do a lot of damage, especially to plastic. The sides of the heating element can easily press against plastic cabinetry when you have to solder in tight places, melting it and ruining the unit's cosmetics.

Finally, be careful where you press your fingers. Most circuitry is fairly hardy, but some components, including video heads, meters, speaker cones, microphone diaphragms, DLP

Latch

FIGURE 3-2   Ribbon connector with flip-up latch.

Latch

FIGURE 3-3   **Ribbon connector with slide-out latch.**

projector color wheels and CD/DVD laser optical heads, just can't stand much stress and
will break if you push on them even with moderate force.

You Fixed It! Is It Safe?

After repair, it's your duty as a diligent device doctor to ensure the product is safe to use.
One common error resulting in an unsafe repair job is neglecting to put everything back
the way it was. If you have internal shields and covers or other items left over after you close
the unit, you'll need to open it back up and put them where they belong. Manufacturers
don't waste a single penny on unnecessary parts, so you know they're important!

It's easy to touch ribbon cables or wires with the side of your soldering iron and melt
insulation while you're concentrating on soldering components. In units with lots of wires
or ribbons, this can happen despite your best intention to be careful. You might not notice
doing it if you're focusing on the action at the tip, but the smoke and smell will alert you.
Should you do this, fix the damage immediately; don't wait until the repair is over. For
one thing, you may have created a short or a lack of insulation that could cause damage or
injury when power is applied. For another, you might forget later and close the unit up in
that condition.

Patching melted insulation can be as easy as remelting it to cover a wire in the case
of low-voltage, signal-carrying wires with only small melted spots. Or, it might require
cutting, splicing and heat-shrink tubing if the wire handles serious voltage or the damage
is too great. If you didn't damage the printed conductors of a ribbon cable, you can cover
exposed copper with electrical tape or some glue. Remember, tape will come off after a
while, so never depend on it for long-term safety.

When the product has an AC cord, take a look at it and run your hand along its entire length, checking for cuts. Naturally, you want to live, so unplug it before doing this! You'll be amazed at how many frayed, cut and pet-chewed cords are out there. Replace the cord or repair it as seems appropriate for its condition, paying extra attention to a good, clean job with proper insulation. If the damage is only to one wire, it's easier to fix than if both sides are involved because at least the two wires can't short to each other. With a damaged AC cord, I like to use both electrical tape and heat-shrink tubing over it.

ATE

Did you ATE? No, I didn't just lose my grammatical mind, and I'm not suggesting you go get a snack! ATE stands for *Always Test Everything*. Have you tested everything? You put in a new battery holder, and obviously that has to work, so why bother testing it, right? Not right! Check it or you might wind up with bite marks on your left butt cheek when you can't figure out why the unit no longer powers up. Even the simplest things have a way of not working when you don't test them.

This is even more of an issue with repairs that aren't so simple. For example, if you've replaced power-handling components like output transistors or voltage regulators, be sure to test the unit for proper operation and excess heat. Stereo amps and receivers, for instance, sometimes require *bias* adjustments when the output transistors are changed. These adjustments, found in the product's service manual, determine how much current flows through the transistors when no signal is being amplified. If you don't set the bias correctly, the unit will work for a while, but it might overheat badly, and the transistors may fail. Set the adjustments as the manual specifies, and then let the unit run on the bench for a few hours at normal listening volume to see how hot it gets. Be sure what you've fixed is really working properly before you close it up. ATE. Seriously. Everything.

We'll revisit ATE periodically throughout this book. It's that important.

Chapter 4

I Fix, Therefore I Am: The Philosophy of Troubleshooting

Imagine if your doctor saw you as a collection of organs, nerves and bones, never considering the synergistic result of their working together, supplying each other with the chemicals and signals necessary for life. No organ could survive on its own, but together they make a living, breathing, occasionally snoring you! Now consider how tough it'd be to solve a murder case without considering the motives, personalities and circumstances of the victim and all potential suspects. The knife is right there next to the body, but anybody on earth could have done the crime. Why was the victim killed? Who knew him? Who might have wanted him dead?

Troubleshooting, which involves skills somewhat like those of doctors and detectives, is a lot like that. You can think of an electronic device as a bunch of transistors, chips and capacitors stuffed into a box, and sometimes that's enough to find simple failures. Taking such a myopic view, though, limits you to being a mediocre technician, one who will be stumped when the problem isn't obvious. To be a top-notch tech requires consideration of the bigger picture. Who made this product, and what were the design goals? How is it supposed to work? How do various sections interact, and what is the likely result of a failure of one area on another?

Like bodies, machines are *systems*. Being built by humans, they naturally reflect our biological origins, with cameras for eyes, microphones for ears, speakers for larynxes and microprocessors for brains. Even the names of many parts sound like us: hard drives and optical disc players have heads, turntables have arms, chips have legs, memory card connectors have fingers and picture tubes had necks. Some products, especially those run by software, exhibit personalities, or at least it can seem that way. Their features and quirks can be irritating, humorous or soothing. Their failures are much like our own, too, with symptoms that may be far removed from what's causing them, thanks to some obscure interaction that nobody, not even the circuit's designer, could have foreseen.

The more you come to understand how devices work at the macro level, the more sense their problems will make. The more you can consider products as metal and silicon expressions of human thinking, the better sleuthing skills you will attain. Before we get to the nitty-gritty of transistors, current flow and signals, let's put on our philosophers' hats and become the Socrates of circuitry, the Erasmus of electrons. Let's look at why products work and why they don't, and how to avoid some of the common pitfalls developing techs encounter. Let's become one with the machines.

Why Things Work in the First Place

When you get a few thousand parts together and apply power to them, they can interact in many ways. The designing engineer had one particular way in mind, but that doesn't mean the confounded conglomeration of components will cooperate!

Analog circuitry has a wider range of variation in its behaviors than does digital, but even today's all-digital gear can be surprisingly inconsistent. I've witnessed two identical laptop computers running exactly the same software, with exactly the same settings, but drawing significantly different amounts of current from their power supplies. I've also seen all kinds of minor variations in color quality between identical digital still and video cameras. I remember a ham radio transceiver whose digital control system exhibited a bizarre, obscure fault in its memory storage operation that no other radio of that model was reported to have, and I never found any bad parts that might explain the symptom. I finally had to modify the radio to get it to work like all the others.

Sure, you string a few digital logic gates together and you will be able to predict their every state. Get a few thousand or more going, run them millions of times per second, and mysterious behaviors can start to crop up. At the level of today's computing products, with tens of millions of microscopic transistors switching on and off, and billions of memory cells, the potential for unpredictability is staggering. Really, it's amazing that they work as intended most of the time.

It's useful to think of all circuitry as a collection of resistors (opposers of electricity) impeding the passage of current from the power supply terminal to circuit ground. As the current trickles through them, it is used to do work, be it switching the gates in a microprocessor, generating laser light for a disc player or projecting your favorite movie on the wall. Electrons, though, are little devils that will go anywhere they can. If there's a path, they'll find it. Malfunctions can be considered either as paths that shouldn't be there or a lack of paths that should.

In essence, when machines work properly, it's because they *have no choice*. The designer has carefully considered all the possible paths and correctly engineered the circuit to keep those pesky electrons moving along only where and when they should, locking out all possible behaviors except the desired one. When choice arises, through failing components, user-inflicted damage or design errors, the electrons go on a spree like college students at spring break, and the unit lands on your workbench.

Products as Art

A machine is an extension of its designer, much as a concerto is an extension of its composer. Beethoven sounds like Beethoven, and never like Rachmaninoff, because Ludwig's bag of tricks and way of thinking were uniquely his, right? It's much the same with products. In this case, however, they tend to have unifying characteristics more reflective of their manufacturing companies than of a specific person. Still, I suspect that an individual engineer's or manager's viewpoints and preferences set the standard, good or bad, that lives on in a company's product line long after that employee's retirement.

Understanding that companies have divergent design philosophies and quirks may help your repair work because you can keep an eye on issues that tend to crop up in different manufacturers' machines. You may notice that digital cameras from one maker have a high rate of imaging chip failures, so you'll go looking for that instead of some other related problem when a troublesome case hits your bench. Or perhaps you've found that tablets from a particular company often develop intermittent charging ports because how they mount them to the circuit board permits too much stress on their solder joints when the plug is inserted and removed, eventually cracking the joints.

When you've fixed enough products, you'll begin to recognize what company made a machine just by looking at its circuit board or mechanical sections. The layouts, the styles of capacitors, the connectors and even the overall look of the copper traces on a board are different and consistent enough to be dead giveaways.

If It Only Had a Brain

Continuing our anatomical analogy, products from the analog and early digital ages were like zombies. Perhaps they had an ear (microphone), some memory (recording tape or a memory card) and a mouth (speaker). Each system did its simple job, with support from a stomach (power supply) and some muscles (motors, amplifiers).

What was missing was much in the brain department. Today's gear is cranium-heavy, laden with computing power. Even 20 years ago, simple mechanical linkages to control sequencing and movement of mechanisms were disappearing. Instead, individual actuators moved parts in a sequence determined by software, positional information got fed back to the microprocessor, and malfunctions might originate in the mechanics, the sensors, the software or some subtle interaction of those elements. Today's designs further that trend and take it to the extreme. You almost certainly won't find potentiometers (variable resistors) to set volume or brightness on a TV; buttons or menu options signal the brain to change the parameters. Heck, most gadgets today don't even have "hard" on/off switches that actually disconnect power from the circuitry. Instead, the power button does nothing more than send a signal to the microprocessor, requesting it to energize or shut down the product's circuitry. This has implications that work against longevity because parts of the product are always on, which wears them out. It's especially the case with TVs and projectors. If you've ever wondered why the old CRT TVs lasted decades, yet you're lucky to get 5 years from a modern LCD set, that's a big factor.

In addition to the brain, modern products have nervous systems consisting of intermediary chips and transistors to decode the microprocessor's commands and fan them out to the various muscles and organs doing the actual work. Failures in these areas can be tough to trace because their incoming signals from the computer chip depend on tricky timing relationships between various signal lines. This is a profound shift from the old way of building devices, and it adds new layers of complication to repair work. Is some part of the circuit not working due to its own malfunction, or is it playing dead because the micro didn't wake it up?

Today's machines are complete electrono-beings with pretty complex heads on their shoulders. Some offer updatable software, while many have the coding hardwired into their chips. Which would you like to be today: surgeon or psychiatrist?

The Good, the Bad and the Sloppy

It's easy for an experienced tech to tell when a repair attempt has been made by an unqualified person. Screws will be stripped, or there will be poorly soldered joints with splashes of dripped solder lying across pads on the board. Wires may be spliced with no solder and, perhaps, covered in cellophane tape, if at all. Adjustments will be turned, insulation melted, and so on. In a word: sloppiness.

That might sound exaggerated, but I used to run into it a lot when I worked in repair facilities. Most shops have policies of refusing to work on items mangled by amateurs, so discovery of obvious, inept tampering was followed by a phone call to the item's owner, who would stubbornly insist that the unit had never been apart and had simply quit working. Um, right, Sony used Scotch tape to join unsoldered wires. Sure, buddy. I remember one incident in which I refused to repair a badly damaged and obviously tampered-with shortwave radio. The owner was so angry that he called my boss and tried to have me fired! The boss took one look inside the set, clapped me on the back, laughed and told the guy to come pick up his ruined radio and go away. Don'tcha wish all bosses were that great?

The key to performing a proper, professional-quality repair job is meticulous attention to detail. Think of yourself as a surgeon, for that's exactly what you are. You are about to open up the body of this electronic "organism" and attempt to right its ills. As the medical saying goes, "First, do no harm." Now and then, repair jobs go awry and machines get ruined—it happens even to the best techs once in a while—but your aim is to get in and back out as cleanly as possible. In Chapters 9 through 13, we'll explore the steps and techniques required for proper disassembly, repair and reassembly.

Mistakes Beginners Make

Beyond sloppy work, beginners tend to make a few conceptual errors, leading to lots of lost time, internal damage to products, and failure to find and fix the problem. Here are some common quagmires to avoid.

Adjusting to Cover the Real Trouble

Analog devices often had adjustments to keep their circuit stages producing signals with the characteristics required for the other stages to do their jobs properly. Once, TVs and radios were full of trimpots (variable resistors), trimcaps (variable capacitors) and tunable coils, and their interactions could be quite complex. With today's overwhelmingly digital circuits, adjustments are much less common. Many are performed in software with special programming devices to which you won't have access, or via hidden service menus, but some good, old-fashioned screwdriver-adjustable parts still exist. Power supplies may have voltage adjustments, for instance, and earlier-generation CD players were loaded with servo adjustments to keep the laser beam properly focused and centered on the track. VCRs had lots of 'em too. Even a modern digital media receiver might have adjustments in its radio and power amplifier sections.

It can be very tempting to twiddle with adjustments or service menu settings in the hope that the device will return to normal operation. While it's true that circuits do go out of alignment—if they didn't, the controls and software tweaks wouldn't be there in the first place—that is a gradual process caused by age and temperature-related drifting of component values. It *never* causes sudden, drastic changes in performance. If the unit won't do something it did fine the day before, it's not out of adjustment, it's *broken*. Messing with the adjustments will only get you into trouble later on when you find the real problem, and now the product really is way out of alignment because you made it that way. Leave those internal controls and settings alone!

Change them only when you're certain everything else is working, and then only if you know precisely what they do and have a sure way to put them back the way they were, just in case you're wrong. Marking the positions of trimpots and trimcaps with a felt-tip marker before you turn them can help, but it's no guarantee you will be able to reset a control exactly to its original position. There's too much mechanical play in them for that technique to be reliable. In some cases, close is good enough. In others, slight misadjustments can seriously degrade circuit performance. At least with service menus, you can write down the values before changing them, and set things back to those values if needed.

I once worked on a pair of infrared cordless headphones with a weak, distorted right channel. After some testing, it was clear that the transmitter was the culprit, and its oscillator for that channel had drifted off frequency. A quick adjustment and, sure enough, the headphones worked fine for a little while. Then the symptom returned. The real problem: a voltage regulator that was drifting with temperature. Luckily, readjusting the oscillator was easy after the new part was installed. When multiple adjustments have been made, it can be exceedingly difficult to get them back in proper balance with each other.

Making the Data Fit the Theory

Most techs have been guilty of this at some time. In my early years, mea culpa, that's for sure. You look at the symptoms, and they seem to point to a clear diagnosis—all except for one. You fixate on those that make sense, convince yourself that they add up, and do your best to ignore that anomaly, hoping it's not significant. Trust me, it is, and you are about to embark on a long, frustrating hunting expedition leading to a dreary dead end. Always keep this in mind: *If a puzzle won't fit together, there's a piece missing!* There's something you don't know, and that is what you should be chasing. Often, the anomaly you're pushing aside is the real clue, and overlooking it is the worst mistake you can make. Many maddening hours later, when you finally do solve the mystery, you'll think to yourself, "Why didn't I consider how that odd symptom might be the key to the whole thing? It was right in front of me from the start!" Ah, hindsight. Nobody needs glasses for that.

Going Around in Circles

Sometimes you think you've found the problem, but trying to solve it creates new problems, so you go after those. Those lead to still more odd circuit behavior, so off you go, around and around until you're right back where you started. When addressing symptoms creates more symptoms, take it as a strong hint that you are on the wrong track. It's incredibly rare for multiple, unrelated breakdowns to occur. Almost always, there is one root cause of all the strangeness, and it'll make total sense once you find it. "Oh, the DVD player's power supply voltage was too low, and that's why the focus wouldn't lock, and the sled motor wouldn't make the laser head go looking for the track." If you're lucky, you'll have discovered that before you've spent hours fiddling with the limit switches and the control circuitry, tracing signals back to the microprocessor. Again, if the puzzle won't fit together, find that missing piece!

That's How It Goes

As with illness in the human body, just about anything can go wrong with an electronic device. Problems range from the obvious to the obscure; I've fixed machines in 5 minutes, and I've run across some oddball, week-eating cases for which a diagnosis of demonic possession seemed appropriate! These digital days, circuitry is much more reliable than in the old analog age, yet modern gear often has a much shorter life span. How can both of those statements be true?

Today's products are of tremendously greater complexity, with lots of components, interconnections and interactions, so there's more to go wrong. Unlike the hand-soldered boards filled with a wide variety of component types we used to have, today's small-signal boards, with their rows of surface-mounted, machine-soldered chips, don't fail that often. But with so much more going on, they include complicated power supplies and a multitude of connectors and ribbon cables. Plus, some parts work much harder than they used to and wear out or fail catastrophically from the stress. And thanks to the rapid pace of technological change, the competition to produce products at bare-bones prices and the high cost of repair versus replacement, extended longevity is not the design goal it once was. Manufacturers figure you'll want to buy a new, more advanced gadget in a couple of years anyway. Contrary to popular myth, nobody deliberately builds things to break. They don't have to; keeping affordable products working for long periods is tough enough. Keeping expensive items functioning isn't easy either! Laptop computers, some of the costlier gadgets around, are also some of the most failure-prone because they're very complex and densely packed, and they produce a fair amount of heat in a small package.

It may seem like electronic breakdowns are pretty random. Some part blows for reasons no one can fathom, and the unit just quits. That does happen, but it's not common. Oh, sure, when you make millions of chips, capacitors and transistors, a small number of flawed ones will slip through quality control, no matter how much testing you do. It's a tiny percentage, though. Much more often, products fail in a somewhat predictable pattern, with a cascading series of events stemming from well-recognized weaknesses inherent in certain types of components and construction techniques. Let's look at the factors behind most product failures.

Infant Mortality

This rather unpleasant term refers to that percentage of units destined to stop working very soon after being put into service. Imperfect solder joints, molecular-level flaws in semiconductors and design errors cause most of these. While many products are tested after construction, cost and time constraints prohibit extensive "burning in" of all but very expensive machines. Typical infant mortality cases crop up within a week or two of purchase and land in a warranty repair center after being returned for exchange. So, you may never see one unless you bought something from halfway around the world, and it's not worth the expense and trouble to return it. Or perhaps the seller refuses to accept it back, and you get stuck with a brand-new, dead device you want to resurrect.

Mechanical Wear

By far, moving parts break down more often than do electronic components. Hard drives, disc trays, laser head sleds and disc-spinning motors are all huge sources of trouble.

Bearings wear out, lubrication dries up, rubber belts stretch, switches get oxidized, nylon gears split, pet hairs bind motor shafts and good old wear and tear grind down just about anything that rubs or presses against anything else. If a device has moving parts and it turns on but doesn't work properly, look at those first before assuming the electronics behind them are faulty. Many new products have no moving parts beyond switches, but anything more than maybe 10 years old probably does. And, of course, DVD and Blu-Ray players, video projectors, hard drives, printers and scanners still have rotating elements and mechanisms.

Connections

Connections are also mechanical, and they go bad very, very often. Suspect any connection in which contacts are pressed against each other without being soldered. That category includes switches, relays, plugs, sockets, and ribbon cables and their connectors.

The primary culprit is corrosion of the contacts, caused by age, oxidation and sometimes, in the case of switches and relays, sparking when the contacts are opened and closed. Also, a type of lubricating grease used by some manufacturers on internal *leaf switches*, whose flexible metal contacts are used to sense the positions of mechanical parts like laser heads in disc players, tends to dry out over time and become an effective insulator. If the contact points on a leaf switch are black, it's a good bet they are coated with this stuff and are not passing any current when the switch closes (Figure 4-1).

A particularly nasty type of bad connection occurs in multilayer printed circuit boards. At one time, a dual-layer board, with copper traces (the printed equivalent of wiring) on both sides, was an exotic construct employed only in the highest-end products. Today, dual-layer boards are pretty much standard in larger, simpler devices, while smaller, more complex gadgets like phones, tablets and laptops may use as many as six layers!

The problems crop up in the connections *between* layers. Those connections, called *vias*, are constructed differently by various manufacturers. The best, most reliable style is with plated-through holes, in which copper plating joins the layers. As boards have shrunk, plated-through construction has gotten more difficult, resulting in a newer technique that

FIGURE 4-1   Leaf switch.

is, alas, far less reliable: holes filled with conductive glue. This type of via is recognizable by a raised bump at the connection point that looks like, well, a blob of glue (see the translucent glue over the holes in Figure 4-2). Conductive glue can fail from flexure of the board, excessive current and repeated temperature swings. Repairing bad glue vias is hard, too. I always cringe when I see those little blobs.

It's uncommon, but a shorted component can pull so much current from the power supply to ground that a copper trace melts! That's not a catastrophe when the trace is on the front or back of the board because you can go around it by soldering on a piece of wire. When the bad trace is part of an inside layer, though, you can't see it. You might find and repair such a failure but it's quite difficult. Typically, a melted inner trace means a ruined product.

Solder Joints

Though they're supposed to be molecularly bonded and should last indefinitely, solder joints frequently fail and develop resistance, impeding or stopping the current. When this happens in small-signal, cool-running circuitry, it's usually the fault of a flaw in the manufacturing process, even if it takes years to show up. Heat-generating components like output transistors, voltage regulators and graphics processing chips on computer motherboards can run hot enough to degrade their solder joints gradually without getting up to a temperature high enough to actually melt them. Over time, the damage gets done and the joints become resistive or intermittent.

Many bad solder joints are visually identifiable by their dull, mottled or cracked appearance. Now and then, though, you'll find one that looks perfect but still doesn't work, because the incomplete molecular bonding lies beneath the surface. Bonding may be poor

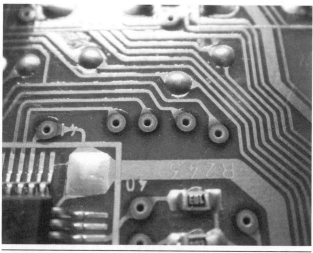

FIGURE 4-2   Conductive glue vias.

due to corrosion on the lead or pad of the soldered components; solder just won't flow into corroded or oxidized metal. When you go to resolder it, you'll have problems getting a good joint unless you scrape things clean first, after removing the old solder.

Heat Stress

Heat is the enemy of electronics. It's not an issue with most pocket-sized gadgets, but larger items like video projectors, TVs and audio amplifiers often fail from excessive heating. So do backlight inverters (the circuits that light the fluorescent lamps behind older LCD screens) and computer motherboards. Power supplies create a fair amount of heat and are especially prone to dying from it.

Overheating from excessive current due to a shorted component can quickly destroy semiconductors and resistors, but normal heat generated by using a properly functioning product can also gradually degrade electrolytic capacitors, those big ones used as power supply filtering elements, until they lose most of their capacitance or their ESR (equivalent series resistance) gets high enough to render them ineffective.

Electrical Stress

Running a device on too high a voltage can damage it in many ways. The unit's voltage regulator may overheat from dissipating all the extra power, especially if it's a *linear regulator*, which functions by acting like a resistor, cutting down the voltage by dissipating current as heat. Electrolytic capacitors can short out from being run too close to, or over, their voltage limits. Semiconductors with inherent voltage requirements may die very quickly.

Overvoltage can be applied by using the wrong AC adapter, a malfunctioning adapter, a bad voltage regulator, or using alkaline batteries in a device made for operation only with nickel–metal hydride (NiMH) rechargeable cells. Those cells produce 1.2 volts each, compared to the 1.5 volts of alkaline cells. So, with four cells, you get 6 volts with the alks compared to the 5 volts the device expects. Most circuits can handle it, but some can't. I've seen digital cameras that were quite fussy about that. Luckily, most new tech runs on lithium batteries. Those range from 4.2 volts per cell down to 3 volts as they discharge, but the products are built for that voltage range, or a multiple of it when there's more than one cell in the battery.

Believe it or not, a few products can be damaged by too *little* voltage. Devices with *switching power supplies* (see Chapters 8 and 15) compensate for the lower voltage by pushing more current through their transformers with wider pulses to keep the output voltage at its required level. This can cause overheating of the rectifiers and other parts converting the pulses back to regulated direct current.

The ultimate electrical stress is a lightning strike. A direct strike, as may occur to a TV or radio with an outdoor antenna that gets zapped, or a hit to the AC power line, will probably result in complete destruction of the product. Now and then, only one section is destroyed, and the rest survives, but don't bet on it. Lightning cases tend to be write-offs; you don't even want their remains in your stack of old boards, lest their surviving parts have internal damage limiting their life spans.

Power surges, in which the AC line's voltage rises to high levels only momentarily, can do plenty of damage. Such surges are sometimes the result of utility company errors, but more often lightning has struck nearby and induced the surge without actually hitting the line, or it has hit the line far away. Typically, the power supply section of the product is badly damaged, but the rest of the unit is unharmed.

When too much current passes through components, they overheat and can burn out, sometimes literally. Resistors get reduced to little shards of carbon, and transistors can exhibit cracks in their plastic cases. The innards, of course, are wiped out. This kind of stress rarely occurs from outside, because you can't force current through a circuit; that takes voltage. When overcurrent occurs, it's because some other component is short-circuiting to ground, pulling excessive current through whatever is connected between it and the power supply.

Nothing kills solid-state circuitry quite as fast as reversed polarity. Many semiconductors, and especially IC (integrated circuit) chips, can't handle current going the wrong way for more than a fraction of a second.

Batteries can be installed backward. Back when 9-volt batteries were the power source of choice for pocket gadgets, all it took was to touch the battery to the clip with the male and female contacts the wrong way around and the power switch turned on. Now that AAA cells and proprietary rechargeable batteries run our diminutive delights, that kind of error occurs less often, because it's routine for designers to shape battery compartments to prevent reversed contacts from touching, and for custom battery packs to fit in only one way, but it still happens on occasion when prudent design practices aren't followed.

By far, the most frequent cause of reversed polarity is an attempt to power a device from the wrong AC adapter. Today, most AC adapters connect positive to the center of their coaxial DC power plugs and negative to the outside, so an automotive cigarette lighter adapter made for the same gadget doesn't present the risk of having positive come in contact with the negative-ground metal car body, which would cause a short and blow the car's fuse. At one time, though, many adapters had negative on the center and positive on the outside instead, and a few still do on items that will never be used in cars. Even from the same manufacturer, both schemes may be employed on their various products.

The train wreck occurs when the user plugs in the wrong adapter, and it happens to have the plug wired opposite to what the device wants. Damage may be limited to only a few parts in the power supply section, or it can be extreme, taking out critical components like microprocessors and the chips that run LCDs.

Not all electrical stress is caused by external factors or random component failures. Sometimes, design errors are inherent in a product, and their resulting malfunctions don't start showing up until many units are in the field for a while. When a manufacturer begins getting lots of warranty repair claims for the same failure, the alarm bells go off, and a respectable company issues an ECO, or *engineering change order*, to amend the design.

Units brought in for repair get updated parts, correcting the problem. A really diligent manufacturer will extend free ECO repairs beyond the warranty period if it's clear that the design fault is bad enough to render all or most of the machines in the field inoperative, or if any danger to the user could be involved.

At least that's how it's supposed to work. Sometimes, companies don't want to spend the money to fix their mistakes, so they simply ignore or deny the problem. That's especially true with the low-cost electronics from obscure brands with which we're flooded today. Or, if only some machines exhibit the symptom, they're treated as random failures, even though they're not. Perhaps it takes a certain kind of use or sequence of operations for the issue to become evident, and the manufacturer genuinely believes the design is sound. And some units aren't used often enough to have experienced the failure, though they will eventually, masking its ultimate ubiquity.

Any of these situations can result in your working on a product with a problem that will recur, perhaps months later, after you've solved it properly. If the thing keeps coming back with the same issue, suspect a defective design. Later in this chapter, we'll look at several examples of them I've run into.

Physical Stress

Chips, transistors, resistors and capacitors can take the physical shock of being dropped, at least most of the time. Many other parts can't, though. Circuit boards can crack, especially near the edges and around screw holes and other support points. Larger parts, with their greater mass, can break the board areas around them. This happens often with transformers (sets of coils wound on an iron core) and big capacitors. On a single- or dual-layer board, you may be able to bridge foil traces over the crack with small pieces of wire and a little solder if the traces are not too small. With a multilayer board, you may as well toss the machine on the parts pile, because it's toast.

Liquid-crystal displays (LCDs), fluorescent tubes and other glass items rarely survive a drop to a hard surface. The very thin, long fluorescent lamps inside older, non-LED-lit laptop screens were particularly vulnerable to breakage. If you run across an old laptop with no backlight, don't be too surprised if it got dropped and the lamps are broken inside the screen. I've seen that happen with no damage to the LCD itself being evident. Newer laptops and virtually all tablets and phones are backlit with LEDs or generate their own light in each pixel, so this sort of failure affects only older products.

If you leave carbon-zinc or alkaline batteries installed long enough, they will leak. Not maybe, not sometimes—they *will*. Devices that take a fair amount of power get their batteries changed often, but those with low current demand, such as digital clocks, remote controls and some kids' toys, may have the same batteries left in them for years. Remotes, especially, are prime candidates for battery leakage damage because most people install the cheap, low-quality batteries that come with them and never change them; their very low current drain ensures those junky cells will be in there until they rot.

Once the goo comes out, you're in for a lot of work cleaning up the mess. They don't call them *alkaline* batteries for nothing! The electrolyte is quite corrosive and will eat the unit's battery springs and contacts. If the stuff gets inside and onto the circuit board, that's where the bigger calamity goes down. Copper traces will be eaten through, solder pads will corrode and those pesky circuit board layer interconnects will stop working. No shop will try to repair such damage, but you might want to give it a go if the device is expensive or hard to replace.

People sit on their phones and digital cameras fairly often, resulting in cracked LCDs, broken circuit boards and flattened metal cases shorting components to circuit ground. It's easy to bend an aluminum case back to an approximation of its original shape, but the mess inside may not be worth the trouble.

Liquids and electronics don't mix, yet people try to combine them all the time, spilling coffee, wine and soft drinks into their laptops and dropping their cameras and phones in the ocean and swimming pools. Good luck trying to save such items. Now and then, you can wash them out with distilled water, let them dry for a long time—some people put them in a container with desiccant packets or rice to absorb the moisture—and wind up with a functional product. Most of the time, and especially with saltwater intrusion, it's a total loss.

Lithium batteries and water make an especially dangerous combination because lithium reacts with water to produce fire! I haven't actually seen it happen from a drenched battery, but I'd be very wary of any lithium cell that got dunked, especially a vented type like the ubiquitous 18650 cells that look like overgrown AAs. Lithium fires are nasty and hard to put out.

Just being near salt water will destroy electronics after a while. Two-way radios, navigation systems, stereos and TVs kept on a boat or even in a seaside apartment get badly corroded inside, with rusted chassis, dull, damaged solder joints and connectors that don't pass current. Very often, you'll see crusty green crud all over everything.

Speaking of the ocean, the beach is a prime killing ground for cameras. Most digital cameras feature lenses that extend when the camera is powered on. Any sand in the cracks between lens sections will work its way into the extending mechanism and freeze that baby up, and it is very hard to get all the grit out. In a typical case, the camera is dropped lens first into the sand, and a great deal of it gets inside. I've taken a few apart and disassembled the lens assemblies, cleaned half a beach out of them and still had little luck restoring their operation. There's always a few grains of sand somewhere deep in those nylon gears, where you can't find them, and even one grain can stop the whole works. If you must take pictures at the beach, you're probably better off using your phone, unless it's an especially expensive one. Phones don't have those protruding lenses, but sand or salt water still can get into the charging or headphone ports.

The Great Capacitor Scandal

Around 1990, a worker at an Asian capacitor plant stole the company's formula, fled to Taiwan and opened his own manufacturing plant, cranking out millions of surface-mount electrolytic capacitors that found their way into countless consumer products from the major makers we all know and love. A few other Taiwanese capacitor makers copied the formula, too.

Alas, that formula contained an error that caused the electrolyte in those caps to break down and release hydrogen. Over a few years, the caps swelled and burst their rubber seals, releasing corrosive electrolyte onto the products' circuit boards, severely damaging them and ruining the units.

This ugly little secret didn't become well known for quite a while, until long after the warranty periods were expired. Billions of dollars' worth of camcorders and other costly small products were lost, all at their owners' expense. Any attempt at having repairs made was met with a diagnosis of "liquid damage—unrepairable." The disaster was so pervasive, and took long enough to show up, that many companies insisted the failures were random and have never to this day admitted any liability for the lost value.

More recently, similar electrolyte problems have continued to plague computer motherboards and the power supplies of various products, affecting even their full-sized capacitors with leads. Caps are dying after just a year or two of use. Even taking into account the rapid charging and discharging of today's power supplies and the high heat of lead-free soldering, they shouldn't go bad that fast.

Lawsuits have been filed, and remedial action has been taken by some manufacturers to purge their product lines of the offending parts. Still, it is highly likely you will run into bulging capacitors in your repair work, perhaps more than any other single cause of failure. Even when they're not bulging, the caps may lose their ability to store energy, showing almost no capacitance on a capacitance meter, or their ESR may be very high. The malady may have started long ago, but bad electrolytic capacitors are still the number one cause of failure in electronic products.

History Lessons

A good doctor understands the value of taking the patient's history before performing an examination. Knowing the factors leading up to the complaint can be very valuable in assessing the cause. How old are you? Do you smoke? Drink? Have a family history of this illness? What were you doing when symptoms appeared?

If you have access to a machine's history, it can provide the same kinds of helpful hints, often leading you to a preliminary diagnosis before you even try to turn it on. Here are some factors worth considering before the initial evaluation:

- *Who made it?* As discussed earlier, products from specific companies can have frequently occurring problems due to design and manufacturing philosophy. Becoming familiar with those differences may help guide you to likely issues, especially if you've seen the problem before in another unit, even of a different model, from the same maker.

It pays to check the internet for reports of similar troubles with the same model product. You may save many hours of wheel reinvention by discovering that others are complaining about the same failure. You might find the cure, too.

- *How old is it?* If made before the 1990s, it shouldn't have the leaking-capacitor problem. It could have a lot of wear, though, with breakdowns related to plenty of hours of use. If it *was* made in the '90s or more recently, those caps are a prime suspect. And we're talking more than 30 years ago! Most of what you'll work on is younger than that.

- *Has it been abused?* Dropped? Dunked? Spilled into? Sat on? Left on the dashboard of a car in the summer? Used at the beach? Had batteries in it for months or years? Had a disc stuck in it, and somebody tried to tear it out? Been in a thunderstorm? Through the washing machine? Kept on a boat? Played with by kids? Cranked up at maximum volume in a club for long periods?

   Each of these conditions can lead you down the diagnosis path. A stereo amplifier used gently at home by 70-year-olds is likely to have a very different failure than one cranked up to high volume levels in a club or one run 40 hours a week in a restaurant for 10 years.

- *What was it doing when it failed?* While devices sometimes quit while in operation, many stop working when sitting idle, and the problem isn't discovered until the next time someone tries to use the product. This is particularly true of AC-powered items that, like most things today, have remote controls. To sense and interpret the turn-on signal from the remote, at least some of the circuitry has to be kept active at all times. DVRs, DVD players and TVs are never truly turned off; some power always flows. A power surge, a quick spike or perhaps just age or—as always—bad capacitors can kill the standby supply, resulting in complete loss of operation. And many devices, like routers and AC adapters for laptops and other digital products, are always on, not just on standby, making the likelihood of age-related failure even higher.

   If the device did crash while being used, it's very helpful to know precisely what operation was being carried out when it quit. If a laptop's backlight went dead while the screen was being tilted, for example, that's a good indication of a broken internal cable rather than a blown transistor in the backlight inverter or LED driver circuit..

- *Did it do something weird shortly before quitting?* Many failing circuits exhibit odd operation for anywhere from minutes to seconds before they shut down altogether. This peculiar behavior can contain clues to the cause of their demise. In fact, it usually does, and it may hold the only hints you have in cases of total loss of function.

- *Was it sudden or gradual?* Some causes of failure, such as drifting alignment, dirty or worn mechanisms and leaking or drying electrolytic capacitors, may manifest gradually over time. Bad caps on computer motherboards are a great example of this as they cause the machine to get less and less stable, with more and more frequent crashes, until boot-up is no longer possible.

   Parts don't blow gradually, though. While it's possible in rare cases for components, and especially transistors, to exhibit intermittent bad behavior, a truly blown (open-circuited) component goes suddenly and permanently, frequently shorting first and then opening a moment later from the heat of all the current passing through its short. So, if the symptoms appeared gradually, it's a safe bet that the problem is *not* blown parts.

Stick Out Your USB Port and Say "Ahhh": Initial Evaluation

Before you take a unit apart, examine it externally and try to form a hypothesis describing its failure. The most potent paintbrush in the diagnostic art is simple logic. Your first brushstroke should be to reduce variables and eliminate as many areas of the circuitry from consideration as you can. Instead of chasing what might be wrong, first focus on what the problem *can't* be. By doing so, you'll sidestep hours of signal tracing and frustration. Before you open the unit, give some thought to these issues:

It's dead, Jim! *Dead* is a word many people use when something doesn't work, but often it's incorrectly applied. If *anything at all* happens when you apply power, the thing isn't dead! A lit LED, a display with something—even something scrambled and meaningless—on it, a hum, a hiss, a little warmth, or any activity whatsoever indicates that the circuitry is getting some power from the power supply, at least. *Dead* means *dead*. Zip, nada, nothing, stone cold. If you do see signs of life, a power supply voltage could still be missing or far from its correct value, but the supply is less likely to be the problem. In a product with a switching supply, you can assume that the chopper transistor is good, as are the fuse and the bridge rectifier (see Chapter 8). You *can't* be sure the supply has no other problem like bad capacitors or poor voltage regulation.

If the device is totally dead, check the fuse. All AC-powered products have fuses, and so do many battery-operated gadgets, though their fuses may be tiny and soldered to the board. A blown fuse pretty much always means a short somewhere inside, so don't expect much merely by changing the fuse. Most likely, it'll blow again as soon as you apply power. Still, give it a try, just in case. Be sure to use the same amperage rating for your new fuse; using a bigger one is asking for trouble in the form of excessive current draw and more cooked parts, and a smaller one may blow even if the circuit is working fine. No matter how tempting it might be, do *not* bypass the fuse, or you will almost certainly do much more damage to the circuitry than already exists. Those fuses are there for a reason, and that reason is protection.

Though nontechnical types tend to think that truly dead machines are the most badly damaged and least worth fixing, the opposite is usually true. Unless the unit was hit by lightning, total loss of activity typically indicates a power supply failure or a shorted part that has blown the fuse. In other words, easy pickings. The really tricky cases are the ones where the thing almost works right, but not quite, or it works fine sometimes and malfunctions only if you turn it facing south during a full moon on a Tuesday. Those are the unruly beasts that may cause you to emit words you don't want your kids to hear.

If the product has a display, is there anything on it?   Although a dead display can be caused by many things, the condition usually indicates that the microprocessor at the heart of the digital control system isn't running. Micros rarely fail, except in cases of electrical abuse like lightning strikes or severe static electricity. The most frequent reasons for a stopped microprocessor are lack of proper power supply (this includes noise on the DC due to high ESR in the filter capacitors) or a clock crystal that isn't oscillating. The clock steps the micro through its instructions, so no clock, no run.

If the display is there but isn't normal, that's a sign that some other issue in the digital system is scrambling the data going to it. If it's a simple system in which the microprocessor directly drives the display, the micro still might be stopped or damaged. If there's a display driver chip between the micro and the LCD, it may be bad. When the unit responds to commands but has a scrambled display, the micro is probably okay. If everything is locked up, suspect the micro or its support circuitry.

Does it work when cold and then quit after it warms up?   Thermal misbehavior can be caused by bad solder joints, flaky semiconductors and bad capacitors. It usually manifests as failure after warm-up, but now and then it's the other way around, with proper operation commencing only *after* the unit has been on for a while. Again, the problem is not a blown part.

Does tapping on it affect its operation?   If so, there's a poor connection somewhere. Typically, it's a cold solder joint or an oxidized connector. Cracked circuit board traces can cause this as well, because the edges are just barely touching across the crack. Those used to be fairly common, but they're quite unusual now, except in cases of physical abuse. Faulty conductive-glue layer interconnects (vias) can make boards tap-sensitive. On *very* rare occasions, the bad connection may be inside a transistor, and I once found one inside a tiny intermediate-frequency (IF) transformer in a radio receiver.

Eliminate variables.   If the device runs off an AC adapter, try substituting your bench power supply, being careful of polarity, as discussed in Chapter 3. If the unit can operate from batteries, put some in and see what happens. The remote control won't turn it on? Try using the front-panel buttons. Even if these attempts don't restore operation, at least you'll know what *isn't* causing the trouble.

Speaking of remotes, they can go wild and emit continuous commands, driving the micro in the product out of its little silicon mind and locking out all other attempts at operation. The situation usually occurs when liquid has been spilled on the remote, causing one or more of the keys to short out. The remote thinks a key is being pressed and sends data ad infinitum. To be sure this isn't the problem, remove the batteries from the remote and see if the product starts working.

Use Your Noodle: Case Histories

Once you've tried these preliminary experiments, think logically about their results, and you will probably have a pretty good sense of where to poke your scope probe first. Let's look at some real-life cases from beginning to end and how this approach helped get me started in the right direction.

Stereoless Receiver

The unit was a fairly high-end stereo receiver with a dead left channel that nobody in the shop could bring back to life. Eventually, they'd given up, and the set had languished on the shelf for 2 years by the time it and I met. The shop's owner handed it to me as an employment test. If I could fix *that* one, I was in. The smug look on his face told me I was in for a challenge.

I saw no evidence of obvious damage or abuse, so I hooked up a pair of speakers, connected a CD player for a signal source and fired it up. My initial evaluation was that the power supply had to be okay because the right channel worked fine. The front panel lit up and the unit seemed to operate normally, other than having a stubborn case of mono. I hooked a clip lead to the antenna terminal and tried FM reception, thinking that the trouble might be in the input switching circuitry feeding audio from the input jacks to the amplifier stages. Nope, FM sounded great, but still from only one channel.

There was no hum in either channel's output, so the power supply wasn't being bogged down by a short someplace. (A loudly humming channel with no audio is classically indicative of a shorted output transistor.) I plugged in headphones because sometimes amplifiers with bad output stages can drive a little bit of distorted signal into headphones, though they can't power a speaker. I kept the cans off my ears, as always, just in case the thing blasted me with punishing volume. There was no difference this time; I couldn't hear a hint of audio from the bad channel, even with the balance control turned all the way to that side. It was as quiet as a mouse. A dead mouse.

I'd eliminated as many variables as I could, so it was time to open up this ailing patient for a little exploratory surgery. I administered anesthesia (I unplugged it), and off came the covers. Several techs had taken their best shots at the poor thing, and evidence of their endeavors was all over the inside. The output transistors had been changed and large components in the power supply resoldered. Other solder work indicated that resistors in and near the bad channel's output stage (the power amplifier that drives the speaker) had been pulled and tested. The focus clearly had been toward the output area, which very often dies in audio amps and is where most techs look first. It made sense, but it hadn't done any good this time.

Thanks to the working channel, I didn't head straight to the power supply. Since the other guys had replaced the output transistors, I didn't bother to check those either. Instead, I stuck my scope probe on the signal line feeding the output stage, and there was no audio signal. Thus, the trouble was farther back in the chain toward the input stages someplace, and everybody had been hunting in the wrong place!

All the World's a Stage

The concept of stages is powerful and useful. Electronic devices are organized into little sections of just a few parts each. Each section, or *stage*, does something to whatever signal the circuit is processing and then feeds its output to the next stage. In a classic audio amplifier like a stereo receiver, the input jacks connect to stages that boost the tiny signals, which then feed the amplified signals to tone control stages that let you adjust bass and treble. Finally, the massaged signal is fed to the power amplifier stages that boost it to where it can drive speakers, which takes a fair amount of power. In modern receivers with digital processing, some of the low-level circuitry consists of specialized chips that decode multichannel sound from HDMI and optical connections and, perhaps, add various effects. In the end, though, the signal still winds up at the power amplifier stages.

This receiver had no digital processing, making the signal path pretty straightforward. I looked closely at a few small-signal transistors and traced their connections between stages. Some amplifiers are capacitively coupled (there's a capacitor between each stage through which the signal must pass), while others are directly coupled, with one stage feeding the next with nothing in between, or resistively coupled, with a resistor between the stages. (See Chapter 7 for details on capacitors and resistors.) The direct and resistive styles are also called *DC coupling* because the DC voltages on one stage get passed to the next, since they're not blocked by any coupling capacitors, which don't pass DC through them. See Figure 4-3 for a view of a signal with and without DC. Direct and resistively coupled amplifiers are tougher types of circuits to design, but avoiding capacitors between stages results in superior sound. So, most good audio gear works this way, and I expected to see that kind of circuitry here.

I wasn't disappointed; this baby had resistors between stages but no capacitors. Thus, the DC voltage levels on one stage could affect those on the succeeding stages. A little light was beginning to glow in the back of my mind, but I needed to take a few measurements before coming to any conclusions.

I went all the way back to the first stage, finding it by tracing the line from the input jack, through the selector switches, and to the amplifier board. I had a known good channel to use as a reference, so I fed the same audio signal to both sides, using a Y-adapter cable. Setting my trusty scope for dual-trace display so I could see two signals on the screen at once, one under the other, and selecting the same voltage range on both of the scope's input channels, I compared the outputs from the receiver's initial left and right channel stages. They looked identical. Same signal levels, same DC voltage. I went to the next stage. The good channel showed 1 volt DC at that stage's output, while the bad side only had about 0.5 volts, with the same audio signal riding on both. Hmmm . . . could such a small difference matter? Half a crummy volt? In a DC-coupled amplifier, you bet it could! Transistors need a *bias*, which is a little bit of DC to keep them turned on, at their bases

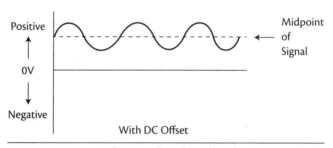

FIGURE 4-3   Audio signal with and without DC.

(input terminals), and not having a high enough bias will make them cut off, unable to pass any signal. The typical silicon transistor requires at least 0.6 volts to turn on. So, that 1volt coming from this stage was crucial to turning on the next one. Half a volt wasn't gonna jump that hurdle. Indeed, the next stage's output was dead, just a sad, flat line on my scope. Without enough bias, the stage was completely cut off. There was the trouble! But why?

I went back to the stage with lower DC output and checked the voltages and signals on the transistor's other terminals. They matched those of the good channel. Only the output was different. So, most likely, that vital half a volt was just getting lost in excessive resistance inside the transistor; there wasn't any other place it could go. The part had to be bad. A whole 25 cents' worth of mysterious mischief that had stymied an entire shop, simply because it wasn't the usual problem. I shut off the receiver, pulled the power plug, desoldered the suspect transistor and popped in a new one. Fired it back up, and voilà! The entire channel came to life and worked perfectly. A check of the previously dead stage's output levels showed that both the signal and DC level matched those of the good channel. Case closed. Of course, I ATE (Always Test Everything, remember?) for an hour, just to be sure there was no overheating or other trouble. I got a few open-mouthed stares from the other techs over that repair, along with an offer of full-time employment at the shop. I decided not to work there, but the episode left me feeling like Sherlock Holmes solving a perplexing crime. All I needed was a pipe and an English accent. "Elementary, my dear Watt-son!"

Silent Shortwave

A friend brought me this set after buying it for very little, knowing it didn't work but badly wanting it to, because he'd always longed for one of these models, and they were hard to find. One of the better digitally tuned *superheterodyne* shortwave receivers, this portable radio had no reception at all. It wasn't dead, though; the display came up normally, and a little hiss came from the speaker. Hmm, where should I start?

The first thing I did was try the various bands. AM, nada. Shortwave, same. FM . . . , hey, the FM worked! Sounded great. The FM band is at a much higher frequency and uses a different kind of signal than AM and shortwave, so all multiband radios have separate stages dedicated to FM reception, and clearly, they were fine. The audio and some other stages are shared, though, so the working FM also confirmed that the power regulation, *frequency synthesizer* (digital tuning control) and audio stages were all functioning. Thus, the trouble had to be in the RF (radio-frequency) or IF (intermediate-frequency) stages of the shortwave section, which also handled AM. Or perhaps the frequency synthesizer was working on some bands and not others. Because the FM worked, I discounted the synthesizer. There still could have been trouble there, but it wasn't suspect number one.

The operation of a radio receiver is a little more complex than we want to get into at this point, but the basic idea of a superheterodyne design is to mix the signals coming in from the antenna with locally generated ones made in the radio itself and then to process the result. The point of doing that is to amplify the resulting signal in a way that picks out a narrow slice of the radio spectrum and rejects everything else. This is called *selectivity*; it's how radios separate stations so you don't hear more than one at a time. Believe it or not, the technique is around 100 years old!

This receiver used two locally generated signals: one whose frequency varied as you tuned the radio, and a second one at a fixed frequency provided by a *quartz crystal*, a component used in most electronic products to provide very stable frequencies that don't vary much with temperature. Just about every piece of tech you own has at least one.

The trouble could have been anywhere along the chain of signal-processing stages, but experience reminded me to check that the fixed signal generator, called the *second local oscillator*, was actually running. It was especially suspicious because it was not used for FM in this radio, but it was for AM and shortwave.

It also rang my déjà vu alarm bell. Back in the 1970s, when I'd worked in the service department of a large consumer electronics chain, tons of CB radios had come in with dead receivers, thanks to a bad batch of oscillator crystals. We'd change 'em and be done in a jiffy, repairing the units without even having to troubleshoot them, since they always had the same problem. I fixed so many of them that the issue of a dead second local oscillator became permanently embedded in my mind. I looked for this set's crystal and touched each end with my scope probe, checking for a nice *sine wave* of a few volts. Nothing. The oscillator was *not* running. Aha!

Crystals actually vibrate at the molecular level when they're working. Sometimes, a weak one can be jolted into operating by putting a capacitor between one of the two leads and the circuit's ground. It unbalances the circuit by making the two connections to the crystal less alike, increasing the electrical stimulation and making the crystal vibrate a little harder. In fact, many oscillator circuits feature such a capacitor because it helps even good crystals start running. So, I touched my finger to each lead of the crystal, one lead at a time, with another finger touching circuit ground via a metal shield, employing my hand as a capacitor. This was all very low-voltage, battery-operated stuff, and it was safe to do that. The first lead, nothing. The second, wham! The radio sprang to life, and the BBC boomed in loud and clear from thousands of miles away. I let go and silence filled the room again. Ah, a bad crystal, and this one, with its oddball frequency specific to this radio, would have to be ordered from Japan. I tried resoldering it, just in case it had a cold joint. No luck. Then, glancing at my friend's glum expression of disappointment that a new crystal would have to be procured from halfway around the world—restoration of the radio would be months away—I decided to grab a magnifying glass and take a close look at the surrounding components. Sure enough, there was a tiny surface-mount capacitor connected from the lead of the crystal I'd touched to ground, performing essentially the same function my finger had. The solder joint on that one looked awfully dull. I resoldered it, and the radio starting playing its little heart out. "This is London calling. And now the news. . . . " Cost: zero. Grin on elated friend's face: priceless.

The Pooped Projector

How about a nice, high-resolution video projector with plenty of lamp life left for $20? Sure, we'd all go for that, right? Oh, there's one small catch: It doesn't work!

I snapped up this craigslist puppy because I knew from the history of its failure exactly what was wrong before I ever saw it. The owner told me that it had started turning itself off randomly and becoming difficult to turn on. Eventually, it stopped responding altogether. Now, what could possibly cause that? Obviously, it couldn't be a blown part. You guessed it: a classic electrolytic capacitor failure. I could picture just what it would look like with its bulging top. I figured it'd be near the DC output end of the switching power supply board, because I'd seen this sort of problem so many times, and that's where it always was. Those kinds of supplies, used in pretty much all modern products, are very hard on their capacitors as they charge and discharge them many times per second, wearing out the parts after a couple of years. It's a gradual process that makes the unit get increasingly flakier until it finally dies. Just like this projector.

Got it home, opened it up, and there was the cap, precisely as I'd pictured it, bulge and all. It was even *where* I'd expected it. I changed the part with an exact replacement I found on one of my scrap boards, a power supply from a computer. Fired up the projector, and it was good to go, with a sharp, bright picture.

While I gave it the ATE work-over, I checked online and found numerous complaints of the same problem in this model, along with various lay diagnoses, including some wacky guesses and the correct answer. The design kept the power supply turned on at all times, stressing that particular cap and causing it to fail after a couple of years, regardless of how much use the projector got. I keep my unit unplugged when I'm not running it, so it should last for a long, long time.

How can you beat a $20 video projector? And *that*, gentle reader, is why repairing electronics is not just fun—it's incredibly economical.

Chatterbox DVD Player

This portable DVD player came from the carcass pile at a repair facility where I worked part time. The machine, one of the better brands, had been a warranty claim, and nobody could fix the thing, so it had been replaced and kept for parts. With its 5-inch widescreen LCD, the player looked kinda cute, and it seemed a shame to cut it up. The shop's owner didn't care if I took it home, so I did. I had no idea what might be wrong with it, but the price was right.

It appeared intact, so I hooked up my bench supply and flipped on the juice. The screen lit up, and the mechanism immediately started making a noise like a machine gun! I killed the power in a hurry because I knew what that rapid-fire sound was.

As I mentioned earlier in this chapter, disc players use leaf switches to sense when the laser head, which rides on a platform called the *sled* (see Figure 9-3), has returned fully to the initial position at the inside of the disc, where it needs to go to begin the startup sequence to play it. The "rat-a-tat" noise was a clear indication that the microprocessor didn't know the mechanical limit had been reached. The unit was cranking its sled motor indefinitely, grinding the nylon gears against each other until they slipped, over and over. I could just imagine the toothless mess it might make of those delicate plastic parts if I let it run for very long. Yikes!

On opening the player up, I looked for the typical leaf switch assembly, with its copper fingers that bend and touch together when pressed on, and couldn't find one! Did this model use optical sensing, with an LED and a *phototransistor* instead of a leaf? I'd seen a few expensive players like that, but never a portable type, so I doubted it. Hmm, there was no trace of optical sensing either. I gently turned the sled motor's gear and moved the head away from the starting position, but I still couldn't see a switch. Finally, I removed the entire spindle assembly, and there it was, a tiny leaf switch hidden underneath the disc motor. It looked fine, though. Why wasn't it being tripped? Or maybe it was, and its signal wasn't getting back to the micro for some reason. Or perhaps the micro was bad.

I forced myself off the trail of wild imagination and back onto a path of pursuit, shaving off the less likely possibilities with *Occam's razor*. The simplest explanation was that the switch must not be getting pressed far enough to work. To protect the microprocessor from stray currents during testing, I disconnected one wire from the leaf switch. Then I connected my digital multimeter (DMM) to its two terminals and set the meter to its ohms scale, watching for the resistance to change from infinite (an open circuit) to near zero (a closed one) as I slowly turned the gear to move the head back toward the switch. The head hit its mechanical limit and would go no farther, but the switch never closed. There was the problem, all right.

After moving the head away again, I could see why, and it was so silly that I couldn't imagine why nobody had caught it. The little metal arm on the laser head that pressed on the switch was bent—not a lot, but just enough to keep it from pushing the leaf far enough to contact its mate. I bent the arm back ever so slightly, and I had a DVD player! Almost. Alas, the disc spindle assembly's three mounting screws also served to align the disc with the laser beam, and I'd had to unscrew them to remove the spindle. Any significant tilt of the disc would cause the reflected beam to miss the center of the head's lens, resulting in poor tracking and skipping. And, with the alignment scrambled, it did. I found the proper test point to use for observing the head's output signal (we'll explore how to do so in Chapter 15), watched it on my scope as I played a disc, and redid the alignment, carefully adjusting those three screws until I got a good, strong *eye-pattern* signal no matter what part of the disc I played. Making me mess up a critical alignment to reach a leaf switch—talk about poor design!

I won't mention the manufacturer's name, but I'd seen flimsy metal parts in some of its other products, so I wasn't terribly surprised to find one here, too. This particular player went on to develop a baffling, chronic problem with the ribbon cable going to its disc motor, causing it to fail to spin the disc fast enough, resulting in an error message and no playback. I kept cleaning the ribbon's contacts and reseating the connector at the circuit board end, and it would work for a few months before failing again. Finally, I checked the other end of the cable, which had looked okay, and found the real trouble; I'd just been wiggling it a little while working on the wrong end, and the movement had helped its connection for a short time. I cleaned and reseated that end, and the unit works to this day. Another mystery solved, another lesson learned in never assuming anything, and another fun freebie.

Goodbye Darkness

A newer projector, this time with LEDs for the light source. I bought this one off ye olde online auction site because the lens on my identical unit had gotten scratched. The replacement was supposed to be "new in damaged box." Whooee, was I surprised when I opened that box and found a dirty, badly worn projector that didn't even light up! A quick note and the seller issued me a prompt and courteous refund—and didn't want the unit back. What was he gonna do with it? It was a useless paperweight. But now it was *my* useless paperweight, and I wondered if it might be reparable.

After looking for all the usual capacitor suspects and finding none, I tossed the carcass in a box and put it on the shelf in my lab as a parts unit. After a few weeks, though, my sleuthing instinct began to gnaw at me. Why didn't this thing work?

I took it apart again and probed a little deeper. It did turn on, but it remained a mysterious pit of darkness. LED-lit projectors use three LEDs, one for each primary color:

red, green and blue. Those super-bright LEDs can fail—I'd run into that plenty of times—but all three? Nah, that didn't add up. Something else was going on.

Because the projector responded to the remote and the fans spun, I figured the power supply, an external adapter, was okay. Just to be sure, I scoped its output while it fed the unit. It was at the stated voltage and nice and clean, with almost no noise riding on the DC. Nothing wrong there.

Since all three colors were out, the LEDs weren't suspect. It had to be whatever supplied power to them. They all connected via wire cables to a board on the side of the optical assembly. Most of the board's components were on the inside, facing the optics, so I couldn't see them. I removed the board and took a look.

The two largest components on it were identical and something I'd never seen before, 1-inch-square metal boxes about a quarter-inch high. Compared to all the other parts, they were gigantic. One of 'em had a brownish, toasted look, while the other didn't. What were they?

The component density on the board was pretty high, but somehow the manufacturer had found room to print *call numbers* on it (see Chapter 8). These parts were labeled "L5" and "L6." You don't find such markings very often anymore because boards are so crowded, so I sure was glad to see 'em. "L" denotes inductors. They were coils! Why would they be so huge? Hmm, maybe they had to handle lots of current—current like the sort used to light really bright LEDs.

I was examining the brownish one when I noticed it was a tad crooked. Today's jam-packed products are built by machine, and the parts placement tolerances are very tight, within a fraction of a millimeter. No way it was askew like that when it was made. Had the coil somehow migrated? I touched it, and the darned thing moved! One end was hanging loose! Now I knew why it looked browned and out of place: It had gotten so hot that it melted its solder joints and slid down the board! In all my years as a technician, I'd never seen anything like it.

A quick resistance check showed that the coil hadn't burned out. Whew. I desoldered the attached end and resoldered both sides properly. Then I did the same treatment on the other coil, just for good measure, figuring that if one could try to run away from home, so could the other. I put everything back together and fired up the projector. I wasn't expecting much, figuring that something else had pulled way too much current through that inductor. Probably a short somewhere. I had some hope, though, because the coil was still good. And, son of an electron gun, it worked! The projector lit up and made a nice, bright picture, all three colors intact.

Now that I could see it, I went to the menu and checked the hours of use. This "new in box" machine had been used for more than 7,000 hours. But thanks to the longevity of LEDs, it made a picture pretty much like a new one. Gotta love it!

So, why did that coil overheat enough to melt solder? Most likely, the unit had run at its maximum brightness setting for every one of those hours, and the coil got pretty darned hot from the high current. Plus, it was soldered with low-temperature solder paste, along with all the small components. This was the real mistake; it didn't have to get super-hot to melt that stuff. Perhaps the resistance in the failing solder joints had caused some of the catastrophic heat, too. I resoldered them with good ol' tin-lead wire solder, and I never run my projectors at full brightness, so it shouldn't happen again. But that there's some seriously shoddy manufacturing, don'tcha think? Defective designs, even from major companies, are very real.

Chapter 5

Naming Names: Important Terms, Concepts and Building Blocks

While many different terms are used to describe electrons and their behavior, you will encounter a core set, common to all electronics, in your repair work. Some deal with electrical units, some with parts and their characteristics, some with circuit concepts, and others with hip tech slang. (Okay, you can stop laughing now!) Others describe frequently employed circuits used as the building blocks of many products. Getting familiar with these terms is crucial to your understanding the rest of this book, so let's look them over before moving on to Chapter 6.

We'll touch briefly on the most vital terms here; for more, and greater detail, check out the Glossary at the back of this book. You'd be doing yourself a favor to read the entire Glossary, rather than just using it for reference. Otherwise, you'll find yourself flipping pages back and forth a great deal as you read on. And believe me, you don't want to miss the definition of *magic smoke*.

Electrical Concepts

Being an intangible essence, electrical energy must be described indirectly by its properties. It possesses quite a number of them, and many famous scientists have teased them out with clever observations throughout the last few centuries. Experiments with electricity have gone on since the 1700s, when Ben Franklin played with lightning and miraculously lived to write about it—talk about *conducting* an experiment! Volta and Ampère built batteries and watched how their mysterious output affected wires, compasses and frogs' legs. Ohm quantified electrical resistance, and many others contributed crucial insights into this amazing natural phenomenon's seemingly bizarre behavior.

Perhaps the most valuable discovery was that electricity is a two-quantity form of energy. Its total power, or ability to do something like light a lamp, spin a motor or push a speaker cone, has two parts: how much of it there is and how strongly it pushes.

The *ampere*, or *amp* for short, is a measure of how much electrical *current* is moving through a circuit. Interestingly, actual electrons don't travel very fast at all and are not what moves through the wires, semiconductors and other parts of a device. Rather, their charge state gets transferred from atom to atom, raising the energy level of each one's own electrons, thus passing the current along. Imagine throwing a stone into a pond and watching the resulting wave. The wave propagates outward, but do the atoms of water at the center, where you threw the rock, actually wind up at the edge of the pond? No, they hardly move at all. They just push against the atoms next to them, transferring the stone's mechanical energy from one to the next.

The amount of current, or number of amperes moving through a circuit, has nothing to do with how hard they push, just as the amount of water in a hose has no relation to how much pressure is behind it. The pressure is what we call *volts*, a measure of how high each electron's energy state rises. Volts tell you how much pressure is pushing the amps through the circuit. In fact, voltage is sometimes referred to as *electrical pressure* or *electromotive force*. It is unrelated to how much electricity there is, just as the pressure in a hose doesn't tell you how much water is present. Volts propel current through the circuit. After all, without pressure, the water will just sit in the hose, going nowhere, right?

The hose isn't infinitely large, of course, and doesn't permit a perfectly free flow. Friction opposes and limits the motion. When the "hose" is a wire, that means it has *resistance*. Resistance is basically friction at the atomic level, and the energy lost to it from electrons and atoms rubbing against each other is converted to heat. The term for resistance is *ohms*, after the man who deduced the relationship between current, voltage and resistance. We call his crucial insight into electrical behavior *Ohm's law*. If you hate math and don't want to memorize formulas, at least get the hang of this rather simple one; it's the most important, useful relation in all of the electrical arts, and grasping its essence will greatly aid your troubleshooting. See *Ohm's law* in the Glossary.

When you put voltage and current together, you get the total picture of the power of the power, so to speak. We call this *watts*, and it describes how much work the energy can do. Determining watts is simple, at least for DC current: Just multiply the volts times the amps. So, 25 volts at 4 amps equals 100 watts, and so does 5 volts at 20 amps. Either arrangement could be converted to the same amount of mechanical work or produce the same light or heat. It's a little more complicated for AC because the peaks of current and voltage don't always happen at the same time.

As it comes from a battery, electricity is in the form of *direct current* (DC), meaning it moves in only one direction. The side of the battery with excess electron charge is called *negative*, while the side with a lack of it is called *positive*. Thus, by definition, current passes from negative to positive as it attempts to correct the imbalance of charges. Why not the other way around? I suppose we could have named either terminal whatever we wanted, but those names were known in Franklin's time and have persisted. And they relate to our modern model of the atom, with the electron's negative charges, so I doubt anyone's going to change them.

Flipping the *polarity*, or direction of current, back and forth turns out to create many useful effects, from easing long-distance power transmission to the magic of radio signal propagation. That's *alternating current* (AC), and you'll see it in just about everything. How fast you flip it is the *frequency*, specified in *hertz* (Hz). The old term was *cycles per second*. It's nice to have one word for it, don't you think? I wonder why we don't have one for speed, instead of miles per hour. We should call them "glorphs." "I'm sorry, sir, you were going 45 glorphs in the 25-glorph zone. License and registration, please."

When you put two conductive plates in proximity to each other and apply voltage, they talk to each other in a peculiar way. A charge builds up on either side of the insulator between them, and that charge can be taken out and turned back into current. We call this phenomenon *capacitance*, and the parts doing the job are *capacitors*. Essentially, capacitors act like little storage wells of electricity.

Electricity and magnetism are very related things, and they interact with each other. In fact, you can turn one into the other quite easily by passing a current through a coil of wire or by moving a magnet in a coil of wire. Passing current through the coil generates a magnetic field, and moving a magnet through a coil generates current. When a coil generates a magnetic field and then the direction of applied current reverses, the field collapses on the coil and generates current in it in the opposite direction to the current that created the field. Essentially, the coil stores some energy in the magnetic field and then puts it back into the circuit, but going the other way. That behavior is called *inductance*, and it has all kinds of important implications in AC circuits. A coil used that way is an *inductor*. Two different-sized coils wound on a common metal core can be used to transform one combination of current and voltage into another, with the magnetic field created by one generating current in the other. That's a *transformer*.

The effect capacitors and inductors have on AC current is called *reactance*, and the combination of capacitive reactance, inductive reactance and resistance is known as *impedance*. This is an especially apt term because it quantifies the amount the circuit impedes the passage of the AC current going through it. Though it doesn't behave exactly like resistance—its effect depends on the frequency of the AC being impeded, for instance—impedance is similar to resistance for AC current and is specified in ohms, just like pure resistance.

Circuit Concepts

When you wire up a bunch of parts such that current can pass through them and return to its point of origin, you create a *circuit*. The circuit concept is central to all electronics, and virtually every device that does anything is part of one. So, naturally, lots of terms are used to describe the functions and characteristics of circuits and the signals that flow through them.

When two or more circuit elements (components) are wired so that the current has to pass through one of them to reach the other, they're in *series*. Examples of things in series are fuses and switches; nothing can reach the rest of the circuit without passing through them first. It makes sense that the current through each element would have to be the same, since the amount of electricity reaching the return end of the power source has to equal what left it in the first place. Indeed, that's true. The current that passes through each element of a series circuit is always equal.

Still, energy has to be used in order for the part to do anything, so something has to give. What changes is the voltage. Each element drops the voltage, essentially using up some of the electrical pressure, until the total drop equals the applied voltage. The elements don't necessarily all drop the same amount of voltage, though. As Ohm so cleverly figured out, the amount dropped is proportional to the element's resistance. If one element has 20 percent of the total resistance of the circuit, it drops 20 percent of the voltage. Another element that has 10 percent of the total resistance drops 10 percent of the voltage, and so on. They'll always add up to 100 percent, right? Thus, all the voltage will be dropped by the time the other side of the power source is reached.

When circuit elements are wired so that multiple components are connected across the power source's two terminals, they are connected in *parallel*. In this case, each one gets the full voltage because nothing is in the way to drop some of it. The amount of current passing through each part is inversely proportional to its resistance, regardless of the other parts also connected. Basically, they have no reason to notice each other. If you measure the total current passing through a parallel circuit, it'll add up to all the currents going through each *leg*, or element. A parallel circuit's conditions are exactly opposite to those of a series circuit: The voltage is constant but the current varies.

Circuits with a path from one end of the power source to the other are said to be complete or *closed*. That's the normal operating state; unless a circuit is closed, nothing flows, and nothing happens. When there's a break in the path, perhaps from a switch in the "off" position or a blown fuse, energy flow stops, and the circuit is *open*. Any failed component no longer capable of passing current is considered open as well.

A condition causing part or all of a circuit to be bypassed, so that current passes straight to the other end of the power source, is called a *short circuit*, and the parts causing the detour are said to be *shorted*. Certain types of components, especially semiconductors (diodes, transistors and chips), often short when they fail.

Although the generation and transport system bringing power into your home provides AC, most electronic circuitry really can't use the stuff. Just as you couldn't drink from a cup swinging back and forth, circuits can't take AC power and amplify or process signals with it; the changes in the power itself would show up in the output. What's needed is a nice, steady cup from which to sip. In other words, smooth DC.

Once AC is *rectified*, or converted into one polarity, it's DC, but it's not steady; it's still a series of waves of power going up and down. To smooth it into a steady voltage, some kind of reservoir needs to store some of it so that as the wave strength approaches zero between waves, the stored energy can fill in and raise the voltage back up. That reservoir is a *filter capacitor*. It's just a big capacitor that can store enough energy to do the job, momentarily emptying itself to power the rest of the circuit until the next power wave fills it back up again.

Smaller filters called *bypass capacitors*, placed close to the part of the circuit pulling current, store some energy to fill in the gaps, just like the big guns do, but a lot less. In the process of accepting a charge, they absorb spikes and other electrical noises, providing a low-impedance AC path to ground in parts of amplifiers where that's needed, without shorting out the DC on the same connection, because steady voltages will not pass through a capacitor.

The circuitry in an electronic product is not just a huge mishmash of components. It is organized into sections and, within those, *stages*, as we discussed in Chapter 4. Each stage performs one function of whatever process is required for the device to do its job. A stage might be an audio preamplifier (a low-level amplifier), a tone control, a video display driver, a demodulator (something that extracts information from a signal), a position detector for a motor, and so on. At the heart of each stage are one or more *active elements*. These are the parts that actually do the work and are generally defined as being capable of providing *gain*, which you'll read about in just a few paragraphs. Supporting the active elements are passive components like resistors, capacitors and inductors. Those can alter a signal, but they can't amplify it. Without them, though, the active elements can't do their jobs. The active elements are the stars of the show, and the passive components are the supporting cast.

Stages feed signals to other stages until the device finally produces whatever output is desired. The components passing the signal from one stage to the next are called *coupling* elements and are usually capacitors, resistors or transformers. We discussed coupling capacitors in the "Stereoless Receiver" example in Chapter 4.

Signal Concepts

Signals are voltages varying in strength, or *amplitude*, to convey some kind of information. *Analog* signals vary the voltage in a pattern resembling the information itself. For instance, the output of an audio amplifier looks like a graph of the original pressure waves of sound in the air that struck the microphone. That's why they reproduce those sounds when applied to a speaker, which moves air in the pattern of the electrical energy. An analog video circuit's signal has varying voltages representing the brightness of each dot on the screen, with a rather complex method of conveying color information and synchronizing the spots to the correct place in the picture. Its graph doesn't look like an image, but it's still an analog signal, with fine voltage gradations portraying the changing picture information. The modern digital equivalent is a lot more complex, but it uses only "ons" and "offs" to represent picture data, like all computing devices. This is why we call digital information a data stream instead of a signal. It has only two states.

The graph of an analog signal is called its *waveform*. Every time the waveform repeats, that's one *cycle*. The number of cycles occurring in 1 second is the waveform's *frequency*, and the amount of time each cycle takes is its *period*. Because the voltage varies over time, it is a mathematical function, meaning that its lines can't cross over themselves. Graphed from left to right, as they are on an oscilloscope, the voltage level, indicated by the vertical position, at each successive moment is to the right of the preceding moment's portrayal.

The purest, most basic waveform is the *sine wave* (Figure 5-1). It is the building block from which all other waves can be created, and it has no *harmonics*, or energy at frequencies that are multiples of the wave's frequency. A sine wave sounds like a pure tone, with no characteristics suggesting any particular musical instrument or tone color. In fact, no nonelectronic musical instrument produces sine waves, though some registers on the flute come close. A tuning fork comes closer.

When a signal switches rapidly between all the way on and all the way off, it assumes a square shape and is called, appropriately, a *square wave* (Figure 5-2). Close enough examination will reveal that the on/off transitions aren't entirely vertical, because it takes time for the state to change. Thus, no square wave is truly square. The time it takes the transition to rise from 10 to 90 percent of its final state is the *rise time*. Going back down from 90 to 10 percent, it's the *fall time*.

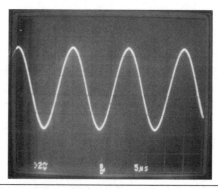

FIGURE 5-1   Sine wave.

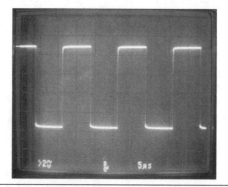

FIGURE 5-2   Square wave.

The percentage of time spent in the "on" state, compared to the "off" state, is called the *duty cycle*, and it can be altered to represent information or control a motor or a voltage regulator using a technique called *pulse-width modulation* (PWM). Unlike sine waves, square waves contain harmonic energy. They include odd harmonics, but not even ones. That is, there is energy at three, five and seven times the frequency, but not at two, four and six times.

A signal used in applications requiring something to move and then quickly snap back is the *sawtooth wave*, so named for its obvious resemblance to its namesake (Figure 5-3). Oscilloscopes employing a CRT (picture tube) use sawtooth waves to sweep the beam across the screen and then have it rapidly return. The old CRT TV sets did as well. Other circuits, including *servos* (motor position controllers) in video tape recorders and optical disc players, use sawtooth waves too. Sawtooth waves include both odd and even harmonic energy.

The relative position in time of two waveforms is called their *phase relationship* and is expressed in degrees. As with a circle, 360 degrees represent one cycle of a waveform, regardless of how long that cycle takes. Two waveforms offset by half a cycle are 180 degrees out of phase. When there is no offset, the waveforms are in phase.

As mentioned earlier, digital streams are entirely different. Represented by a series of pulses resembling square waves, digital information is always in one of two states, on or off, indicating the binary numbers 1 and 0. Each number is a *bit*, and a group of them is a *byte*. This has tremendous advantages over the analog method because keeping track of those two states is a lot easier than accurately moving and processing a voltage with infinitely fine gradations. Noise in a digital channel has no effect at all until it's so strong that the two states can't be determined, while noise in an analog channel is very hard to separate from the desired signal and corrupts it badly. That's why scratches on an analog LP record create clicks and pops in the audio, while scratches on a CD don't. All circuits introduce some noise, so the digital method is less susceptible to degradation as it moves through various processes. A digital data stream is also much easier to store and manipulate, again because it has only two states to worry about.

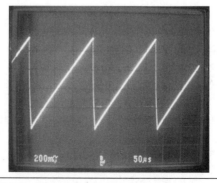

FIGURE 5-3   Sawtooth wave.

The world of sound and light is inherently analog, though; nothing in nature exists only as ons and offs! To *digitize* natural phenomena like sounds and images, an *analog-to-digital converter* (ADC) is used to chop the analog information into a rapid series of *samples*, or measurements, that are then encoded, one by one, into the binary 1s and 0s of digital data. Conversion is a complicated process that introduces quality limitations of its own, so digital is no more perfect than is analog. Digital's imperfections are different, though, and generally less objectionable. And, by using lots of data to describe an analog phenomenon, digital can come much closer to perfect than any analog system ever could.

Building Blocks

There's one heck of a variety of circuits out there! For any given function, a designer can find lots of ways to build something that works. While the circuitry "wheels" get reinvented all the time, they're all round and they all spin, so common circuit configurations are found in pretty much all products. Sometimes they have significant variations, but they're still basically the same old thing and can be recognized easily once you get familiar with them. Let's look at some common circuits you're likely to find.

The basic circuit at the heart of most stages is an amplifier of some kind. Amplifiers have *gain*, which means they take steady DC from the power supply and shape it into a replica of an incoming signal, only bigger. The "bigger" can be in terms of voltage, current or both. Most voltage amplifiers also flip the signal upside down, or *invert* it. Sometimes that's important to the circuit's operation, but much of the time it's just an irrelevant consequence of how voltage amplifier stages are configured.

A common type of current amplifier is the *emitter follower*. This one takes its output from the emitter (one of the terminals) of a transistor, and the amplified signal mimics, or follows, the input signal without inversion or change in voltage swing. Only the amount of current the signal can pump into a *load* is increased. Current amplifiers are used to drive things that do a lot of physical work, like lamps in video projectors, motors and speakers.

When amplifying analog signals, *linearity*, the ability to mimic the changes in the input signal faithfully without distortion, is a critical design parameter. The term comes from the graph that results if you plot the input signal against the amplified output. In a truly linear circuit, you get a straight line. The more gain, the more the line points upward, but it's still straight, indicating a ratio of input to output that doesn't change as the signal's voltage wiggles up and down. If the amplifier is driven past the point that its output reaches the power supply voltages, the transistors will be all the way on or all the way off during signal extremes, resulting in an output that no longer accurately follows the input. That's serious *nonlinearity*, also known as *clipping* because it clips off the tops and bottoms of the waveform; the amplifier simply can't go any farther (Figure 5-4). If you've ever turned up a stereo loud enough to hear ugly distortion, you've experienced clipping. Even a small amplifier driven to clipping can burn out the tweeters (high-frequency speakers) in a pair of speakers that normally could take the full power of a bigger amplifier without harm.

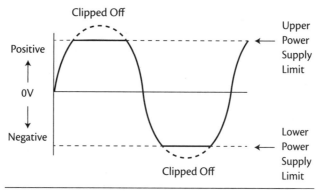

FIGURE 5-4   Clipping.

The high-frequency content of a clipped waveform is much greater than that of a linearly amplified signal, thanks to the steep edges of the clipped area, so it drives disproportionate power into the tweeters. I've seen it happen. It can injure your ears, too.

Most high-fidelity analog audio amplifiers use a *complementary* design, referred to as *class AB*. Complementary amplifiers split the audio waveform into its negative-going and positive-going halves and amplify each half separately. Then, they combine the two halves at the output, rebuilding the waveform. Why do that? The technique allows for excellent efficiency because almost no power is dissipated when the waveform is near the zero-voltage level. Only when the input signal gets big does the amplifier draw a lot of power, keeping power usage proportional to output.

In a non-complementary design, the amplifier has to be *biased* to set its output halfway between ground and the power supply voltage when no signal is applied so the negative and positive peaks of signals can make the output swing both higher and lower in step with them. It sits there eating half the available power at all times, even when it's not amplifying anything. Some high-end amplifiers, called *class A*, actually do it this way to avoid certain subtle distortions associated with splitting and recombining the waveform. Those amplifiers sound especially pure but run very hot and waste a lot of power.

Digital amplifiers work differently. We'll get into those in Chapter 15.

Tuned amplifiers use *resonant* circuits to select a particular frequency or range of frequencies to amplify, rejecting others. They're used in classic superheterodyne radio and TV receivers to separate and boost incoming signals. See *resonance* in the Glossary. The very first amplifiers in a receiver that strengthen the weak signals from the antenna are called the *front end*. Some are tuned to amplify only the frequency to which the radio is set, and some aren't, depending on the design. Later amplifier stages, operating at a fixed frequency to which all incoming signals are converted, are called *intermediate-frequency* (IF) amplifiers. Those are always tuned, and they provide most of a receiver's *selectivity*, or ability to separate stations.

Even digital gates in integrated circuits (ICs) are amplifiers. They take in digital pulses and amplify them enough to ensure that they swing all the way to the supply voltage and ground, making up for any losses that may have occurred and preventing them from accumulating until the pulses can no longer be reliably processed. These amplifiers are deliberately nonlinear; the only desired states are fully *saturated*, or all the way on, and fully *cut off*. Their *linear region*, where small changes in the input signal would be faithfully amplified, is made as narrow as possible. Such small wiggles in digital signals only represent noise, and you don't want to amplify that.

Oscillators generate their own signals. They have many uses, including providing timing pulses to, or *clocking*, microprocessors, generating tones, mixing with radio signals to convert their frequencies, and lots more. An oscillator is basically an amplifier with its output fed back to its input in phase, reinforcing the input and sending the signal around and around again indefinitely.

Oscillators can be designed to produce any of the basic waveforms. In analog signal-processing circuits, sine wave oscillators are often the most useful, thanks to their purity. As I mentioned, sawtooth wave oscillators are used to drive electron beams across CRTs and any time something needs to be swept and then quickly returned to its starting point. With digital circuits, which operate with pulses, square waves are the order of the day.

The significant challenge with most oscillators is setting the frequency and keeping it constant. When only a single frequency is needed, a quartz crystal or ceramic resonator can keep the oscillator very accurate. These parts resonate mechanically at the molecular level, and they're dimensionally stable, so they drift very little with temperature. The tradeoff is that a given crystal can generate only one frequency.

When frequency variability is required, as in a radio tuner, simple resonant circuits like inductor/capacitor combinations work but are not terribly stable, especially regarding thermal drift. While early radios could tolerate some drift, today's high-precision systems simply can't. Do you really want to get up to fine-tune your digital HDTV every 20 minutes as the receiver drifts off frequency and the picture drops out? Of course not! The receiver has to sit on its tuned frequency the moment you turn it on and stay there all day long.

The *digital frequency synthesizer* solves the stability problem by providing frequency agility while still being referenced to the unvarying frequency of a quartz crystal. Pulling that off isn't simple. There are two basic techniques. In a classic hybrid analog/digital synthesizer, an analog oscillator's frequency is controlled by a voltage from the synthesizer. The resulting frequency is digitally divided or multiplied until it matches that of a reference crystal. The two are then compared, and the controlling voltage is adjusted to keep them at the same frequency. So, even though the analog oscillator really isn't at the same frequency as the crystal, the comparison circuit thinks it is. This kind of circuit is called a *phase-locked loop* (PLL).

Tuning the oscillator is accomplished by changing the ratio of division and multiplication, forcing the control voltage to change the true frequency to match the result of the division to that of the unvarying crystal. Many receivers have been built this way, but the method has a serious downside. In order for the comparison and correction process to work, there has to be a little error to correct. The oscillator's frequency wobbles ever so slightly as it drifts and gets corrected, resulting in a phenomenon called *phase noise*. With careful circuit design, the noise can be kept low, and some really nice radios worked just fine with this style of synthesizer, but it caused enough signal distortion that better methods were sought.

As digital technology advanced, chips got fast enough that the oscillator could be done away with altogether, and the required output signal could be built directly from digital data and converted to analog, in much the same way a CD player rebuilds the audio waveform of a music disc from samples. In most applications, this *direct digital synthesis* (DDS) technique has replaced hybrid analog/digital synthesizer designs. The tipoff is if you see a component on the schematic that looks like a combination of a capacitor and a diode. That's a *varactor*, or voltage-variable capacitor, and it's what tunes the analog oscillator with the control voltage from the digital part of the circuit. Where there's a varactor, there's a PLL.

Varactor diode symbol

PLLs are used in digital systems too. Digital data is recovered from hard drives and optical discs using a PLL to synchronize the data rate read off the platter to the circuit processing it.

Servos are a lot like PLLs, except that they slave a motor's rotational speed and phase (position at a given moment), or some other moving part's behavior, to a reference signal. A servo regulates a DVD player's disc rotation, keeping it at whatever speed is required for a given data rate. The rapid lens motions required to focus the laser beam on the microscopic pits and track them as they whiz by are also controlled by servos. In VCRs, servos adjusted capstan motor and head drum rotation, locking them to a reference signal recorded on the tape so that the rotating heads traced over the recorded tracks correctly. More and more, though, servos have gone all-digital, emulating in software the comparison and adjustment functions once provided by varactor-controlled oscillators. Analog PLLs and servos, with their sometimes-finicky adjustments, are quickly becoming relics of history.

Voltage regulators keep power supply voltages constant as current demand varies with circuit function. Linear regulators use a *series pass transistor* as a variable resistor, automatically changing the resistance to set the voltage, and dissipating the unwanted extra power as heat. They're simple and effective but also inefficient. Linear regulators handling serious power get quite hot.

Switching regulators use pulses to turn a transistor on and off like a switch, and then reconstitute the pulses passed through the transistor back into steady DC power with a filter capacitor. Varying the pulse width permits more or less power to get through the transistor over a given period of time. Because the transistor is always completely on or completely off, except for the short moments when it switches states, little power is converted to heat, and there's no excess to waste because only as much energy as needed is allowed through. Switching regulators are more complicated, and they have the potential to generate electrical noise that can interfere with radio and TV reception, but their cool-running efficiency makes them very desirable, especially in battery-operated devices, where conserving power is critical.

Now that you've seen some basic terms, core concepts and circuit building blocks, go read the Glossary and see lots more. Seriously. Do it now, before going on to Chapter 6. I know, "Grrr, this guy is such a nag." You'll thank me later. Really, you will. And hey, there's some cool stuff there. You'll enjoy it, I promise.

Chapter **6**

Working Your Weapons: Using Test Equipment

E ffective application of test gear is key to your sleuthing success. Especially with the oscilloscope, the settings you make while probing around a circuit determine what elements of the signals you will see and which ones you may miss. Soldering, too, can be more or less effective, depending on your technique. Let's look at each piece of basic test gear and how to use it to your best advantage.

Digital Multimeter

Digital multimeters (DMMs) are great for measuring things that don't change quickly. Battery and power supply voltage, along with resistance and current, are prime candidates for being checked with a DMM. The instrument is less effective for observing changing voltages and currents, which look like moving numbers and are tough to interpret.

Overview

The DMM's great advantages over other instruments are its precision and accuracy. Even a digital scope has fairly limited resolution; you can't tell the difference between 6.1 and 6.13 volts with one very easily, if at all, and measuring resistance and current is impossible with normal scope setups.

All that detail in the meter's display can get you into trouble, though, if you take it too seriously. When interpreting a DMM's readings, keep in mind that real life never quite hits the specs. Don't expect the numbers you see to be perfect matches for specified quantities. If you're reading a power supply voltage that's supposed to be 6 volts, a reading of 6.1 probably isn't indicative of a circuit fault. The same is true of resistance; if the reading is very close, the part is most likely fine. And if the rightmost digit wanders around a little bit, that's due to normal noise levels or the DMM's digitizing noise and error inherent in any digital sampling system. Remember, when a part goes bad, it's not subtle! Real faults show readings far from the correct values.

Most DMMs run on batteries, and that's a good thing because it eliminates any ground path from the circuit you're testing back to your house's electrical system. The instrument "floats" relative to what's being tested (there's no common ground), so you can even take measurements across components when neither point is at circuit ground. If your DMM has the option for an AC adapter, don't use it. Always run your DMM on battery power. The batteries will last for hundreds of hours anyway.

DC Voltage

To check a circuit point's voltage, first you must find circuit ground. Usually, it's the metal chassis or metal shields, if there are any. *Don't* assume that heatsinks, those finned metal structures to which are attached larger transistors, voltage regulators and power-handling integrated circuits (ICs), are connected to ground! Sometimes they are, sometimes they're not. In switching power supplies (see Chapters 8 and 15), the chopper transistor's heatsink may have several hundred volts on it. You sure don't want to hook your meter there.

In some devices, especially small ones like digital cameras, you may find no shields, and there's no metal chassis either. So where is ground? In most cases, the negative terminal of the battery will be connected to circuit ground, and you can use that. Particularly if you can trace it to a large area of copper foil on the board, it's a fairly safe bet. Also, look for electrolytic capacitors in the 100-microfarad (μF) and up range with voltage ratings lower than 50 volts or so (see Chapter 7). Those are most likely power supply filter capacitors, even in battery-operated gear, and their negative terminals will be connected to ground. If you see two such identical caps close to each other, the device may have a split power supply, with both negative and positive voltages. Trace the caps' terminals and see if the negative lead from one is connected to the same point as the positive from the other. Where they meet is probably circuit ground.

If all else fails, you can use the outer rings of RCA jacks on audio and video gear. The only way to get an alligator clip to stay put on one of those jacks is to push half of it into the jack, with the other half grabbing the ground ring. It's better to use an input jack, rather than an output, so that the part of the clip sticking inside can't short out an output, possibly damaging the circuitry. You can't hurt an input by shorting it to ground.

Don't use the negative terminal of a speaker connection as a ground! In some designs, it is in fact ground, but in others, neither connection to the speaker is grounded. You can cause all kinds of damage by assuming that a speaker terminal goes to ground.

Turn on your DMM, set its selector switch to measure DC voltage, and connect its negative (black) lead with a clip lead to circuit ground, regardless of whether you intend to measure positive or negative voltage. A DMM will accept either polarity; measuring negative voltage simply adds a minus sign to the left of the displayed value.

With power applied to the circuit under test, touch the positive lead's tip to the point you want to measure, being careful not to let it slip and touch anything else. Many DMMs are autoranging and will read any voltage up to the instrument's ratings without your having to set anything else. Keep the probe in place until the reading settles down; it can take 5 or 10 seconds for the meter to step through its ranges and find the appropriate one.

If your DMM is not autoranging, start at the highest range, and switch the range down until a proper reading is obtained. If you start at the lowest range and the voltage you're measuring happens to be high, you could damage the DMM.

If you see a nice, steady number somewhere in the voltage range you expect, it's safe to assume you have a valid measurement. If, however, you see a moving number at a very low voltage, you're probably just reading noise on a dead line, and you may have found a circuit problem. You'll also see this if you accidentally set your DMM to read AC volts instead of DC. If you see a voltage in the proper range but it won't settle down, that indicates noise on the line, riding along with the voltage. Such a reading can suggest bad filter capacitors, but only when the point you're measuring is supposed to have a clean, stable voltage in the first place. Regulated power supply output points should be steady, but some other circuit points may carry normal signals that fool the DMM, causing jumping readings. To see what's going on with those, you'll be using your scope. Generally, electrolytic caps with one lead going to ground shouldn't have jumpy readings, since their reason for being in the circuit is to smooth out the voltage.

AC Voltage

You'll usually use this as a go/no go measurement. Is the voltage there or not? DMMs are optimized to read sine waves at the 60-hertz AC line frequency, so the reading doesn't mean much if you try to measure an audio signal or the high-frequency pulses in a switching power supply. Measurements are taken across two points, as with DC voltage, but in many circuits, neither point will be at ground.

DMMs indicate AC voltage as root-mean-square (RMS), which is a little bit more than the average voltage in a sine wave when taken over an entire cycle. It's a useful way of describing how much power an AC wave will put into a resistive load, which is a pure resistance that would convert the power to heat, with no capacitance or inductance to alter its behavior. Essentially, it's a description of the amount of work the AC can do, compared to steady DC power, but it is not a measurement of the actual total voltage swing. The RMS value is much smaller than the peak-to-peak voltage—the total swing between the positive and negative peaks—you'll see with your scope. *That's* what tells you the maximum voltage of the AC as it reaches its peak in each cycle. American AC line voltage, for example, is 120 volts RMS and reads about 340 volts peak to peak on a scope. (If you want to keep your scope, don't try viewing the AC line with it unless you have an isolation transformer!)

The reason DC and AC can't be evaluated the same way is *time*. The AC wave spends only part of its time at maximum voltage, unlike DC, which stays at the same voltage continuously. For a sine wave, RMS is 0.3535 times the peak-to-peak value. For other waveforms, it can be quite different because the time they spend at various percentages of their peak values varies with the shape of the wave. DMMs are calibrated to calculate RMS for sine waves, so the reading will be way off for anything else, at least with hobbyist-grade meters.

Resistance

When measuring resistance, *turn off the power to the circuit!* The battery in your DMM supplies the small voltage required to measure resistance, and any other applied power will cause unfortunate consequences ranging from incorrect readings to a damaged DMM. In addition to removing the product's batteries or AC adapter, or unplugging AC-operated products from the wall, it pays to check for DC voltage across the part you want to measure and to discharge any electrolytic capacitors that could be supplying their stored-up voltage to the area under test before you take a resistance reading.

Some resistances can be checked with the parts still connected to the circuit, but many cannot because the other parts may provide a current path, confusing the DMM and resulting in a reading that is lower than the true value. For most resistance measurements, you will need to unsolder one end of the component. When one side of it goes to ground, leave that lead connected, and connect your DMM's negative lead to the ground point; it's just more convenient that way. When neither side is grounded, it doesn't matter which lead you disconnect.

Set the DMM to read resistance (ohms or Ω). If it's autoranging, that's all you need do. Let it step through its ranges, and there's your answer. If it isn't autoranging, start with the *lowest* range and work your way up until you get a reading so you won't risk applying the higher voltages required to get a reading on the upper ranges to sensitive parts. DMMs with manual ranges have an "out of range" indicator to show when the resistance being measured is higher than what that range can accept, usually in the form of the leftmost digit's blinking a "1." (If you're on too high a range, you'll see all 0s or close to it.)

With a manually ranging DMM, you can get more detail by using the lowest range possible without invoking the out-of-range indicator. Let's say you have a typical 3½ digit hobby-grade DMM. This means there are three digits plus a "1" at the left. If you are reading a 10-ohm resistor on the 20-kilohms (20,000-ohms) scale, you'll see 0.01. If you switch to the 200-ohms scale, you'll see 10.0 or thereabouts. If the resistor's measured value is too high by, say, 20 percent, which is a significant amount possibly indicating a bad part, it might show 12.0, critical data you'd miss by being at too high a range, where there aren't enough digits on the right to show the overage. Also, the farther to the right the digits lie, the less accurate the reading because they're at the limit of the instrument's resolution, or ability to determine small values. Autoranging meters always use the lowest possible range so they can get the most detailed and accurate reading.

Resistance has no polarity, so it doesn't matter which lead you connect to which side of a resistor. If you're checking the resistance of a diode or other semiconductor, it does matter, and you must swap the leads to see which polarity has lower resistance. The essence of a semiconductor is that it conducts only in one direction, so a good one should have near-infinite resistance one way and low resistance the other. Checking semiconductors for resistance with a DMM can yield unpredictable results, though, because the applied voltage may or may not be enough to turn the semiconductor on and allow current to pass, depending on what you're testing and on the meter's design. There are better tests you can perform on those parts, but a reading of zero or near-zero resistance pretty definitively indicates that the component is shorted. To be sure, swap the test leads. If you still see zero, you've found a short.

Continuity

Continuity simply shows whether a low-resistance path exists, and is intended as a yes or no answer, rather than as a measurement of the actual resistance. It's exactly like taking a resistance measurement on the lowest scale, except that many meters have a handy beeper or buzzer that sounds to indicate continuity, so you don't even have to look up. Use this test for switches and relay (electromagnetically operated switch) contacts or to see if a wire is broken inside its insulation or a connector isn't making proper contact.

In many instances, you won't need to pull one side of the component to check continuity, because the surrounding paths will have too much resistance to fool the meter and invalidate the conclusion. There are some exceptions, however, involving items like transformers, whose coil windings may offer very little resistance and appear as a near-zero-ohm connection across the part you're trying to test. If you're not sure, pull one lead of the component. And, as with resistance measurement, make sure all power is off when you do a continuity check!

DC Current

Most DMMs can measure current in amps or milliamps. To measure current, the meter needs to be connected between (in series with) the power source and the circuit drawing the power so that the current will pass through the meter on its way to the circuit. The easiest way is to break the connection between the battery or AC adapter's positive terminal and the device, and connect the meter between the two points that were previously attached. Thus, neither of the DMM's leads will be connected to ground. Never connect your DMM *across* (in parallel with) a power supply's output, from its voltage output to ground, when the meter is set to read current! Nearly all the supply's current will go through the meter, and both the instrument and the power supply may be damaged. At the least, the meter will blow its internal fuse.

Even with the meter properly connected, it's imperative that you not exceed its current limit or you will damage the instrument or, if you're lucky, blow its fuse. For many small DMMs, the limit is 200 milliamps (mA), or 0.2 amps. Some offer higher ranges, with a separate terminal into which you can plug the positive test lead, extending the range to 5 or 10 amps.

In estimating a device's potential current draw, take a look at what runs it. If it's a small battery, as you might find in a GPS unit or a digital camera, current draw probably isn't more than an amp or so. For many such devices, it's much less, in the range of 100 to 200 mA. If the unit uses an AC adapter, the adapter should have its maximum current capability printed on it somewhere, and it's safe to conclude that the product requires less than that when operating properly. Some gadgets state their maximum current requirements on the backs of their cases, too. When they do, they indicate the maximum current needed under the most demanding conditions—for example, when a disc drive spins up or a Bluetooth speaker is playing at its loudest—and normal operating current should be less.

When taking a current measurement, don't worry about test lead polarity; all you'll get is a minus sign next to the reading should you attach the leads backward. If you want to know the current consumption of an entire product, connect the meter between the positive terminal of the battery or power supply and the rest of the unit. If you want to measure the current for a particular portion of the circuitry, disconnect whatever feeds power to it—perhaps a resistor or a wire on a connector—and insert the meter there.

The DMM measures current by placing a low resistance between the meter's leads and measuring the voltage across it. The higher the current, the higher that voltage will rise. With a big current, the resistance can be very low, and there will still be enough voltage across it to get a reading. With smaller currents, the resistance needs to be higher to obtain a significant, measurable voltage difference. Thus, the meter's higher ranges place less resistance between the power and the circuit. Start with the meter's highest range and work your way down. Using too low a range may impede the passage of current enough to affect or even prevent operation of the product you're trying to test. It also may heat up the DMM's internal resistor enough to damage it. If you're taking current measurements and the meter suddenly goes from a reasonable reading down to zero, you've probably pulled too much current through it and blown the meter's fuse. Most of those are in the 250- to 500-mA range. Take off the back and you'll see the fuse in its holder. Some meters keep a spare fuse inside for just these occasions.

Current is perhaps the least useful measurement and, consequently, the one most infrequently performed. Now and then, it's great to know if excessive current is being drawn, but heat, smoke and blown fuses usually tell that story anyway. The more revealing result is when current *isn't* being drawn; that tells you some necessary path isn't there or the unit isn't being turned on. Especially because breaking connections to insert the meter is inconvenient, however, you won't find yourself wanting to measure current very often.

Diode Test

Some DMMs offer semiconductor junction tests, making them handy for checking diodes and certain types of transistors. The measurement is powered by the meter's battery, as with resistance measurements, but it's taken somewhat differently. Instead of seeing how many ohms of resistance a part has, you see the voltage across it. And to complete the test, you must reverse the leads and check the flow in the other direction.

Kill the power and disconnect one end of the component for this test. A good silicon diode should show around 0.6 to 0.7 volts in one direction and no continuity at all (an open circuit) in the other. This lack of flow will be shown as the maximum voltage being applied, typically around 1.4 volts. (You can check your meter's open-circuit value by setting it to the diode test without connecting the leads to anything.) If you see 0 volts or near that, the part is shorted. To verify, switch the leads, and you should see 0 volts in the other direction too. If you see 1.4 volts (or whatever your meter's maximum is) in both directions, the part is open, a.k.a. blown. If the meter indicates the normal 0.6 volts in the conducting direction but also shows even a slight voltage drop the other way around, the diode is leaky and should be replaced. If you see such a reading, make sure your hands aren't contacting both test lead tips, because even a small current through you will affect the reading, making a good diode look leaky.

Some DMMs perform capacitance, inductance, frequency and other measurements, but most don't. If yours does offer these readings, see the sections on those kinds of meters, and the principles will apply.

ESR Meter

With changes in technology have come changes in what goes wrong with it. By far, the biggest change I've seen is the epidemic of high ESR (*equivalent series resistance*, as discussed in Chapter 2) in electrolytic capacitors.

Today's circuits require squeaky-clean DC power. By *clean*, I mean free of fluctuations. Fast ones, a.k.a. noise, spikes or glitches, can really foul up a digital system, causing everything from erratic operation to total system shutdown. In analog circuits, the noise may show up in the output, manifesting as a hum, buzz or whine. The telltale symptom of ESR trouble in digital devices is that operation of the system gets less stable with time. Flat-panel TVs and computer monitors are particularly prone to these problems. So are computers themselves. At first, a TV begins having trouble turning on, but it'll do it if you try several times. Then, the set starts turning itself off randomly. Finally, it won't run at all. Head straight for your ESR meter!

Using an ESR meter is easy. Because it employs very low test voltages in the tens of millivolts, semiconductors (transistors, diodes and ICs) connected to the capacitors won't turn on, so they effectively disappear from the measurement. This lets you test for ESR without having to unsolder the capacitors, unless there's a coil or another cap connected across (in parallel with) the part under test. Beware: Computer motherboards often have lots of electrolytic caps in parallel in order to achieve extra-low ESR. With those, you'll have to unsolder one lead of each cap while you test it or you'll be testing the whole herd at the same time, which tells you nothing about an individual capacitor. That can be a real problem! I fixed a wireless router with some bad electrolytics, and one showed an in-circuit ESR of 0.72 ohms, which seemed pretty good. When I took it out and retested it, it measured 40 ohms! Why such a vast difference? There were three other caps on the board in parallel with it, masking how bad it was.

If your ESR meter offers a zeroing function, use it before testing. Just connect the test leads together and hit the "zero" button. The meter will adjust itself to compensate for the resistance in its own test leads to obtain the most accurate reading. Then, unhook the leads and get ready to connect them to the capacitor under test. Be sure to *discharge the test cap completely* before hooking up the meter! I can't stress this enough. A charged cap can ruin your ESR meter. Even if you're lucky and it doesn't, you won't get a valid reading from a charged capacitor.

To discharge the capacitor, short across it for at least a few seconds. For small ones, that'll do the trick. It can cause trouble with big ones used in power supplies, though, because they hold so much energy that a huge surge can occur, causing a spark or even damaging other components. To discharge those, it's best to use a 10-ohm resistor across the terminals. Just keep the resistor connected for about 20 seconds, and the part will be discharged safely, with the resistance limiting how much current can flow at once.

Connect the meter's leads, being sure to hook + to the cap's + and − to its −. Connect them directly across the capacitor. Even if one side of the cap goes to ground (as many do), *don't* use a ground point somewhere away from the part; the resistance of the circuit board trace between it and the cap will distort the reading, making the ESR look higher than it is. Fractions of an ohm of resistance matter here.

The meter will display the ESR in ohms. Be prepared to see a fraction of an ohm to perhaps a few ohms. Remember, the lower the better.

Unlike with many other measurements, interpreting ESR isn't a hard and fast science. Caps from various manufacturers may have somewhat different ESRs. Also, there are several chemical and mechanical formulations of electrolytic capacitors, all with different ESRs.

How do you know what's okay and what isn't? If the name of the manufacturer is visible on the cap, go online and look up the data sheet. That will show the ESR of a new part. Is yours close? Don't expect it to be the same, but if it's higher by less than 20 percent or so, it's still good. Any more than that means the cap is starting to go, even if it hasn't caused problems yet.

If you can't find specs for your particular cap, compare the measured ESR against that of a new capacitor of the same capacitance and voltage rating, regardless of who made it. If the ESR values are close, the test cap is okay. A capacitor with high enough ESR to cause circuit malfunctions will read at least two or three times what a good one does. If you don't have a new cap to compare with, try looking up an ESR chart online to see what typical parts with the same voltage and capacitance values should read. Some meters come with them. Mine did.

Switching power supplies (which we'll cover in Chapter 15) and most computer motherboards use polymer electrolytic caps, which have especially low ESR ratings. If you compare a bad low-ESR cap with a good standard one, the bad cap may seem to compare well when it's really too far gone to work in a circuit requiring low ESR. Be sure to compare similar parts.

Finding Short Circuits

ESR meters have a valuable off-label use: They can serve as ohmmeters with a very low scale. That, combined with their super-small test voltages that make most components disappear from the readings, lets you find elusive short circuits. Even more important, you can rule out what *isn't* shorted without having to remove components.

Now and then, you'll work on a device with a short across the DC power supply line. This happens frequently with automotive electronics, due to the harsh electrical environment in cars. The problem seems simple enough, but where is the short? Everywhere you poke your DMM's probe, it's the same: 0 ohms.

Ah, but it's not! Really, there's some fraction of an ohm that's too low to see with a DMM. It's exactly the range measured by an ESR meter, though. Use it like any other ohmmeter, with power disconnected from the circuit under test and power supply capacitors discharged. Hook the negative test lead to the negative power connection where it joins the circuit board. In nearly all circuits, that's ground. Be sure to connect it right at the board, not on the metal chassis, because fractions of an ohm tell the story here, and you can't predict how much resistance is between the chassis and the board or how close the connection between the two is to the short you're hunting.

Probe around the traces connected to the positive side of the power input, looking for the lowest resistance. If the value goes up everywhere you look relative to the reading directly across the power input points, either the short is right there at the input points or, more likely, your ground isn't near the short. Try other ground points on the board. Find the point of lowest resistance, and you're as near to the short as you can get. The bad part is almost in your hands!

When you get close, you may find several components connected together that could be the culprits. Which one is it? Check each of them with the ESR meter, with both of the test leads directly across the suspected part. Good electrolytic capacitors will show low values, but they shouldn't be as low as the value of the short. No matter what kind of part you test, if the reading isn't as low as or lower than the one that led you there, that part is not shorted. Keep looking. When you find that lowest reading, you've found the shorted component.

This technique led me to a shorted diode in a mobile two-way radio that its owner had given up for dead. I don't know how else I could have found it. ESR meters rock!

Oscilloscope

Back in Chapter 2, I emphasized that the oscilloscope is your friend. Now it's time to get acquainted with your new best buddy. This is the most important instrument, so learning to use it well is absolutely vital to successful repair work. There's no need to be intimidated by all those knobs and buttons; we'll go through each one and see how it helps you get the job done. Various scope makes and models lay out the controls differently, and some call them by slightly different names, but they do the same things. Once you get used to operating a scope, you can figure out how to use any model without difficulty.

While the functions of analog and digital scopes are basically the same, each type offers a few features unique to its species, along with some characteristic limitations. These days, most new scopes are digital, with LCD screens. But all scopes are based on the analog style that ruled for decades, and they are a bit easier to use and understand. So, let's look at a scope's functions and operation using an analog instrument as an example, with digital-specific differences noted along the way. Then we'll review some important items to keep in mind when working with a digital unit.

Overview

You cannot harm your scope by misadjusting its controls, outside of possibly damaging the CRT (cathode-ray tube) with an extremely high brightness setting. But even that doesn't happen in an instant. So have no fear as you play with the knobs, and feel free to experiment and learn as you go. Just keep the brightness to reasonable levels and you'll be fine. This isn't an issue at all with an LCD screen, though. There's no way to hurt one of those with any settings you can make.

The purpose of a scope is to plot a graph of electrical signals, with horizontal motion representing time and vertical motion representing signal voltage. Various controls adjust the speed and the voltage sensitivity so you can scale a wide range of signals to fit on the display and perform rough measurements based on how they line up with the grid, called the *graticule*, overlaid on the screen. Others help the scope trigger, or begin drawing its graph, on a specific point in the signal for a stable image. Still more let you perform special tricks that are helpful in viewing complex signals. Now you know why a scope has so darned many knobs!

To get signals on the screen, first you will connect the probe's ground clip to circuit ground of the device you're examining. If circuit ground is connected to the wall plug's round ground terminal, that's fine, but *never* connect the clip to any unisolated voltages— that is, points connected to the AC line's hot or neutral wires. Doing so presents a serious shock hazard, along with the distinct possibility of destroying your scope. (This issue crops up mostly when you're working on switching power supplies, so read up on them in Chapter 15 carefully before you try to connect your probe to one.)

Next, you'll touch the probe's tip to the circuit point whose signal you want to see, and set the vertical, horizontal and trigger controls to scale the signal to fit on the display and keep it steady. Really, that's all there is to it. The rest is just details.

Graticule and Measurements

The graticule boxes on the face of the screen are used for visual estimation of the signal's voltage and time parameters, and the vertical and horizontal controls are calibrated in *divs*, for divisions (Figure 6-1). One box equals one division, and the boxes are subdivided into five equal parts for an easily visible resolution of 0.2 divisions, or 0.1 divisions if you want to count the spaces in between the subdivision lines. Compared to a digital meter, a scope's graticule calibration is neither especially accurate nor precise. The graticule is still useful as all heck, though, as you'll discover as we continue exploring scope operation.

So, if the vertical input control is set to 0.5 volts/division and your signal occupies two divisions from top to bottom, it's a 1-volt signal, *peak to peak* (from the bottommost point to the topmost). Similarly, if the Time/div control is set to 0.5 μs (microseconds, or millionths of a second) and one cycle of your signal occupies two divisions from left to right, it has a period of 1 μs and is repeating at a rate of about 1 MHz. (1/period = frequency; something that occurs every millionth of a second happens a million times a second, right?).

Notice I said "*about* 1 MHz." As I mentioned, scopes are not intended for measurements requiring tremendous precision or accuracy. A few percent is the best they do. Newer designs offer more accurate time calibration, thanks to digital generation of their internal timing clocks, but the precision is still low compared to that of a digital frequency counter's many digits. Plus, the vertical input specs don't approach those of the horizontal, even on digital scopes, because scaling the incoming voltage so it can be digitized is still an analog process, with all the attendant drift and error inherent in such processes.

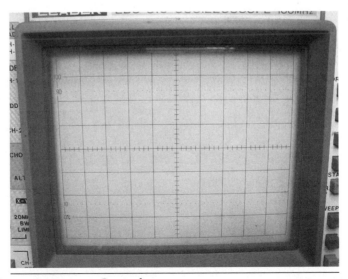

FIGURE 6-1   Graticule.

Though the layout of scopes varies by manufacturer, vertical controls are usually near each other, with horizontal controls grouped together somewhere else. Typically, vertical stuff is on the left and horizontal is on the right. Other functions, like triggering and screen controls, may be anywhere, although screen settings are usually under the screen, with triggering near the far-right edge of the control panel. Some digital scopes, especially the pocket variety, make extensive use of menus and don't have control layouts anything like the traditional scheme. Most of the functions are the same, but you'll have to locate the settings in the menus. So, whenever I talk about pressing buttons, take it to mean selecting menu options, if that's how your scope works.

To begin, find the power button, most likely near the bottom of the screen. Turn the scope on and adjust the following controls.

Screen Settings

This group of controls, found only on CRT-type scopes, is pretty much always located beneath the screen (Figure 6-2). Most scopes with CRTs are analog, but some early digital units used them too.

- *Brightness or intensity.* Set it to midrange. If it seems rather bright, turn it down a bit. If you see nothing on the screen, *don't* turn the brightness way up. Other controls may need to be set before you'll see a line, or *trace*, on the display. If your scope has dual brightness controls for A and B, use A. The B control is for delayed sweep operation, an advanced option we will explore a little later on.
- *Focus.* If you can already see a trace on the screen, adjust the focus for the sharpest line. Otherwise, set it to midrange.
- *Astigmatism or Astig.* Set it to midrange. Some scopes don't have an Astig control, but most do. It may be on the back panel, since it doesn't get adjusted very often.

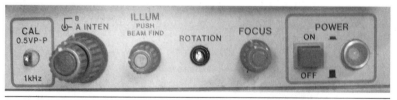

FIGURE 6-2 Screen controls.

Do It All For Me, Please

Many modern digital scopes, and a few analog types, offer a one-button "auto setup" feature. If yours does, push that button! It should get you a trace on the screen. But keep reading anyway. Automatic settings are a handy shortcut but don't always show you what you want to see. You'll still need to know how to set things manually for effective troubleshooting with your scope.

Vertical Settings

- *Vertical mode.* This control or set of buttons may be anywhere on the scope, but will usually be found near the rotary controls of the vertical channels. Set it to channel 1.
- *Channel 1's Volts/div or Attenuator knob.* Set it to 0.5 volts. Make sure its center knob is fully clockwise. On most scopes, it'll click at that position.
- *Input coupling or AC/DC/GND.* Set it to DC.
- *Channel 1's Vertical position* (Figure 6-3). Set it to about one-third of the way up.

FIGURE 6-3 Vertical controls.

Horizontal Settings

- *Sweep rate or Time/div.* Located to the right of the screen, this probably will be the biggest knob on the scope, and it may have a smaller knob inside, with an even smaller one in the center of that. Set the *outer* knob to 2 milliseconds (ms) per division. Make sure the innermost knob at the center of this control is all the way clockwise.
- *Sweep mode.* Look for "Auto," "Normal" and "Single." Set it to Auto.
- *Horizontal display.* Look for a knob or buttons labeled "A," "A intens B" and "B." Set it to A (Figure 6-4).

FIGURE 6-4   Horizontal controls.

Trigger Settings

To find the trigger controls, look for a knob labeled "Level." Also look for switches for "Coupling," "Source" and "Slope." If your scope has separate trigger sections labeled "A" and "B," use A (Figure 6-5).

- *Source.* Channel 1.
- *Coupling.* AC.
- *Level.* Midrange; on most scopes, the knob's indicator line will be straight up.

These settings should result in your seeing a horizontal line across the screen. If not, turn the channel 1 vertical position knob back and forth. You should see the line moving up and down.

If you still don't see anything, try turning up the brightness control pretty far. Still nothing? Turn it back down to medium. If you saw a spot on the left side when you turned it up, then your scope is not sweeping across the screen. (On an LCD scope, you can't do this because there is no brightness control in the traditional sense. There might be a setting to adjust the contrast of the screen or the brightness of the backlight, but it won't show the dot on the left.)

Check that the Sweep mode is set to Auto. If it's on Normal or Single, you will not see anything when no input signal is applied. If you still have a blank screen, either you have

FIGURE 6-5   Trigger controls.

made an error in this initial setup, your scope is not getting power or it's not functional. Go back and check all the settings again.

Viewing a Real Signal

Assuming that you do see the line, connect a scope probe to the channel 1 (or A) vertical input by pressing its connector onto that channel's input jack and then turning the sleeve clockwise about a quarter turn until it locks. If the probe has a little "10X/1X" switch on it, set it to 10X. Usually, the switch is on the part of the probe you hold, but some types have the switch on the connector, at the scope end. Touch your finger to the probe's tip, and you should see about one cycle of AC on the screen. It might wobble back and forth a little, but it should be fairly stable. If all you see is a blur, try adjusting the trigger level knob back and forth until the image locks. You can also step the vertical attenuator's range up and down so the image occupies most of the screen. If your scope has auto setup, press the button again while you're touching the probe tip, and it should scale everything to show you the signal. You're looking at the voltage induced into your body from nearby power wiring! Pretty startling, isn't it?

What All Those Knobs Do

Now that you have the scope running, let's look at what each control does and how to use it.

Screen Controls

Screen controls adjust the electron beam for optimal tracing, accounting for changes due to drift, writing speed, and so on. Again, these controls are specific to CRT displays; you won't find them on scopes with LCD screens.

Brightness or Intensity   This sets the brightness of the trace. It should be adjusted to a medium value. Don't crank the brightness way up for very long or you may burn the trace into the tube's phosphors, and there's no undoing that. Depending on the speed of the signal you're examining, you might have to turn the brightness up so high that the trace will be way too bright when you remove the signal or slow the scope's sweep rate back down, and you'll have to back off the brightness again. Be sure to do so without much delay.

You may find two brightness controls: one for the main A sweep and another for the delayed B sweep. The extra brightness control is there because the beam will be sweeping at two different speeds, and the faster sweep might be too dim to see without a brightness boost.

If you turn the brightness up very high to see a fast signal, the beam's shape may distort or go out of focus a bit, even if the displayed brightness remains low. That's normal and is nothing to worry about, but it means the scope's circuits are being driven to their maximum levels, so it's a good idea to keep the brightness down below the point at which distortion becomes significant.

Focus   This focuses the beam. Turn it for the sharpest trace. When adjusting the astigmatism control, you may have to alternate adjusting Astig and Focus for maximum sharpness.

Astigmatism, or Astig   This sets the beam's shape and should be adjusted for the sharpest, thinnest trace when displaying an actual signal. You can't set it while observing a flat line. An easy way to adjust it is to touch the probe's tip to the scope's cal (calibrator) terminal, which outputs a square-wave signal useful for calibrating several of the instrument's parameters. You can let the probe's ground wire hang, since it's already grounded to the scope through the cable. Adjust the channel 1 input attenuator, the A trigger level and the horizontal Time/div control to get a few cycles of the square wave on the screen. Turn the Astig and Focus controls until the waveform looks sharpest. If your scope offers measurement features and displays alphanumeric info on the screen along with the trace, you may have to compromise a little to keep everything looking reasonably sharp. Normally, you won't have to mess with Astig again. Now and then, you might touch it up when viewing very fast signals with the brightness control cranked up. If you do, you'll need to reset it afterward to keep slower waveforms looking sharp.

Rotation   This is usually a recessed control under the screen, accessible with a screwdriver. It compensates for ambient magnetic fields that can cause the trace to be tilted. Get a flat line, use the channel 1 vertical position knob to center it right down the middle, and adjust the rotation to remove tilt. Unless you take the scope to another locale, you'll probably never have to touch this control again.

Illumination   This adjusts the brightness of some small incandescent bulbs around the edge of the screen so that you can see the graticule better, especially when shooting photos of the screen. On a used scope, if the control does nothing, the lamps may be burned out. Their functionality has no effect on the operation of the scope. I always leave mine turned off anyway. The days of oscilloscope cameras are long gone.

Beam Finder   Activating this button stops the sweep and puts a defocused blob on the screen so you can figure out where the beam went, should it disappear. If the blob is toward the bottom of the screen, the trace has gone off at the bottom. If it appears near the top, the trace is above the top. If it's in the middle, the trace is within normal viewing limits but the horizontal sweep is not being triggered to move the beam across the screen.

On some scopes, the sweep continues to run, but all dimensions get smaller and everything gets brighter so you can see a miniature version of what might otherwise be off the screen. That can be more informative than just a blob.

Vertical Controls

The vertical controls scale the incoming signal to fit on the screen from top to bottom. They also allow you to align the image to marks on the graticule for voltage-measurement purposes.

Probe Compensation   This makes the probe match the input channel's characteristics to ensure accurate representation of incoming signals. It is also a screwdriver adjustment, but it's not on the scope itself. Instead, you'll find it on the probe, and it can be at either end. It won't be marked, so look for a hole with a little slotted screw. Make sure the channel 1 input coupling is set to DC. If the probe has a "10X/1X" switch, set it to 10X. Touch the probe to the scope's cal terminal, and adjust the input attenuator so that the square wave uses up about two-thirds of the vertical space on the screen. Set the Time/div control so you can see between two and five cycles of the square wave. Look at the leading edge of the waveform, where the vertical line takes a right turn and goes horizontal at the top of each square. Adjust the probe compensation for the squarest shape. In one direction, it'll make a little peak that sticks up above the rest of the waveform. The other way, it'll round off the corner. It may not be possible to get a perfect square, but the closer you can get, the better (Figures 6-6 and 6-7).

Once the probe is matched to the input channel, it'll have to be recalibrated if you want to use it on the other channel or on another scope. This takes only a few seconds.

Vertical Input Attenuator or Volts/Div   The outer ring of this control scales the vertical size of the incoming signal so it will fit on the screen. The voltage marking on each range refers to how many volts it will take to move the trace up or down one division, or box, on the graticule when you're using a 1X probe, which passes the signal straight through to the scope without altering it.

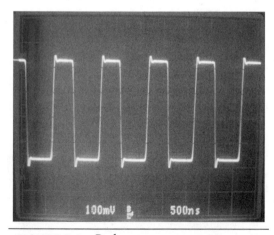

FIGURE 6-6   Probe overcompensation.

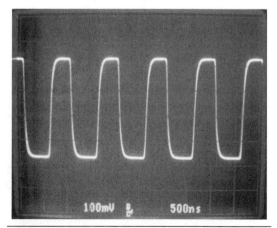

FIGURE 6-7   Probe undercompensation.

Here's where things get interesting. Remember that "10X/1X" switch on the probe? When set to 10X, it divides the incoming signal's voltage by 10. If your probe has no switch, it still does the same thing, as long as it's a 10X probe, which most are. You have to multiply the attenuator's setting by 10 to make up for that.

For example, if you measure a voltage that makes the trace rise 3.5 divisions and your attenuator is set to 0.1 volts/div, multiply 3.5 × 0.1 × 10 to get 3.5 volts. This might sound overly complicated, but there's an easy way to do it: Just move the decimal point one space to the right when looking at the attenuator. If it's set to 0.1 volts/div, remember it's really reading 1 volt/div with a 10X probe. Then multiply that by what you see on the screen, and you're all set.

Some fancier scopes automate the 10X factor when used with their own brand of probes. The probes alert the scope to the scaling factor, and the input attenuator's markings are illuminated at the correct spot, or the corrected attenuator value is shown on the screen, so you don't have to do the arithmetic.

When you're interpreting the displayed signal, keep in mind that the precision to which you can measure things depends on the setting of the attenuator. If it's set to 1 volt/div (after accounting for the 10X probe factor, of course), you can visually estimate down to about 0.1 volts using the subdivision lines and the spaces between them. If it's set to 10 volts/div, you can estimate only down to about 1 volt, since the same graticule box now represents 10 volts instead of 1, so each subdivision line represents 2 volts.

You may be wondering why probes divide the signal by 10 and why some have switches. To display a signal, the scope has to steal a tiny amount of that signal from the circuit you're testing. It's like a blood test: You have to take a little blood! The object of the probe's division is to present a very high *impedance* (essentially, resistance) between the scope and the circuit to avoid loading it down—that is, stealing enough current from it to alter its behavior and give you a false representation of its operation. Because the 10X probe needs to present to the scope only a tenth of the signal's actual voltage, it has internal voltage-dividing resistors that give it an impedance of 10 MΩ, (megohms, or 10 million ohms). That's very high compared to the resistances used in any common circuit of the sort you'll want to measure. Extremely little current passes through such a high resistance, so the circuit under test doesn't notice the loss. Also, using a 10X probe lets you measure voltages 10 times higher than what the scope normally can accept without damage.

If your probe offers 1X, that switch position removes the voltage divider, passing the signal directly to the scope and resulting in an input impedance of 1 MΩ. That's still pretty high, but it can affect some small-signal and high-frequency circuits. Usually, you'll keep your probe at 10X unless the signal you want to see is so small that you can't get enough vertical deflection on the screen even with the attenuator set to its most sensitive range. Now and then, you may scope a circuit that generates enough electrical noise to get into the probe through the air like a radio signal gets into an antenna. When you're at 10X, the impedance is so high that it takes very little induced signal to disturb your measurement. Switching to 1X may make the extraneous noise disappear or at least get much smaller. That kind of thing happens mostly when probing CRT TV sets, LCD fluorescent backlighting circuits and switching power supplies, all of which use high-voltage spikes capable of radiating a significant short-range radio signal.

Using 1X will let you see rather small signals as long as it doesn't load them down too much—which depends on the circuit you're probing. Also, some scopes let you pull out the center knob on the attenuator to multiply the sensitivity of the selected range by a factor of 10, letting you view really tiny signals of the sort you might see from a phono cartridge or in low-level stages of radio and TV receivers. Doing so causes some signal degradation, so use this only when you really need it. You probably never will.

Ah, that center knob. It's called the *variable attenuator*. Normally, you keep it in the fully clockwise, calibrated position so the volts/div you select will match the graticule, allowing you to measure voltage values. Sometimes, you may want to do a relative measurement—that is, one whose absolute value doesn't matter, but you need to know if it's bigger or smaller than it was before, or its size relative to another signal. To make such measurements easy to read, it's very helpful to line up both ends of the signal with lines on the graticule. Turning the center knob counterclockwise gradually increases the attenuation, reducing the vertical size of the signal and letting you align its top and bottom with whatever you like. It's crucial to remember, though, that you can't take an actual voltage measurement this way; the vertical spread of the signal has no absolute meaning whenever the variable attenuator is engaged. To remind you, many scopes have a little "Uncal" light near the attenuator so you'll know when the variable attenuation is on and the channel is uncalibrated.

Input Coupling (AC/DC/GND)   This determines how the signal is *coupled*, or transferred, into the vertical amplifiers, and it's one of the most important features on a scope. As you will see, the choice of coupling enables a neat trick for examining signal details and is not limited to being used in the obvious way, with DC for DC signals and AC for AC signals. You'll find yourself switching between the two settings quite often when exploring many types of signals.

GND   The GND setting simply grounds the input of the scope, permitting no voltage or signal from the probe to enter, and discharging the coupling capacitor used in the AC setting. (More about this shortly.) It does *not* ground the probe tip! You don't need to remove the probe from the circuit under test to switch to the GND setting. This selection is used to position the trace at a desired reference point on the screen, using the input channel's vertical position control, with no influence from incoming signals. Not all scopes have a GND setting. Some probes offer it on their 10X/1X switches. If you don't have one in either place, it's no big deal, because you can always touch the probe tip to the ground clip. Having a GND setting just makes getting a clean, straight line of 0 volts a bit more convenient.

DC   In the DC position, the signal is directly coupled, and whatever DC voltage is present will be plotted on the screen. If you want to measure the voltage of a power supply or the bias voltage on a transistor (a DC voltage applied to it to turn it on, as described in Chapter 4), or see when a signal is riding on a DC voltage, use this position.

To measure a DC voltage, first remove the probe from the circuit and touch it to the ground clip, or switch to the GND setting, and turn the vertical position knob to wherever on the screen you want to call 0 volts. If you're measuring positive voltage, the bottommost graticule line is a good place to put the trace. If the voltage is negative, set the trace at the topmost line because it will move down when the signal is applied. Then, touch the probe to the point you want to measure, or switch back from GND to DC if you kept the probe touching the test point, and observe how many graticule divisions the trace rises or falls. Unless your scope does it for you, don't forget to multiply the attenuator's marking by 10 to compensate for your 10X probe!

If you touch a voltage point with the probe and the trace disappears, you have probably driven it off the screen with a voltage bigger than can be handled by the range you've selected with the vertical attenuator control, so set that to higher voltages per division until you can see the trace. After switching the attenuator to a different range, perform the zero setting again before trying to estimate a measurement from the screen, because sometimes the resting position of the trace drifts a little bit when you change the range. If the zero line doesn't line up with the graticule anymore, your measurement won't be correct. This can happen with both analog and digital scopes.

AC   As we discussed in Chapter 4, many signals contain both DC and AC components. They have a DC *voltage offset* from 0 volts, but they're also not just a straight line; there's variation in the voltage level, representing information or noise. So, how is a varying DC voltage AC? Isn't it one or the other?

As I mentioned in Chapter 5, voltage level and polarity are entirely relative. Any voltage can be positive with respect to one point and negative with respect to another. Take a look at Figure 6-8. That signal is above ground in the positive direction, with none of it going negative, so it's a DC signal with respect to ground, and its height above ground is

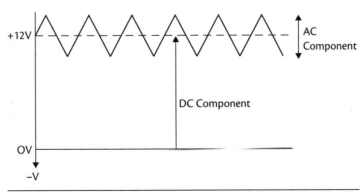

FIGURE 6-8   AC and DC components of a signal.

its *DC offset* or *DC component*. The top of it, however, is wiggling up and down. Imagine if you could block the DC offset and that dotted line became 0 volts. The signal would then be AC, with half above the line and half below.

It looks nice on paper, but you can't actually do that, right? Sure you can! Passing the signal through a capacitor will block the DC level, but, as the signal wiggles, those changes will get through, resulting in a true AC signal swinging above and below ground, with polarity going positive and negative. Hence, those wiggles are known as the *AC component* of the DC voltage. (And now you know a quick-and-dirty way to turn a positive voltage into a negative one! See, I wasn't kidding! There truly is no absolute polarity.)

When you select AC coupling, the scope inserts a *coupling capacitor* between the probe and the vertical amplifier, blocking any DC voltage from deflecting the beam. This changes everything! By blocking the DC component of a signal and passing only the AC component, you can examine that component in great detail. Let's see how.

Cleaning Up the Dirt   Suppose you have a 12-volt power supply in an LCD monitor with erratic operation. The backlight doesn't like to turn on. When it finally does, sometimes it shuts itself off. You suspect the supply might not be putting out clean power. In other words, some noise could be riding on top of the voltage, perhaps caused by weak filter capacitors, confusing the microprocessor and turning it off. You fire up your scope, set it to DC coupling and check that 12-volt line, but it looks okay. Hmmm . . . there might be a little blurriness on the line, but it's hard to tell for sure, so you crank up the sensitivity of the vertical attenuator to take a closer look. Oops! The trace is now off the screen. You turn the vertical position control down to get it back, but now that control is as low as it will go, and the only way to get the line on the screen is to turn the attenuator sensitivity down again. You can see the line once more, but you're back where you started. You can't check for any small spikes or wobbles in it because it keeps going off the screen every time you up the sensitivity enough to examine the small stuff.

If only that same noise were centered around 0 volts instead of 12 volts, then the trace would stay put, and you could fill the screen with even the tiniest changes. No problem. Switch to AC coupling to block the DC component of the signal, and you can crank the attenuator for maximum sensitivity without budging the trace. This is a very powerful technique for examining signals with both DC and AC components. Many signals are of that form, and you'll use AC coupling quite often, regardless of whether the signal is really AC or not. In fact, a symmetrical AC signal (one whose positive and negative halves are exactly the same, with no DC offset) will look identical with either DC or AC coupling, so switching between the two is an easy way to see if an offset exists. If the trace doesn't shift vertically when you flip the coupling switch, there's no offset.

Tilt   If the signal's changes are slow enough, the coupling capacitor charges up and eventually begins to block the slowly changing signal voltage until the other half of the cycle discharges it. This is called *low-frequency roll-off*; the coupling cap acts as a high-pass filter, permitting high frequencies to pass through while gradually rolling off (attenuating)

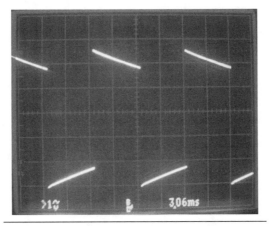

FIGURE 6-9   Tilt due to low-frequency roll-off.

lower ones, passing nothing when the frequency reaches zero. The effect will distort low-frequency signals, causing their flat areas to head toward the center of the waveform as the coupling capacitor charges (Figure 6-9).

To see this in action, view about four cycles of the square wave from the cal terminal, and switch between AC and DC coupling. (Adjust the vertical position if a DC offset moves part of the waveform off the screen when you change the coupling.) The flat tops and bottoms look tilted with AC coupling, losing amplitude and heading toward the middle of the waveform as the coupling cap charges. As the signal frequency rises, the cap has less time to charge, so this effect fades away, for a truer representation of the signal. When you're viewing low-frequency signals with AC coupling, always keep in mind that long, sloping areas may in fact be flat, and the scope's coupling cap might be causing the apparent slope. The easy way to verify the presence of roll-off is to switch to DC coupling (unless a DC offset drives the trace too far off the screen for the vertical position control's range to bring it back). If the sloping lines become flat, you know the cap was fooling you.

After using AC coupling, always ground the probe tip or momentarily switch the input coupler to GND to discharge the coupling capacitor before probing another point. Otherwise, whatever is stored on it will discharge into the next point you touch, confusing your reading and possibly damaging the circuit under test, although there's only a remote chance of that. When you ground the probe, you should see the trace jump for a fraction of a second and then return to where it was as the cap discharges.

AC coupling can be inconvenient when you're working with signals that change amplitude (vertical size) in an asymmetrical fashion. Analog video signals, for instance, were "clamped" to a fixed voltage, with the sync pulses not deviating from their position at the bottom as the video information at the top changed with the TV picture's content. Using AC coupling with such a signal will cause both ends of the waveform to bounce around as video content rises and falls, because the midpoint of the signal is constantly shifting. The effect is disconcerting and difficult to interpret. Video signals and others with similar asymmetry, like digital data streams, are best viewed with DC coupling so that the unchanging end of the signal stays put.

The example was from a real case of an LCD monitor I fixed using the AC coupling on a DC signal technique. Sure enough, there were narrow 0.5-volt spikes on the power supply line confusing the microprocessor and tripping the unit off. Once I switched to AC coupling and saw the spikes on the filter capacitor's positive terminal, I didn't even have to check the cap, because I knew a good one would have filtered them out. I just popped in a new part. A quick check confirmed that the spikes were gone, and the noise on that line was now well under 100 millivolts. The micro was happy, and so was I. The monitor worked great. Without the AC coupling trick, I'd never have known those spikes were there.

Vertical Position   This moves the trace up and down. Use it to set the zero reference point to one of the graticule markings when taking DC measurements. Otherwise, set it wherever you like for examining the signal you're viewing. When using the scope in Dual-Trace mode, you'll usually set the two input channels' vertical position controls to keep the waveforms separated. It's conventional to put channel 1 in the top half of the screen and channel 2 in the bottom, but you can place them wherever you want, even on top of each other. Now and then, this is useful when comparing two signals.

Bandwidth Limit   This limits the frequency response of the vertical amplifiers. In 100-MHz scopes, the limit is typically 20 MHz. Switching in the limit removes noise bleeding in from external sources, particularly FM radio stations. With 100-MHz bandwidth, the scope can pick up the lower two-thirds of the FM band, resulting in a noisy-looking signal if you happen to live near an FM broadcast station. Some other radio services may get into your measurements, too, if they're strong enough. Also, the product you're servicing might induce objectionable noise if it uses fast pulses, like a switching power supply does, or contains radio-frequency sections that emit signals strong enough to get into your scope probe through the air. Leave the bandwidth limit turned off unless noise problems are making your trace blurry.

Vertical Mode   On a dual-trace scope, this selects which channels you will view, and how. Most scopes offer these options:

- *Ch 1*. You will see only channel 1. Input from channel 2 can still be used to feed the trigger (the circuit that keeps the waveform steady on the screen) if you want, even though you can't see it.
- *Ch 2*. The same, but in reverse. You'll see only channel 2.
- *Add*. The voltages of the two channels will be added together and shown as one trace.

Adding two signals together is pretty pointless, so why is this here? One or more of your channels should have a button marked "Invert" or "Inv." Pressing it makes the channel flip the waveform passing through it upside down, with positive voltages deflecting downward and negative ones upward. If you flip one channel upside down and then select the Add mode, the signals will be *subtracted*, and that is very useful in certain circumstances.

What's the Difference?   Let's say you have an audio amplifier with some distortion. Or, perhaps, the high-frequency response is poor. You want to find which stage of the circuit is causing the problem. Feed some audio to the amplifier's input. Set both scope channels to AC coupling. Connect one channel to a stage's input and the other to the stage's output. If the amplifier hasn't already inverted the signal, do the invert-add trick. Then, use the vertical attenuator of the channel with the bigger signal (usually the stage's output) to reduce its displayed amplitude until the signals are the same size and cancel each other out, so the resulting trace is as flat as possible. What's left is the difference between the two signals: the distortion. Using this technique, you can easily see what a circuit is doing to a signal, and the result may be enlightening, leading you to a diagnosis. Most amplifier stages will show some difference, but it should be minor. When you see something significant, you've found the errant stage.

Chop   This puts the scope into Dual-Trace mode, displaying two signals at once. The two will appear to be independent, and you can set the vertical attenuation and position of each channel at will. The horizontal sweep speed set with the Time/div control will apply to both channels, and the scope will trigger and start the sweep based on the timing of whichever of the two channels you choose for triggering. However, the scope is not truly displaying two simultaneous events; it just looks that way. In reality, it is switching rapidly between the two channels, with the beam bouncing between them, drawing a little bit of one channel and then a little bit of the other. It is chopping them up.

This method works very well, ensuring that the time relationship between the two signals is well preserved, since they are really being made by the same beam as it traverses the screen from left to right. Chop mode has a few serious limitations, however. If you crank the sweep speed way up, you can see the alternating segments of the two channels and the gaps between them. So, the mode is not useful at high sweep speeds. Also, if the two signals are *harmonically unrelated* (their frequencies are not a simple ratio, so the cycles don't coincide), only the channel chosen for triggering will be visible; the other will be a blur.

Alternate, or Alt   This is the other Dual-Trace mode, and it works rather differently. Instead of chopping the waveform, it draws an entire channel across the screen and then goes back and draws the other one, in two separate, alternating sweeps. Alt mode leaves no gaps in the traces, so it's suitable even for very fast sweep speeds. Plus, under certain circumstances, you can view harmonically unrelated signals, with separate triggering for each channel making them both look stable.

Alt mode has its own limitations, though. At slow sweep speeds, you'll see the two sweeps occur, one after the other, with the previous one fading away, resulting in a rather uncomfortable blinking or flashing effect. Also—and this one is much more serious—the timing relationship between the two signals may be disturbed because of when the scope triggers and how long it takes to sweep the screen before it has a chance to draw the second waveform. In Alt mode, the displayed alignment in time of one channel to the other *cannot be trusted*. Always choose Chop mode when you need to compare or align the timing of two signals. In fact, use Chop mode as much as possible, and select Alt mode only when chopping interferes with the waveform, or the two signals you're viewing are unrelated in time so that each one needs its own trigger.

X-Y   This mode stops the sweep and lets you drive the horizontal deflection from a signal input to channel 2. Be careful when pressing this button, because the stopped beam will create a *very* bright spot on the screen that can burn the tube's phosphors quickly. Turn the brightness all the way down before trying it, and then gradually turn it up until you see the spot. As with other brightness concerns, this isn't an issue with LCD screens.

At one time, X-Y mode was a useful way of detecting nonlinearity (when an amplifier's output doesn't follow its input faithfully), comparing two frequencies and observing some other parameters. Today, it has little or no application, at least not for general service work. I have never needed to use X-Y mode even a single time. It can be fun to play with, though. Feeding various signals into the two channels in X-Y mode will make your scope produce some really interesting patterns.

Trigger Controls

To present a stable waveform, the scope must *trigger*, or begin each sweep, at the same point in each cycle of the incoming signal so the beam will draw the same signal features on top of the last ones. Otherwise, all you'll see is a blur. The trigger locks to a specific voltage level in the signal that you set with the controls. Stable triggering is one of the most critical features on a scope, and it's important that you get good at using the trigger controls to achieve it.

Trigger Lock Light   This indicator tells you when the trigger is locked to a feature of the signal. When it's on, you should see a stable waveform. If not, some other control is improperly set, or the trigger may be locking to more than one spot in each cycle of a complex waveform that has more than one occurrence of the same voltage level per cycle (Figure 6-10).

Source   This selects which channel will be used to trigger the sweep, along with some other options:

- *Ch 1*. Channel 1's signal will feed the trigger. Use this mode for single-channel operation or for dual-channel work when you want channel 1's signal to control timing of the sweep.

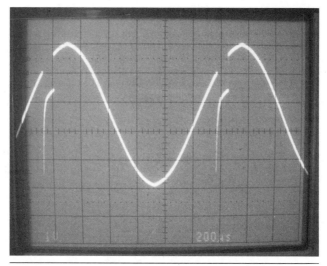

FIGURE 6-10   Waveform with multiple trigger points.

- *Ch 2.* Channel 2's signal will feed the trigger.
- *Alt.* This is different from the Alt vertical mode, but it shares the same limitation. In Alt trigger mode, each channel will feed the trigger, one after another. This completely invalidates the timing relationship between displayed signals because you have no idea how much time has elapsed between when the first channel's sweep finished and the second channel's sweep began. This is useful when you want to look at two signals whose periods are not related, and you want them both to display stably. Just remember that it's like having two separate scopes; no time relationship exists between the displayed signals. Not all scopes offer this mode.
- *Line.* This uses the 60-Hz AC power line as a timing reference, generating 60 sweeps per second. It's useful when viewing signals at or very near that frequency whose own features make for difficult triggering. Now and then, it can be handy when troubleshooting AC line-operated gear, especially linear power supplies.
- *External, or Ext.* Many scopes have extra inputs that can be used as vertical channels and/or trigger inputs. External triggering is great for locking very complex signals the normal trigger can't get a grip on.

For instance, when adjusting VCR tape paths while viewing the radio-frequency (RF) waveform from the video heads, there was no stable way to trigger at the start of each head's sweep across the tape by using the signal it produced. Instead, you had to drive the scope's trigger from another signal in the VCR that was synced to the rotation of the heads. The external trigger input provided a place to feed it in, keeping channel 2 (from which you could accomplish the same thing) free for viewing other signals.

When servicing something modern like a DLP video projector, you might want to lock the scope to the position detector on the color wheel so you can see if it is properly synchronized to the circuit driving it.

External trigger inputs are primitive, with nowhere near the flexibility of a normal scope channel. If the signal you want to trigger on is complex or especially small in voltage, feed the triggering signal to your second vertical input channel, and trigger off that. You can choose not to display it by selecting only channel 1 for display even though you're triggering off channel 2.

Coupling   This is somewhat like the input coupling on the vertical amplifiers, but it offers a few more choices specific to triggering needs:

- *DC.* The trigger will lock to a specific signal voltage relative to ground. As with the DC coupling option for the vertical amplifiers, it's most useful with asymmetrical waveforms.
- *AC.* The trigger will lock to a voltage above or below the midpoint of the signal, regardless of its voltage relative to ground. It works just like the AC coupling option on the vertical amplifiers, placing a coupling capacitor in line to block the signal's DC component. You can use AC trigger coupling even when you choose DC coupling for the vertical channel. You'll use AC triggering most of the time because it makes triggering easy as you look at various signals with different DC components (Figure 6-11).
- *HF reject.* This feeds the signal through a low-pass filter, smoothing out high-frequency noise or signal features that might confuse the trigger and cause it to trip where you don't want it to. If your waveform has high-frequency components causing jittery display, try this option. If you *want* to trigger on a high-frequency feature, however, selecting HF reject will prevent triggering. Using this setting affects only the trigger operation; the signal going to the vertical amplifier is not filtered.
- *LF reject.* This feeds the signal through a high-pass filter, rolling off low frequencies and preventing them from tripping the trigger. Use it to help trigger on high-frequency

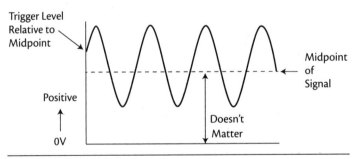

FIGURE 6-11   **AC-coupled triggering.**

signal components when lower-frequency elements are causing triggering where you don't want it. Again, the filtering affects only the trigger.

- *TV-H.* This is a specialized trigger mode for use with analog TV signals. It is optimized to help the scope trigger on the horizontal sync pulses in the TV signal. It comes from a time when much service work was on TV sets, and not all scopes offer it, especially modern LCD types, because analog video is rarely encountered anymore.
- *TV-V.* This enables triggering on the vertical sync pulses in an analog TV signal.

Slope   This selects whether the trigger locks on signal features that are rising or falling. With many kinds of signals, like audio and oscillators, it doesn't matter. Normally, leave it on +. When you want to trigger on the falling edge of a waveform, switch it to −. To see the slope feature in action, connect the probe to the cal terminal, get a locked waveform and then switch between the slopes. Look at the leftmost edge of the screen to see what the waveform was doing when the sweep triggered.

Level   This sets the voltage level at which the trigger will trip. As mentioned earlier, when the trigger is set to DC coupling, that level is absolute relative to ground. When the trigger is set to AC coupling, it's relative to the midpoint of the signal. On older scopes, the level control had no calibration. Newer analog scopes, and most digital units, show a cursor on the left side of the screen illustrating the trigger voltage. Some even display the voltage numerically. For the most part, you don't care; just use whatever setting provides the most stable display. Turn the level control back and forth until the trigger locks. If the trigger stays locked through a wide swath of this control's range, that indicates a solid trigger lock, and you should see a very stable display. If you can get trigger lock only over a very narrow range of the level control, the lock is not great, and you can expect the waveform to jump around if the signal level or shape changes even a little bit. To get a better grip on the signal, try the various coupling options, especially HF reject and LF reject, rotating the Level control back and forth for each one.

Once in a while, having a trigger level display is handy. If you're running into trouble getting stable triggering on a complex waveform, seeing the trigger level can help you understand why, and can aid in finding a more suitable spot in the signal from which to trigger.

Hold-off   This keeps the trigger held off, or unable to trip, for an adjustable period of time after its last trigger event. It's used on complex signals with irregularly spaced features of similar voltages to avoid having more than one in each cycle trip the trigger and blur the displayed waveform. If you can't get a waveform to stabilize any other way, try turning this control. Otherwise, keep it at the "Normal" position, which reduces hold-off to a minimum. Forgetting and leaving it on can make triggering confusingly difficult because the hold-off will make the trigger miss consecutive cycles if it's set for too long a period, resulting in a jumpy mess.

Horizontal Controls

The horizontal axis represents time, and the scope's drawing of the waveform from left to right is called the *sweep*. The settings controlling it determine what you'll see, even more than do those for the vertical parameters.

Horizontal Position   This positions the trace left and right. Set it to fill the screen, with the left edge of the trace just off the left side. Now and then you may wish to line up a signal feature with the graticule to make a rough measurement of period or frequency, and you can use this control to do so. Some scopes have both coarse and fine horizontal position controls.

Sweep Mode   This control offers three options: auto, normal and single.

- *Auto.* The trigger will lock to the incoming signal, and sweep will begin. When there's no signal or the trigger isn't locking on it for some reason, the sweep will go into free-run mode, triggering itself continuously so you can see a flat line when there's no signal at all or, in the case of trigger unlock, a blur. This is the mode you will use most of the time.
- *Normal.* Sweep will begin when the trigger locks on a signal but will stop when the signal ceases. This can be handy for observing when rapid interruptions occur in intermittent signals because the trace will blink at the moment the signal disappears. It's a little disconcerting sometimes, though, because if the screen goes blank, you don't know why. Maybe the trace is off the screen, maybe the brightness is too low or maybe the trigger isn't locked. Normal mode is more useful on analog scopes than on digital types because analog units display the incoming signal instantaneously. On digital scopes, there's a delay between signal acquisition and display, making it hard to find intermittent problems with them.
- *Single.* This is for Single-Sweep mode, in which the trigger will initiate one sweep and then halt until you press the Reset button (which may be the Single button itself). This helps you determine when a signal has occurred, because you'll see the flash of one sweep go by. Look for a little indicator light near the Single button labeled "Ready" or "Armed." When it's lit, the sweep can be fired one time. After that firing, the light will go out, and you must press the Reset button to rearm the sweep. You won't use this mode very often on an analog scope because the displayed signal disappears immediately. Digital scopes can freeze the waveform so you can see it after it's gone, which can be handy for examining transient events.

Time/div   The outer ring of this large control sets the *time base* (sweep speed) at which the beam will travel across the screen from left to right. It is calibrated by time in seconds, milliseconds (ms, or thousandths of a second) and microseconds (μs, or millionths of a second). The calibration number refers to how long the beam will take to traverse one division, or graticule box.

For a signal of a given frequency, the faster you set the sweep, the more horizontally spread the display will be, and the fewer cycles of the signal you will see at one time. Often, you will want to set it as fast as possible to see the most detail, but not always. In some instances, the aggregate effect of many displayed signal cycles (its *envelope*) can be more revealing than a singular signal feature. If you have low-frequency variation, such as AC line hum, on a fairly fast signal, you can't see the hum if the sweep is set fast enough to display individual cycles of the signal. But, when you turn the sweep speed low enough to view a 60-cycle event, you will see the hum riding on the signal clearly, even though the signal's waves themselves will be too crammed together to be resolvable.

To see the effects of various sweep rates, look at the calibrator's square wave, and click the Time/div knob through its ranges. On most scopes, the number of displayed waves will grow or shrink with the sweep rate. If it doesn't, your scope is changing the frequency of the calibrator to match the time base. Some of them do that.

Variable Time   The center knob uncalibrates the time base, slowing it down as you turn the knob counterclockwise. This is the horizontal equivalent of the vertical channels' variable attenuators, and it can be useful for lining up events with the graticule during relative time measurements of two signals. Normally, you'll leave it in the fully clockwise position. As with the variable attenuators, it has an "Uncal" light to remind you that the time base no longer matches the graticule, and you can't make valid time measurements.

Pull X10   On most scopes, pulling out the Variable Time knob speeds up the sweep rate by a factor of 10. The sweep does remain calibrated in this mode. It's a quick-and-dirty way to spread out a signal, and it also lets you get to the very fastest sweep rate by turning the Time/div knob all the way up and then pulling this one out. Normally, keep this knob pushed in, and pull it only when you really need it. Don't forget to push it in again, or your time measurements will be off by a factor of 10. I haven't seen this feature on a digital scope; the menu just goes up to the fastest available sweep rate.

Delayed Sweep Controls

Delayed sweep is an advanced scope function you won't need for basic repair work, but it's worth learning for those more complex situations, like servicing camcorder motor control servos, in which it's essential. My apologies for any neck injuries caused by making your head spin while reading this section! Once you actually play with delayed sweep a few times, you'll discover it's really not difficult, and it's quite nifty.

With delayed sweep, your scope becomes a magnifying glass, allowing you to zoom in on any signal feature, even though it's not the one on which you're triggering the main sweep. Why do this? It's very powerful, providing a level of signal detail you couldn't otherwise examine.

Let's say you have a sine wave from an oscillator, but it doesn't look quite right. A spike or something is distorting its shape at a particular spot. It's hard to tell what it is, but you can see a thickening of the trace at that point. You want to get up close and personal with that spot so you can really see the details of the distortion and determine what's causing it. You scope the sine wave and crank up the sweep rate, but when you get it going fast enough to spread out the mystery spot, it has already gone off the right side of the screen. You turn the sweep rate down a little bit, but now the signal is too crammed together to permit a good look at the spot.

Cue superhero music. This is a job for . . . Delayed Sweep, Slayer of Stubborn Signals, Vanquisher of Villainous Voltages! In this mode, the scope triggers on the waveform as usual, as set by the A trigger. It begins sweeping at the rate selected with the main Time/div knob. After a period of time you set with the Delay Time Multiplier knob, the B time base takes over and the beam finishes the sweep at the speed set by that time base's knob, the one inside the main sweep's knob, between the main knob and the variable one used for uncalibrated operation, as described above under "Variable Time."

The result is a compound view of the signal, with a lower sweep rate for the events leading up to your spot of interest, followed by a stretched-out, detailed view of the spot! Even cooler, you can rotate the Delay Time Multiplier knob and scan through the entire signal, examining any part of it. It's practically a CAT scan for circuitry, and you don't even need a litter box. Let's look at the controls involved with setting up the delayed sweep mode.

Delay Time or B Time Base   This control, located inside the main Time/div knob, sets the speed of the B sweep, which stretches out the waveform for close examination. Think of it like a zoom lens on a camera: The faster you set it, the more you're zooming in for a closer look at a smaller area. This can be set equal to or faster than the main time base, but not slower.

On some scopes, notably those made by Tektronix, the same knob is used for both the A and B sweeps. To engage the B sweep, you pull out the knob, mechanically separating the two time base controls. In this position, the outer ring, which sets the A sweep rate, will not move when you turn the knob clockwise to speed up the B sweep. The Horizontal Display modes discussed in the next section also may be controlled by this knob.

Horizontal Display Mode   This selects which time base (sweep generator) will drive the beam across the screen, and is how you choose between normal and delayed sweep modes.

- *A.* The main time base, which is set by the big Time/div control, will control the sweep. Delayed sweep mode will not be engaged.
- *Alt.* Both time bases will be displayed, one on top of the other. The A sweep's display will be highlighted over the area that the B sweep is stretching out. The faster you make the B sweep rate, the narrower the highlight will get, and the more stretched the displayed B sweep will be (Figure 6-12). Look for a knob called "Trace Separation." It lets you position the B sweep's display above or below that of the A sweep so you can see them better. It has no calibration or meaning in voltage measurement terms; it's just a convenience.

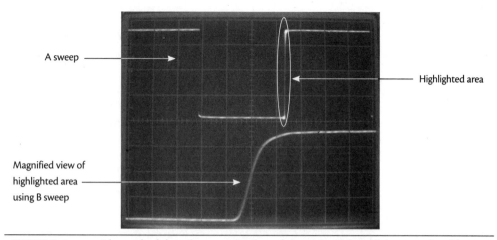

FIGURE 6-12   Alt mode delayed sweep display of the rising edge of the scope's calibrator square wave. The A trace is on the top, with the B trace below it. Notice the highlighting of the portion of the waveform being stretched out by the B trace.

If your scope has two brightness controls, you may need to increase the brightness of the B sweep to keep it visible at high sweep rates. The faster the beam sweeps, the dimmer it will appear. You won't find this on LCD scopes because the brightness of their pixels is independent of what's shown on them.

- *A intens B.* This stands for "A intensified by B" and shows you the A sweep, highlighting the portion of the signal the B sweep will cover, just as is shown in Alt mode. It's useful for zeroing in on the spot you wish to examine without cluttering up the screen with the B trace, before you switch to a mode that displays B. On CRT scopes, in both Alt and A intens B modes, the highlighting of the A sweep may not be visible if you have the A brightness control set too high. Try turning it down for a more prominent highlight.

- *B.* This shows only the B trace, at the sweep rate selected by the B or Delay Time setting (the knob inside the big, outer Time/div knob). It turns off the A trace, but the settings that result in the delayed sweep B view remain in effect, including triggering, A sweep rate and delay time. Once you've zoomed in on your spot of interest, you can switch to this setting to see only the detailed area, without the main waveform from which it is derived. In this mode, the Trace Separation control doesn't do anything.

- *Start After Delay/Triggered.* This determines what happens after the A sweep has reached the beginning of the intensified area where it will switch to B. In Start after Delay mode, the B sweep begins as soon as the A sweep hits that point. This lets you scan continuously through the signal with the Delay Time Multiplier control, and is my preferred way of using delayed sweep. It has one drawback, though. Any jitter (instability) in the triggering of the A sweep will get magnified by the faster sweep rate of the B sweep, resulting in wobbling back and forth of the stretched waveform. Sometimes, the wobble is bad enough to make examining the signal difficult.

To remove the wobble, select the Start Triggered mode. Then, when the A sweep reaches the intensified portion, the B sweep will wait for a trigger before beginning. This is what those B trigger controls are for. What's great about having a separate trigger for this function is that you don't have to use the same trigger settings for the B sweep that you selected for the main trigger. So, you can treat the detailed area as a separate signal, setting the trigger for best operation on its particular features.

The downside to this mode is that you cannot scroll smoothly through a signal, because the B sweep waits for a trigger. You can see only features on which it has triggered. Keep your scope in Start after Delay mode unless you run into wobble problems or are examining fine details in a complex signal with many subparts, like an analog video signal, where triggering on the subpart you're trying to view is essential for stability.

Delay Time Multiplier This sets how many horizontal divisions the delay will wait before starting the B sweep (or looking for a new trigger for it, in the Start Triggered mode), and is referenced to the A sweep rate. For example, if the A sweep is set to 0.1 millisecond and the Delay Time Multiplier knob is at 4.5, it'll wait 0.45 milliseconds before the highlighted area begins. For most work, you can ignore the numbers and just keep turning the knob to scan through the signal, using the highlight to pick your expansion area. Most scopes provide a vernier (gear-reduction) knob for fine control of the delay time.

B Ends A This turns off the A sweep after the B sweep starts. On some scopes, it presents both sweeps in one line, switching from A to B when B begins. On others, the two sweeps are still on separate lines, and A just disappears after the point where B takes over.

Try It!

How are those neck muscles doing? Ready to try some of this? Let's use delayed sweep to get a close-up look at the rising edge of your scope's calibrator square wave, as shown in Figure 6-12.

Square waves aren't really square! If they were, they would rise and fall instantly. But there is no *instantly*! Everything takes time to occur. Each jump up or down takes a tiny moment, meaning that what looks vertical on the screen really slants a little toward the right. It's just too fast to slant enough that you can see it when viewing entire cycles of the square wave. Here's how to see it and even measure it if you want.

Connect the probe to the cal terminal, and set up the scope for normal, undelayed operation by selecting the A horizontal display mode. Adjust the vertical, trigger and sweep rate controls to see a couple of cycles on the screen, and use the input channel's vertical position control to put them on the top half.

Disengage B ends A mode if it is currently turned on. Make sure Start after Delay is engaged, not Start Triggered. Select the A intens B mode and move the Delay Time Multiplier knob to center the highlight over the leading (rising) edge of the square wave, with the start of the highlight just before (to the left of) the edge. On a CRT scope with two brightness controls, set them as desired so you can see the highlight clearly. The A brightness control will adjust the bulk of the waveform, with the B brightness setting the highlight's intensity.

If more than one cycle of the waveform is on the screen, any of the leading edges will do. For optimum trigger stability, though, it's best to keep the A sweep as fast as possible without losing the feature you want to examine off the right side. So, try not to have more than one or two cycles visible.

Adjust the B sweep rate so that the highlight is a little wider than the square wave's edge, covering a smidge of the waveform before and after it. Note that the faster you make the B sweep, the narrower the highlight gets, meaning that you will be zooming in on a smaller area, enlarging it proportionally more.

Now switch to Alt Horizontal Display mode (*not* the Alt Trigger or Alt Vertical mode). You should see the zoomed-in rising edge of the waveform superimposed on the original wave. Use the Trace Separator knob to move it down so you can see it clearly. Increase the B sweep rate one step at a time. As you crank it up, the magnified edge will move off the right side of the screen, and you'll have to turn the Delay Time Multiplier knob clockwise just a tad to bring it back. If the magnified waveform gets too dim, crank up the B brightness.

When you get the B sweep going pretty fast, you can see the edge's slope clearly, and you may also notice some wobbles or other minuscule features at its bottom and top, details you can't see at all without the magic of delayed sweep! For fun, scan through the waveform with the Delay Time Multiplier knob, and take a look at the falling edge too. Pretty awesome, isn't it? If you want to measure the time each rise or fall in the wave takes, count graticule boxes and multiply by the speed of the B sweep. On a CRT scope, just don't forget to turn the A brightness down when you switch back to normal, undelayed operation, if you turned it up.

Cursor Controls

If your scope offers numerical calculation, it will have cursor controls that let you specify the parts of the waveform you wish to measure. The results of the measurements will be shown as numbers on the screen, along with the waveforms.

The layout of controls can vary quite a bit in this department, but the principles are pretty universal. For amplitude measurements, you move two horizontal cursors up and down to read the voltage difference between them. For time measurements, you move two vertical cursors left and right to measure the time difference between them or calculate approximate frequency (Figure 6-13).

On some scopes, you can lock one cursor to the other after you've set them, so you can move one to the start of a waveform feature and the other will follow, letting you see how the signal aligns against the second one.

Always keep in mind that these measurements provide nowhere near the accuracy or precision of those you'll get from your DMM or frequency counter! Still, you can't measure the voltage or frequency of features within signals with anything but a scope.

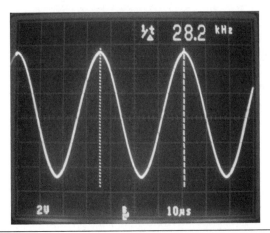

FIGURE 6-13 Cursors and frequency (1/time) measurement on a Tektronix 2445 analog scope.

Digital Differences

Operating a digital scope isn't that much different from using an analog type, but there are some items to keep in mind. First, many of the controls on an analog instrument are replaced by menus on its digital counterpart. This approach unclutters the front panel, but it's slower and more awkward to have to step through nested menus than it is to reach for a knob. As long as you keep the basic functions of vertical, horizontal and triggering in mind, though, you should have no trouble remembering where to find the options you need.

All the screen controls are gone. Because the display is an LCD screen and is not being swept at varying rates by an electron beam, you don't need astigmatism, focus, illumination and separate A and B brightness. You'll find a main brightness or contrast adjustment in a menu. Once set, it won't require any changes for different modes or sweep speeds.

The display will show various operating parameters like trigger status, volts/div and time/div, so you don't have to take your eyes off the screen and interpret a bunch of controls.

Digital scopes are generally more accurate, especially in the horizontal (time) domain. They also tend to have more stable triggering and very little drift.

Delayed sweep may be handled a bit differently. On the Tektronix TDS-220, for instance, the equivalent of the analog scope's highlight is called the *window zone*, and you set the delay time multiplier and window width using the Horizontal Position and Time knobs after selecting the mode from a menu. Instead of a highlight, you get a set of long, vertical cursors. Then, you select "Window" to see the magnified waveform. The principles are the same, of course, but there's no equivalent to the analog instrument's Alt mode, in which both the undelayed and delayed sweeps are shown simultaneously.

More than likely, the digital scope will include various acquisition modes, so you can grab waveforms, store them and display them along with live signals. It'll probably also have measurement options for frequency, period, peak-to-peak voltage, RMS voltage, and so on. Especially nice is the auto setup function we discussed earlier, which sets the vertical, horizontal and trigger for proper display of a cycle or two of most waveforms without your having to twist a single knob. Hook up the probe, hit the Auto button and there's your signal. It's the oscilloscope equivalent of autoranging on a DMM, and it saves you a lot of time and effort.

Reading the screen on a digital scope requires more interpretation. The limitations imposed by the digital sampling process, the finite resolution of the dot-matrix display and the slower-than-real-time screen updating have to be kept in mind at all times. For one thing, curved areas can have jagged edges, and it's important to remember that they are not really in the signal. Also, lines may appear thicker and noisier than they really are, due to digital sampling noise. Aliasing of the signal against the sampling rate and also against the screen resolution can seriously misrepresent waveforms under certain circumstances, as discussed in Chapter 2. Finally, the slow updating causes some details to get missed. All scopes, analog or digital, show you snapshots, but with digitals there's much more time between snaps.

When the input signal disappears, many digital scopes freeze the waveform on the screen for a moment before the auto sweep kicks in and replaces it with a flat line. This makes it hard to know exactly when signals stop. If you're wiggling a board while watching for an intermittent, the time lag can hinder your efforts to locate the source of the dropout.

Overall, a digital instrument offers more stability, options and conveniences, but an analog scope gives you a truer representation. As with just about everything, digital is the future, like it or not, and analog scopes will probably disappear from the marketplace in the next few years. There are still some being manufactured, but not many, and they're a lot more expensive than digitals of similar bandwidth because analog units cost more to make, and the market is smaller. If you snagged a good one, hang onto it! If you chose a digital instrument, just keep these caveats in mind and you'll be fine.

Soldering Iron

Soldering is probably the most frequent task you'll perform in repair work. It's also one of the easiest ways to do damage. Competent soldering technique is essential, so let's look at how to do it.

Never solder with power applied to the board! The potential for causing calamity is tremendous. First, you may create a path from the joint, through your iron, to ground via the house wiring, resulting in unwanted current that burns something out. Second, it's very easy for the iron's tip to slip off the joint and touch other items nearby, causing a potentially damaging short circuit if there's any current to short. Make sure all power sources are truly disconnected, remembering that many products don't actually remove all power with the on/off switch. Unplug the item or remove the batteries to be sure. It's good practice to check the point you want to solder for voltage with your DMM. If there's still some voltage there, even with all power disconnected, discharge the power supply capacitors as we discussed earlier until the voltage is gone.

A good solder joint is a molecular bond, not just a slapping of some molten metal on the surface. The solder contains a core of rosin, an organic material that helps it flow into the metal of the component's leads and the copper circuit board traces. When it doesn't flow well, the result is called a *cold solder joint*, and it will fail fairly quickly, developing resistance or, in some cases, completely stopping the passage of current.

To get a good joint, first *tin* the iron's tip. Warm up the iron to its full temperature, and then feed a little bit of solder onto the tip. It should melt readily; if not, the tip isn't hot enough. Coat the tip with solder—don't overdo it—and then wipe the tip on the moistened sponge in the iron's base. If you have no sponge, you can use a damp (not dripping wet) paper towel, but strictly avoid wiping the tip on anything plastic. Melted plastic contaminates the tip badly and is tough to remove.

Once the tip is nice and shiny, put another small drop of solder on it. Press the tip onto the work to be soldered, being sure it makes contact with both the circuit board's pad and the component's lead (or contact point in the case of surface-mounted, leadless parts). Then, feed some solder into the space where the lead and the pad meet, until you have enough melted solder to cover both without creating a big blob (Figure 6-14). To get a good idea of how much solder to use, look at the other pads (Figure 6-15).

 Before trying to solder tiny, surface-mount parts in a device you're trying to repair, practice on a scrap board. Experience really helps in developing successful soldering technique with these minuscule components.

FIGURE 6-14   Proper soldering technique.

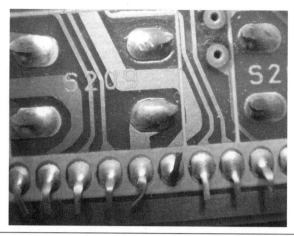

FIGURE 6-15   Good solder joints.

As the solder feeds, it should flow into the metal. Check around the edges for smooth integration into the joint. If you see a ring of brown rosin, gently scrape it away with an X-Acto knife or very small screwdriver so you can get a good look. Also, check for proper flow around the component lead. Sometimes, the flow is fine to the board's pad, but the solder is pooled around the lead without having bonded to it because the iron's tip didn't make good enough contact to get it adequately hot, or the lead had a coating of oxidation that blocked the necessary chemical bonding. In fact, that style of cold solder joint is a major cause of factory defects resulting in warranty claims. At least it was, back when most components had leads. It doesn't occur nearly as often with leadless, surface-mount parts, but I've seen a few with cold joints. Frequently, though, the problem is where the solder meets the circuit board's pad (Figure 6-16).

FIGURE 6-16   A cold solder joint. Note how the edge fails to flow into the surrounding pad.

If the joint looks like a bead sitting on top of the pad, you have not created a molecular bond and will need to reapply the iron. To get enough heat, the wattage of the iron has to be appropriate to the size of the joint. Also, you have to apply some pressure to the tip for effective heat transfer; a very light touch won't do it. Don't press really hard, though; it won't improve transfer and could cause damage.

If the iron's tip is contaminated, heat transfer will be limited. It should look shiny. Especially if it has come in contact with plastic, it could have a coating blocking the heat. Although plastic contamination is most easily removed by scraping the tip when the iron is cold, tinning and then wiping a hot tip may cut through the coating.

Contamination can occur on the leads of replacement components, too, especially with parts that have sat around for years in your parts drawers. If the leads look dull, apply some fine sandpaper or scrape them clean before attempting to solder them. They should be shiny for good solder flow.

Once you have a nice, properly flowed joint, remove the supply of solder and then the iron, in that order. If you do have to reflow the joint, add a small amount of new solder so you'll have fresh rosin on the joint to help facilitate bonding.

Soldering leadless, surface-mount components with an iron is tricky, mostly because they are so small that it's hard to keep them in place while applying the heat. Make sure the board's pads are completely flat, with no solder blobs on them, and then put the part in place. Hold it down with a small screwdriver placed in the middle of the component while you solder one end. Unless you're quite unusual anatomically, you won't have an extra hand to feed solder to the joint, so put enough solder on the tip to make a crude joint. Don't even worry about molecular bonding. Just tack that side down, even if it's with a bad joint. Then let go of the component and properly solder the other side, taking care to make a good joint. Finally, go back to the first side and do it right. You might have to wick off your first attempt before trying again. (We'll discuss wick shortly, in the section on desoldering tools.) Be extra careful not to heat the part too much, or the good side will come unsoldered; those tiny parts conduct heat much faster than larger components with leads. Also, a lot of heat can delaminate and destroy the component's solderable platings. What works best is adequate heat applied quickly. Get on and off the part with minimum delay.

After soldering, the board will be left with a coating of rosin on and around the new joint. Some techs leave it on, but it's not a good idea because it can absorb moisture over time, causing some unwanted conductivity between solder pads, as if there were high-ohm resistors there. In many circuits, it won't matter, but it can. I just repaired a two-way radio that was eating its memory backup battery in a few months, rather than the 5 years it was supposed to take, for that very reason! The tiny current flow across the rosin drained the battery slowly but surely. Loosen the rosin by gently scraping with the tip of an X-Acto knife or a small screwdriver. Wipe up what's left with a swab wet with contact cleaner or isopropyl alcohol.

To join wires, first twist them together for a solid mechanical connection. If the wires are stranded, try to separate the strands a bit and intertwine them when twisting the wires together (Figure 6-17). If you're using heat-shrink tubing, keep it far from the soldering

FIGURE 6-17   Stranded wires twisted together.

work or it'll shrink before you get a chance to slip it over the joint. And don't forget to slide the tubing onto one of the wires before entwining and soldering them! It really helps you avoid the expletives from having to cut the wires and start over.

Full of Hot Air

Back in Chapter 2, I mentioned the *hot-air rework station*. Once an expensive instrument found only in industrial settings, hot-air rework stations are now easily available and inexpensive. These things can blow hotter air than a politician in an election year! See Figure 2-24.

They're called rework stations because they can completely rework a solder joint, from removing the old solder to applying the new solder. While rework stations can be used with traditional solder and through-hole parts, they're really made for use with *solder paste* and surface-mount components, and they work best with those.

Solder paste melts at lower temperatures than regular solder, and it has a peculiar, very useful property: It migrates toward connection pads and doesn't stick to the rest of the board. To use it, you just paint it on the board's pads—no need to worry if it slops over the edges of a pad—place your surface-mount component, and apply the air stream. If you're soldering only one pad at a time, use a fine tip on the rework station to avoid heating up nearby components and possibly desoldering them inadvertently. But you can solder multiple pads simultaneously, as long as they're close to each other, by using a somewhat broader tip.

As the paste melts, you'll see it shrink toward the joints without forming bridges between them, a crucial characteristic for installing chips with lots of leads very close together. It'll even cause small parts to align themselves with the pads! It's really quite amazing to watch.

Solder paste is soft and not very strong. Some of it has flux mixed in, while some requires the use of separate flux on the pads to be soldered, such as when performing reballing on a ball-grid-array chip. (We'll get to those a little later.) Paste is great for tiny components, but I recommend against using it to solder anything that will bear physical stress, such as laptop power jacks and USB ports. Some manufacturers' use of paste for affixing USB ports is a major cause of failure. I can't count how many of those things I've resoldered.

While I highly recommend you get a rework station and learn to use it, it's not a substitute for a standard soldering iron, because not everything is suited to hot-air soldering. You'll still want a normal iron for soldering wires, switches and any other sizable parts like power transistors and voltage regulators. See Chapter 12 for how to use an iron to solder leads too close together to get the tip onto one at a time, as you'll find on a micro-USB port or a high-density chip.

Desoldering Tools

You'll use desoldering tools almost as often as your soldering equipment. The two basic types are wick and suckers.

Wick

For small work, wick is the best choice. It's easy to control, doesn't splatter solder all over the area and doesn't run the risk of generating a static charge capable of damaging sensitive components. Its only real drawbacks are that it can't pick up a lot of solder at once, it's a tad expensive and it's not reusable.

To wick the solder off a joint, place the wick on the joint and heat it by pressing the iron to the other side or by hitting it with the hot air from your rework station. With a regular soldering iron, applying a little pressure helps. In this case, *don't* put that extra drop of solder on the tip first or it'll flow right into the wick, wasting some of the braid's capacity to soak up the joint's solder.

When the wick saturates with solder, pull it and the iron or hot-air tip away at the same time. If you remove the heat source first, the wick will remain soldered to the joint. If desoldering is incomplete, clip off the used wick and try again.

When desoldering components with leads, it can help to form the end of the wick into a little curve and press it against the board so that when it's heated, it'll push into the hole in which the lead sits and soak up the solder stuck inside. Be extra careful to keep the wick up to temperature until you pull it away or you could lift the copper off the board, creating a significant problem.

Not all desoldering is for component removal. Quite often, you'll need to clear a hole so you can insert a new component. On thick, multilayer boards, this can be remarkably difficult because you can't get enough heat through the hole to melt the solder on the other side. Even worse, component density on that side might be too high to allow putting your iron or hot-air tip to the hole there.

To clear such holes with an iron, you need one that's a bit hotter than one you'd use to solder in new parts. Form the wick into a point, and push the point against the hole. Press on it with the hot iron, and the wick should push into the hole pretty deeply. Hold it there for a few seconds, and then remove the wick and iron together. It may take a few tries, but that ought to clear even the deepest holes.

The big risk in clearing holes with an iron on thick multilayer boards is tearing the copper out of the inside of the hole. Doing so can ruin the board and your product. An iron that's way too hot can do it, but usually the cause is just the opposite—the iron is too cold, and the copper gets torn out when you pull on the wick, which is stuck to it.

You may be able to clear stubborn holes with your hot-air station by pressing the tip against the hole. In this case, be certain to use a fine tip, and expect solder to dribble out the other side of the board. Keep an eye on that, because you don't want it dripping onto nearby connections and causing a short. Also be careful to avoid burning the board with the tip, as it can delaminate the copper traces and make a hard-to-fix mess.

Sometimes, you are left with a film of solder the wick refuses to soak up. If this occurs, try resoldering the joint with a minimum of solder, just enough to wet it down a little. Then wick it all up. The fresh rosin of a new joint can help the wick do its job, carrying the old solder to the wick with it.

Wick has another handy, off-label use. Now and then, you do everything right but still can't get a good joint. Even with leads and pads looking nice and clean, the solder simply won't flow into the metal like it should. Apply a little too much solder to the joint. Then, use the wick as you would for desoldering, but pull it away before all the solder has flowed into it. Quite often, this will leave you with a lovely, well-bonded solder joint with just the right amount of solder.

Suckers

Solder suckers come in several varieties. The most common are bulbs, bulbs mounted on irons, and spring-loaded suckers. Bulbs work well when there isn't a lot of solder to remove; they tend to choke on big blobs of it. To use a bulb, squeeze the air out of it, heat up the joint with your iron, position the bulb with its nylon tube directly over the molten solder, get the tube into the solder and relax your grip on the bulb. Although the end of the bulb is plastic, it won't contaminate your iron's tip because the plastic used is a high-temperature variety that doesn't melt at normal soldering temperatures.

After a few uses, the tube may clog with solder. Just push it inside with a screwdriver, being careful not to pierce the bulb. If it's so clogged that you can't budge the solder, pull the tube out and expel the plug from the other end. Eventually, the bulb will fill up and you'll have to remove the tube to empty it anyway.

When you have a large joint with lots of solder, a spring-loaded sucker is the only thing short of a professional, vacuum-driven desoldering station that will get most or all of the solder in one pass. Cock the spring, and then use it like a bulb.

The fast snap of a spring-loaded sucker can generate a static charge reputed to be capable of damaging sensitive parts, especially MOSFETs (metal-oxide–semiconductor field-effect transistors) and IC chips of the CMOS (complementary metal-oxide–semiconductor) variety. To be on the safe side, don't use one on those kinds of parts.

Vacuum-driven desoldering stations work like the other suckers, except that the action is continuous. To use one, you just press the hot, hollow tip against the solder pad and activate the vacuum. They're great for clearing holes and removing larger parts, but many of these tools are too big for use at the size scale of today's components.

Power Supply

When powering a device from your bench power supply, you need to consider several factors for successful operation, and to avoid causing damage to the product.

Connection

Many battery-powered items also have AC adapter jacks offering a convenient place to connect your supply. The usual style of plug that fits into these is the *coaxial plug*, also called a *barrel connector* (Figure 6-18). These plugs come in many sizes, and both inside and outside diameters vary. Most have a sleeve for the center terminal, but a few feature a pin inside it, so be careful to use the same type. If you can find one in your parts bins that fits, perhaps from an old AC adapter or car adapter cord, you're in luck! Sometimes, you can use a plug with a slightly different diameter, but don't force things if it isn't a good fit. You may find one that seems to fit but doesn't work because the inner diameter is too large, so the jack's center pin won't contact the inner ring of the plug.

The polarity of the plug is paramount! Don't get this backward or you'll almost certainly do severe damage to the product as soon as you turn on the power supply. Coaxial

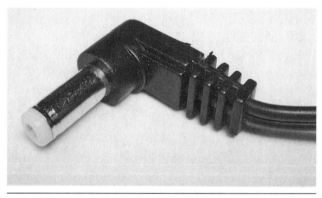

FIGURE 6-18 Coaxial power plug.

plug polarity is usually printed on the device somewhere near the jack, and it will be on the AC adapter as well. Most modern products put positive on the tip and negative on the outer sleeve, but not all. Always check for the polarity diagram. It should look like one of these:

Positive Tip   Negative Tip

If there's no adapter jack, or you choose not to use it, you can connect clip leads to the battery terminals of most devices. This should work fine with anything using standard, off-the-shelf cells like AAs. The convention is to use a red lead for positive and a black one for negative, and I strongly suggest you do so to avoid any possible polarity confusion, which could be disastrous.

In a typical case, you open the battery compartment and find a bunch of springs and contacts, one set for each cell. Most of them link one cell's positive terminal to the next one's negative terminal, forming a series string. One spring (the negative terminal) and one positive terminal (usually a flat plate or wire) feed the circuitry from each end of the string. Which are the two you need?

Sometimes, the positions of the terminals offer mechanical clues. If you see two connected directly to the board, or if they're placed such that they could be, those are probably the right ones. Also, if one set of terminals is on a flip-out or removable door, that pair is almost certainly *not* what you're looking for. (The exception is if you find a single terminal there. That may indeed be the end of the string. It's unusual, but I've seen a few like that in devices that didn't have batteries next to each other.) If you find no such hints, use your DMM's continuity feature to determine which terminals are connected to adjacent ones. Whatever's left should be the two magic terminals.

If you can reach the terminals with probes while the batteries are installed, pull the cells out, use your DMM to measure each cell, and then reinstall them. Add up the voltages to get the total series voltage, and then look for it between two of the terminals. Make sure the product is turned off so its current draw won't pull the voltage down. Only the two correct terminals will show your calculated voltage; any other combination will be significantly lower.

If the device uses a square 9-volt battery, the larger, petal-like terminal of the product's snap-on connector is positive. Nine-volt gadgets aren't common these days, but you will encounter this battery style if you work on old transistor radios or tape recorders. Some alarms and smoke detectors are powered by 9-volt batteries, and many clock radios use them for memory backup in case of power failures.

In products with proprietary lithium-ion batteries, it's often possible to provide power through the device's terminals, but not always. Some of them, especially camcorders and laptops, use "smart" batteries containing their own microprocessors, and the device won't recognize power applied to the terminals without the data those micros provide.

Many smaller items, like digital cameras, may have three terminals. Two are for power, and the third one is for a temperature sensor to prevent overheating during the charge cycle. Usually, these devices can be powered from a power supply, with the third terminal left unconnected.

To determine which terminal does what, look at the battery. You probably won't find any polarity markings on the product, but they are nearly always printed on the battery, and you can place it in the orientation required for insertion and see which terminals line up with the ones in the unit. You want the + and – terminals, of course. The other one may be marked "C" or have no marking at all.

If the battery is also unmarked, try measuring its voltage with your DMM. Most commonly, the voltage output is from the two outer terminals, with the sensor terminal between them. Once you find the right terminals, you'll also know the polarity. This assumes, of course, that the battery has at least a little charge on it; there's no way to read a dead battery, and don't even *think* of trying to apply a momentary charge from your power supply without knowing the polarity. Lithium-ion batteries do not like reverse-polarity charging one bit, and they're nasty when they burst. They can start seriously hot fires, and you can get hurt. Even when you do know the polarity, putting too much current through lithiums too fast can make them swell up or go boom.

Voltage

Set the supply's voltage before connecting it to the device, and try to get it pretty close to what the unit expects. Most products list their battery voltages either on the back of the unit or on the battery in the case of proprietary cells. Lithium-ion cells are 3.6 or 3.7 volts apiece, so their packs' voltages will be multiples of those values. In AA- or AAA-driven units made for alkaline cells, just multiply the number of cells by 1.5. Remember, though, that NiMH (nickel–metal hydride) rechargeable cells are only 1.2 volts each. If you run into the rare item that is made for use *only* with NiMH cells, multiply by 1.2 instead. Some older digital cameras fall into this category; they will not function correctly with alkalines or other throw-away cells, and the higher voltage may damage them.

Most traditional hobby-grade power supplies had analog meters, and they could be off by quite a bit. Even newer supplies with digital meters should be checked for accuracy with your DMM. Just 'cause it's digital doesn't mean it's right! If the supply's voltmeter doesn't agree with your DMM within maybe 100 millivolts, trust the DMM and use it to set the voltage. There's no guarantee that a DMM is telling the truth either, but its primary function is measurement, so it's more likely to be reasonably calibrated than the meter on a typical hobby-type power supply.

When matching the voltage to the product's requirements, don't worry about millivolts; just stay within a quarter of a volt or so and you should be fine. For example, 3.7-volt lithium-ion batteries start out at 4.2 volts when fully charged, with the voltage dropping as the charge is drained, so devices that use them can tolerate a small range of voltage variation. The curve is a lot flatter than with other battery technologies, though, which is why it's a good idea to match the rated voltage the best you can.

Also keep in mind that a volt of variation might be tolerable at the battery terminals but not at the DC input jack. Unlike old products, where those two sources typically wound up at the same place, devices running on lithium batteries have quite a bit of circuitry between the jack and the cell. USB-powered products, especially, expect a pretty accurate 5 volts at the jack. I had a poor-quality USB AC adapter that put out 5.2 volts, and it caused problems with some gadgets I tried to power with it.

Once you've set the voltage, turn your power supply back off and connect the leads, double-checking the polarity. Then, hit the switch and pray. No smoke? Great! You're in business.

Current

The current drawn by a device will vary, depending on what the unit is doing. Especially with any product employing moving parts like a hard-drive platter or laser optical head, current demand goes way up during mechanical motion, dropping again when movement is reduced or ceases.

As long as your supply has sufficient current capacity, it doesn't matter. If, however, the supply has enough for some modes of a device's operation but not others, the results can be unpredictable.

This issue crops up during service of camcorders, disc players and hard-drive-based devices. The drive spins up or the tape loads, and suddenly the device shuts down or its micro gets scrambled due to the lowered voltage from an overloaded supply. If your supply has a current meter (ammeter), keep an eye on it to be sure you're never pulling the supply's maximum current. If the meter does show maximum, the product's demand is probably exceeding what the supply can offer, and the voltage is dropping. To work on the device, you'll need a supply that can provide more current.

If you suspect a current overload but your supply has no ammeter, measure the current draw with your DMM's current function, as described in the section on DMMs. As with a supply's internal meter, if it shows as much current as your supply can provide, you're probably overloading the supply and the voltage is dropping.

Transistor Tester

Using a transistor tester requires taking the transistor out of the circuit. Doing so ranges from easy, with through-hole small-signal transistors, to a hassle, with tiny surface-mount parts or power transistors mounted on heatsinks. Transistors have three leads (see Chapter 7), so you'll have to disconnect at least two of them, though it's typically easier just to desolder all three and pull the part off the board. Big power transistors with metal cases employ the case as one terminal, usually the *collector*. With those, you may find it more convenient to leave the part on the board or heatsink and disconnect the other two terminals, especially if they're connected with wires instead of soldered directly to circuit board traces.

There are several basic types of transistors, and testing procedures vary. With some transistor testers, you need to know which terminal is which, while others will try out all the combinations automatically and recognize when the correct configuration has been found. Most modern testers will do that.

Fancy transistor checkers and some newer cheapies can evaluate a transistor's characteristics in actual operation by using the transistor as part of a functioning circuit built into the checker. They can show you the part's gain and leakage. Gain specifies how much current has to be at the input of the part to produce a certain amount at the output, and is part of a transistor's design specification, which you can look up. Leakage means the part is passing current in the wrong direction, or it's letting current flow even without the required input, like a faucet with a drip. It's not common, but it happens.

For most service work, such sophisticated measurement is unnecessary. Usually, you just want to know if the part is open or shorted, because most of 'em fail in those ways. To test a transistor, connect its three leads as specified in your tester's instructions, and read whatever info it gives you. There are too many types of testers to detail their operation here, but, as I said, mostly you're interested in opens and shorts.

Some DMMs include transistor test functions. If yours has a little round socket marked "E," "B" and "C," you have a transistor tester!

Capacitance Meter

Checking capacitors requires disconnecting at least one of their leads because other circuit elements will distort the reading. Be absolutely sure to discharge the capacitor before testing it, especially with electrolytic caps, which can store a lot of energy capable of trashing your tester.

Turn the meter on, and connect the capacitor to the input terminals. Some meters have terminals into which you can press the cap's leads. You can use those or the normal clip leads, whichever is more convenient. If the capacitor is polarized, which most electrolytics are, connect it the right way around, + to + and − to − ! See Chapter 7 for polarity marking styles, because some varieties of capacitors mark the negative lead and some mark the positive. Also, the circuit board may be marked, usually with a + sign adjacent to the positive terminal of the capacitor.

If your meter is autoranging, it'll step through its ranges and read the cap's value. If it isn't, begin at the most sensitive range, the one that reads picofarads (pF, or trillionths of a farad) and work your way up until the "out of range" indicator goes away and you get a valid reading. When you're finished, unhook the capacitor and discharge it. Very little energy is put into the component to test it, so you can short across its terminals without worry.

The meter will show you the value of the capacitor in fractions of a farad. Some types of capacitors, especially electrolytics, have fairly wide tolerances, or acceptable deviations from their printed values. Typically, an electrolytic can be off by 20 percent of its stated value even when new. Manufacturers deliberately err on the high side to ensure that filtering will be adequate when the caps are used in voltage-smoothing applications, as many are. If the cap reads a little high, don't worry about it. If it reads a little low, that may be okay too. If it reads more than 20 percent low, suspect a bad cap. And if you can't get rid of the out-of-range indicator on any scale, the cap is probably shorted or very leaky. Open caps will read as extremely low capacitance.

Remember, a cap can show the correct value on a capacitance meter and still have unacceptably high ESR (equivalent series resistance), rendering it ineffective. But if a capacitor reads significantly below its intended value or shows a short or an open, it is bad, regardless of the ESR. Bad electrolytics are the most common problem in electronics service work, so always be on the lookout for them.

Inductance Meter

You'll need to measure inductance only rarely. Typically, a bad inductor (coil) will show as an open circuit, which you can test with your DMM's ohms scale. Often, too much current gets pulled through an inductor by some other shorted part, and the wire inside the inductor melts, rendering it open. Shorts in inductors occur when the wire heats up enough to damage its enamel insulation but doesn't melt, and the bare wire from one winding touches the bare wire of another. The short bypasses part of the coil and reduces the inductance, but the coil still passes current. This can also happen in coils and transformers that handle high voltages, which can break down the enamel even without excessive heat. The windings arc and get welded to each other. You'll find this sort of problem in switching power supplies and fluorescent backlight inverters.

Another cause of reduced inductance is a cracked core in coils that are wound on metallic cores, as opposed to air-core coils or those wound on non-inductive forms. Metallic cores are used to increase the inductance for a given number of turns of wire. In larger inductors that handle significant current, the cores can crack from excessive heating. Low-inductance-value coils usually don't employ metallic cores unless they're adjustable types, where moving the core in or out of the coil changes its inductance. You'll find such coils in the tuned amplifier stages of radio equipment, but not much else. Small adjustable cores are delicate and can be broken by using the wrong tool to adjust them.

It's impossible to tell with an ohmmeter if you have a shorted turn because the resistance, which is what an ohmmeter measures, will vary so little that you can't see it. Besides, in most cases you had no way to know what it was supposed to read in the first place. This is where an inductance meter is handy. To measure inductance, you'll need to disconnect one end of the coil from the circuit, or just remove the coil if that's easier. Unlike capacitors, coils can store energy only for a fraction of a second, so there's nothing to discharge before measurement. Connect the meter to the coil's two terminals. Inductors have no polarity, so it doesn't matter which way around you connect the meter. It also doesn't matter which range you select. If it's too high, the coil will read something close to zero, and if it's too low, the meter will show its out-of-range indicator. When you hit the right range, the reading will be evident, with at least a couple of digits showing something.

Some inductors are marked for their values using a scheme similar to that used with capacitors, as described in Chapter 7. Instead of farads, though, the unit for inductance is *henries*. Typical coils range from microhenries, for tiny coils used in radio gear, to millihenries, for bigger ones used in power supplies.

Don't expect the inductance to read exactly what's marked. As with all components, there's a tolerance within which the part is good. Unfortunately, you almost certainly won't be able to look this up because the manufacturer and part number usually aren't marked on the coil. If the measurement is within around 10 percent of the marked value, the coil is probably okay. If it's much lower, the part might have a short between windings. That's about the best you can do here. There's no way for a coil to fail by having too much inductance. So, if your reading is significantly high, something is wrong with your measurement. If all the scales read zero, the coil is open. Verify with your DMM's ohms scale.

Multi-component Tester

I mentioned these back in Chapter 2. You plug a component into the tester's socket and the tester figures out what it is, what lead does what, and shows you the type of part and its characteristics. At least that's what is supposed to happen. Sometimes, it actually does! I've had mixed results with these devices, though. I still own one, but I'd never get rid of my dedicated meters.

The displayed parameters on a multi-component tester are the same as you'll see from your meters, as described in this chapter.

Signal Generator

A signal generator is used to replace a suspected bad or missing signal temporarily so you can see what its insertion will do to a circuit's behavior. It also gives you something for a circuit to process, providing a steady input. Inserting a signal is a very handy technique when working on audio and radio circuitry. Plus, it can help you check clock oscillator function in digital gear or sub for a missing oscillator signal in radio equipment.

For audio testing, it's best to use a sine wave somewhere in the lower middle of the audio spectrum, at around, say, 1 kHz. Using a sine wave prevents the generation of *harmonics* that could damage the amplifier under test, the speakers or your ears. It also lets you see distortion from a bad circuit stage clearly on your scope. If a pure sine wave goes in and some other shape comes out, you've found the trouble!

For clock oscillator substitution, set the generator to the same frequency as the missing oscillator (it should be marked on its crystal), and use a square wave. Set the peak-to-peak voltage of the generator just below whatever voltage runs the chip normally doing the oscillating. Don't exceed it, or you could *latch* the chip, ruining it. Also, the bottom of the waveform coming from the generator shouldn't go below 0 volts. Most generators have a DC offset control that lets you adjust this. Use your scope to set it so it's all positive, with the bottom at or very near 0 volts, making sure that the scope's vertical channel is set to DC coupling, as we discussed earlier in this chapter.

For radio oscillator substitution, use a sine wave at the frequency of the missing oscillator. It's best to feed the signal from the generator through a capacitor of around 0.01 μF to avoid loading down the radio's circuits, and also to block any DC offset from the generator; unlike clock oscillators, radio signals do swing both positive and negative. Set the peak-to-peak oscillator voltage to much less than the power supply feeding the radio's stage. Usually, a small fraction of a volt is plenty in this kind of experiment.

Frequency Counter

Frequency counters are used to adjust a device's oscillators to a precise, accurate frequency or to verify a frequency. Radio and TV receivers, video recorders and even some all-digital devices can require carefully set oscillators for proper operation. Frequency measurement also may aid in troubleshooting optical disc players.

A counter works by totaling up how many cycles of an incoming waveform go by in a period of time specified by the instrument's *gate period*. The gate opens, the waves go by, it counts 'em and puts the count on the display. That's it.

Ah, if only real life were so straightforward! Sometimes this works and gives you a correct count; sometimes it doesn't. For one thing, how does the counter know when a cycle has occurred? Unlike the trigger on a scope, the counter's trigger is very simple: It looks for *zero crossings*, or places where the signal goes from positive to negative, and counts every two of them as one cycle. For simple waveforms with little or no noise, that does the job adequately.

A lot of signals have noise or distortion on them, however, that can confuse the zero-crossing detector, resulting in too few or too many counts. If you connect a counter to a test point using a scope probe and then switch between 10X and 1X on the probe, the count will probably change a little bit, with a lower count in the 10X position. Which one is the truth? It's hard to say for sure. Usually, the count is more accurate when the input voltage is lower because noise on the signal is less likely to trip the zero-crossing detector. If it gets too low, though, the detector may miss some cycles altogether, resulting in an incorrect, low count. If you get counts that seem very low for what you were expecting— say, a 10-MHz oscillator reads 7.2 MHz—the input voltage is likely too low, and the detector is missing some cycles. If the count seems too high, the signal may be noisy and also could be too strong, adding false cycles to the count.

Complex, irregular signals like analog audio, video and digital data streams cannot be counted in any useful way with a frequency counter. What comes in during each gate period will vary, so the display won't settle down. Also, correct tripping of the zero-crossing detector is impossible. Use this instrument only for simple, repeating waveforms such as those produced by oscillators.

Back in Chapter 2, we looked at precision and accuracy and how they affected each other. Nowhere does this issue come up more often than with frequency counters. Most counters have lots of digits, implying high precision. Accuracy is another matter.

The count you get depends on how long the gate stays open. That is set by a *frequency reference*, which is an internal oscillator controlled by a quartz crystal. In a very real sense, the counter is comparing the incoming signal's frequency with that of the instrument's crystal, so variation in the crystal oscillator's frequency will skew the count. Crystals are used in many oscillator applications requiring low drift, but they do wander a bit with temperature and age. Most counters have internal trimmer capacitors to fine-tune the crystal's frequency, but setting them requires either another, trusted counter for comparison, an oscillator whose frequency is trusted, or some cleverness with a shortwave radio that can receive WWV, the National Bureau of Standards atomic clock's time signal originating from Fort Collins, Colorado. That station broadcasts its carrier at a highly precise, accurate frequency, and it is possible to compare it audibly with your counter's oscillator by putting the counter near the radio and using the radio as a detector. When the WWV signal and your counter's oscillator *zero-beat*, or mix without generating a difference tone, your counter is spot on frequency. Performed very carefully with a counter that's been fully warmed up, zero-beating against WWV can get you within 1 Hz, which is darned good. This trick works only if your counter has a 10-MHz reference crystal, though. Luckily, many do. (WWV broadcasts at other frequencies, too, but not at those commonly found in frequency counters' reference oscillators.)

If you sprang for a GPS-disciplined oscillator, you can use its output in place of a WWV signal and zero-beat your counter's oscillator to it. Or, you can set your counter to read exactly 10 MHz when it's connected to the GPSDO, no radio required. This method isn't quite as accurate, though, because even slight differences not noticeable at 10 MHz may cause some errors when you use the counter at much higher frequencies. Still, it's pretty good.

If your counter offers a 10-MHz reference input, you can connect your GPSDO to it. You might have to push a button or set a menu option on the counter to switch it from using its internal oscillator to being clocked by the GPSDO. Once your instrument is locked to GPS timing , you never have to calibrate its frequency accuracy. It'll always be dead on.

On many counters, the gate time can be selected with a switch. Longer gate times give you more digits to the right of the decimal point, but their accuracy is only as good as the counter's reference oscillator. Don't take them terribly seriously unless you are certain the reference is correct enough to justify them. If your rightmost digit specifies 1 Hz but the reference is 20 Hz off, what does that digit mean? For audio frequencies, you'll need a fairly long gate time to get enough cycles to count. For radio frequencies, a faster gate time is more appropriate.

Connecting a counter to the circuit under test can be tricky in some cases. As with a scope or a DMM, some signal has to be stolen to measure it. Loading of the circuit's source of signal generation can be a real problem, pulling it off frequency and affecting the count significantly. Especially when you touch your probe directly to one lead of a crystal in a crystal oscillator, the capacitance of the probe and counter can shift the oscillator's frequency lower by a surprising amount, making your measurement fairly meaningless. To probe such a beast, look for a *buffer* (an amplifying stage) between the oscillator and everything else, and take your measurement from its output. Because the circuits following the oscillator could load it down and affect its frequency, it's highly likely that a buffer is present right after it. These days, most crystal oscillators are formed with IC chips rather than transistors, with the buffer being on the chip and the frequency output coming from a pin *not* connected to the crystal. Use your scope to find it, and then connect the counter to that pin.

Analog Meter

Using an analog VOM (volt-ohm-milliammeter) or FET-VOM (field-effect transistor–volt-ohm-milliammeter) requires interpreting the position of a meter needle rather than just reading some numbers off a display. Why would you bother with this? Well, that meter needle can tell you some things a numerical display can't. Specifically, how it moves may give you insight into a component's or circuit's condition.

Taking most kinds of measurements with a VOM is pretty much like taking them with a DMM. VOMs are not autoranging, so you have to match the selector knob's scale to the markings on the meter movement. Also, you need to zero the ohms scale with the front-panel trimmer knob every time you change resistance ranges. Select the desired range, touch the leads together and turn the trimmer until the meter reads 0 ohms.

Unlike FET-VOMs and VTVMs (vacuum-tube voltmeters), VOMs have no amplification, so they load the circuit under test much more when reading voltage than do other instruments. This is of no consequence when measuring the output of a power supply, but it can interfere significantly with some small-signal circuits.

VOMs can pull a few tricks not possible with their digital replacements. Slowly changing voltages will sway the meter needle in a visually indicative way instead of just flashing some numbers. Also, a little noise won't affect the reading, thanks to the needle's inertia. A lot of hum on a DC signal can vibrate the needle in a very identifiable manner. Some old techs could read a meter needle almost as if it were a scope!

If you don't have a capacitance meter, you can gauge the condition of electrolytic capacitors with the meter's ohms scale. This works pretty well for caps of about 10 μF or more; it doesn't work at all for anything under 1 μF or so. Set the meter to its highest range and touch the test leads together. When the needle swings over, use the trimmer on the front panel to adjust it to read 0 ohms. (If the needle won't go that far, the meter needs a new battery!)

Connect the leads to the *discharged* cap, + to + and − to −, and watch what happens. The meter should swing way over toward 0 ohms and then gradually fall back toward infinity. The greater the capacitance, the harder the needle will swing, and the longer it'll take before it finally comes to rest. If the needle doesn't fall all the way back, the cap is leaky. If it doesn't swing toward zero, it's open or of low capacitance.

The meter applies voltage from its battery to the capacitor, so be sure the voltage rating printed on the cap is higher than the voltage of the battery in the meter or you could damage the cap. Some meters use 9-volt batteries for their higher resistance ranges. If yours does, it's wise not to try this test on caps rated lower than that. If your meter uses only an AA cell, there's nothing to worry about because all electrolytic capacitors have voltage ratings higher than 1.5 volts.

The ohms scale can also be used to check diodes and some transistor characteristics, just as with a DMM. You won't find a transistor tester function built into an analog VOM, though.

Contact Cleaner Spray

Cleaner spray is handy stuff, especially for use with older, analog equipment. Volume controls, switches and sockets can all benefit from having dirt and oxidation cleaned away. When controls exhibit that characteristic scratchy sound, it's time to get out the spray.

For spray to be effective, you have to be able to get it onto the active surface of the control or switch. Sometimes, that's not easy! If you look at the back of a *potentiometer*, or variable resistor, you may find a notch or slot giving you access to the inside, where the spray needs to be. Always use the plastic tube included with the spray can! Trying to spray the stuff in with the bare nozzle will result in a gooey mess all over the inside of your gear. Once you get some spray into the control, rotate it back and forth a dozen times. That'll usually cure the scratchies.

Switches can be a bit tougher. Some simply have no access holes anywhere. If you can't find one, you'll have to spray from the front, into the switch's hole. Never do this where the spray may come into contact with plastic; most sprays will permanently mar plastic surfaces. As with potentiometers, operate the switch a bunch of times after spraying.

Trimmer capacitors (little variable caps adjustable with a screwdriver) should not be sprayed. They have plastic parts that are easily damaged by the spray and may lose function after contact with it.

Do your best not to splatter spray onto other components. Wipe it off if you do. Also, it could shatter a hot lamp, so don't use it near projector bulbs. And, of course, avoid breathing it in or getting it in your eyes. Spray has a nasty habit of reflecting back out of the part you're blasting, right into your face. Keep your kisser at least 12 inches away. Farther is better.

To avoid making a mess or getting soaked, try controlling the spray by pressing gently on the nozzle until only a gentle mist emerges. Some cans have adjustable spray rates, but many don't. Some brands are more controllable than others, too. Experiment with this before you attack expensive gear.

Component Cooler Spray

Cooler spray can be incredibly useful for finding thermal intermittents. If a gadget works until it warms up or it works only *after* it warms up, normal troubleshooting methods can be hard to implement, especially in the second case. How are you going to scope for trouble in something that's working?

Before hitting parts willy-nilly with cooler, decide what might be causing the trouble. The most likely candidates for cooler are power supply components like transistors and voltage regulators, output transistors and other parts with significant temperature rises during normal operation.

As with cleaner, use the spray tube, and try controlling the spray rate. Also, the same caveats regarding breathing it in and getting hit in the face hold here. This stuff is seriously cold and can damage skin and eyes. A small amount hitting your hands won't do you any harm, but I shudder to think of your getting a single drop on your cornea.

If spraying a suspected component reverses the operational state of your device (it starts or stops working), most likely you've found the trouble. After spraying, moisture will condense on the cold component. Be sure to kill the power and wipe it off.

Sometimes, you spray a part and the behavior change suggests you've found the culprit. You might have, but the real trouble could be next to it, and some spray has splattered onto that. It's not easy to cool just one component, especially in today's equipment, with its crammed-together construction. Let the part warm back up again, and then spray from a different angle so the adjacent components that get hit won't be the same ones. Eventually, you'll narrow the cooling to the right part.

Also, the real problem could be the part's solder joints. Before replacing the suspect component, try resoldering it.

Chapter 7

What Little Gizmos
Are Made of: Components

Although there are hundreds of types of electronic components and thousands of subtypes (transistors with different characteristics, for example), a small set of parts constitutes the core of most electronic products. Let's look at the most common components and how to recognize and test them. In this chapter, we'll cover out-of-circuit tests, the kind you perform after removing the part from the board. We'll get to in-circuit testing in Chapter 11 when we explore signal tracing and diagnosis.

Safety Components

Anything involving electricity offers the potential for danger. There's enough energy coming out of a wall socket to toast bread, microwave your dinner, run a fridge or power a radio transmitter that can be received around the world. Even small batteries will deliver quite a wallop—especially the modern lithium type. A direct short circuit across a phone's lithium battery can heat up a wire and make it glow like a light bulb in less than a second, for more than long enough to start a fire.

To keep products safe, various components dedicated to electrical safety are incorporated into pretty much all of them. Some are designed to self-sacrifice as part of their function, but they can fail even with no provocation, stopping a device's operation. Let's look at the major types of safety components, starting with the most common of all.

Fuses

Fuses protect circuitry and prevent fire hazards by stopping the current when it exceeds the fuses' ratings. Though simple in concept, fuses have a surprising number of parameters, including current required to blow, maximum safe voltage, maximum safe current to block and speed of operation.

The primary criterion is the current required to melt the fuse's internal wire and blow it. If you're not sure of anything else, be certain to get that right when replacing a fuse.

The speed at which the fuse acts is also important in some devices. Time-delay, or *slow-blow*, fuses are used for applications like motors, which may require momentary high startup currents that would blow a faster fuse. Ultra-fast-acting fuses are used with especially sensitive circuitry that must be protected from even transient overcurrent conditions. Most consumer electronics gear uses standard fuses, which are considered fast-acting but not ultra-fast.

Fuses come in many shapes and sizes, from the ubiquitous glass cylinders with metal end caps to tiny, rectangular, surface-mount parts hardly recognizable as what they are (Figure 7-1). You'll find fuses in holders and also soldered directly to circuit boards. Be on the lookout for glass fuses with internal construction including a spring and a little coil; those are the slow-blow variety (second from the left in the figure).

Is It Really a Fuse?

Especially in dense, surface-mount products, 0-ohm resistors are sometimes used as fuses. They look exactly like all the other tiny resistors, but they have no resistance. Sometimes they have a zero written on them, but don't count on it. They also have no other fuse characteristics such as speed or a particular current rating. A short will pull enough current through them to burn them out, and the damage will be obvious.

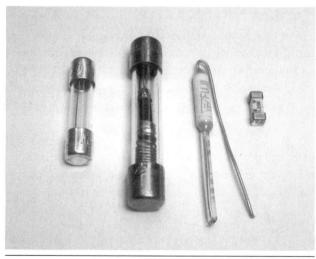

FIGURE 7-1   Fuses.

But hang on a minute! Such resistors may also be used as jumpers to cross over other conductors on the circuit board layout, so you can't assume the fuse function when you see a 0-ohm resistor. If it's charred, though, it's highly likely the part was employed as a fuse, and it functioned as intended.

These resistors also get installed as configuration jumpers for setting product characteristics. For instance, the same microprocessor might be used in a family of products, with the jumpers specifying which product the chip is in so that it can properly respond to the front-panel buttons. Jumpers are common in shortwave and two-way radios to set the permitted frequency ranges, which vary by country, so the radios can be sold all over the world without violating government regulations. In these cases, the resistors are used as jumpers instead of solder blobs or pieces of wire because they can be placed and soldered by automatic placement machines along with all the other parts when the board is made.

Symbol

Markings

At least four marking systems are used on fuses. The primary marking is the melting current, shown by a number followed by an "A." You may also see "3AG" or "AGC," both of which indicate standard-speed glass fuses. You must look up other markings to get the speed rating. Many online catalogs offer this information.

Uses

Just about everything that plugs into the wall will have a fuse on the hot side of the AC line, usually right as it enters the device, before it even gets to the power switch. The standard color-coding scheme for 3-wire grounded plugs is black for hot, white for neutral and green for ground. Follow the black wire and it should lead straight to the fuse. Battery-operated gadgets often have fuses too, typically between the battery's positive terminal and the rest of the unit. Some products have multiple internal fuses protecting various parts of the circuitry. I've seen as many as eight of them in one camcorder!

Some audio amplifiers and receivers use fuses in line with the speakers to protect the amp, should a speaker's *voice coil* overheat and short out. Such a fuse can also protect the voice coil if the amp develops a shorted output transistor that sends the power supply's entire current capacity straight to the speaker, which would fry it in a hurry were it unprotected.

What Kills Them

Most fuse failures aren't failures—they're successes caused by the fuse doing its job. A short in the circuitry pulls too much current through the fuse, so it blows. Now and then, you may run across a fuse that has fatigued with age and use. If it's a glass fuse, take a look at the inside. When the two wires are almost touching and there's no discoloration on the glass, the fuse blew gently, and there's a possibility that the circuitry isn't shorted. That happens sometimes with speaker fuses when the amplifier is played at high volume for extended periods. The fuse's wire gets just warm enough to fracture, but there's no malfunction in the circuitry. If you see a wide gap between the wires and spattering on the inside of the glass, the fuse blew hard, indicating a lot of current and certain circuit failure.

Very rarely, fuses can develop resistance, continuing to pass current but interfering with the full flow. I've seen it a couple of times, and my first such case drove me bonkers trying to figure out why a power supply's voltage was low and erratic. I assumed a fuse was either good or bad, so I never considered it a possible culprit until I'd wasted hours looking at everything else. Incredulous, I pulled it and discovered that it had become a 10-ohm resistor! I haven't trusted the little buggers since.

Out-of-Circuit Testing

Check fuses with your DMM's ohms scale. They should read 0 ohms or pretty close to it. If you see 1 or 2 ohms, try touching the meter's leads together; you may get the same reading, thanks to the resistance of the leads themselves. If not, and the resistance is definitely in the fuse, replace it. But if it's an especially low-value fuse in the 200-mA range or under, it might show an ohm or so without being bad, so look it up online to see what the maximum resistance should be. A blown fuse will read completely open, of course.

Polarized AC Plugs

These two-prong plugs have one blade wider than the other so they will fit into the wall socket only one way around (Figure 7-2). Given that AC current alternates back and forth in polarity, it might seem odd that a plug could be polarized. After all, there's no positive or negative that stays put for more than 1/60 of a second, so what are we polarizing?

It really isn't about polarity. I think nobody could come up with a better word, so that one stuck. It's actually about electrical symmetry.

Were there only two wires used to power homes, there'd be no difference between them. But there's a third current path, even in products that have no third wire. It's *ground*, and it is not symmetrical with the other two. Ground is literally that: It goes to a rod in the ground, which you will find near your electrical panel. Neutral is normally connected to it, so there can't be a voltage between them. The voltage should read zero or close to it when measured between neutral and ground.

The action takes place between hot and neutral. But because neutral and ground are connected together at your panel, the same voltage will be present between hot and ground.

FIGURE 7-2   Polarized AC plug.

In a device with a three-wire plug, ground is typically connected to the product's metal chassis. This keeps it safe. No matter what happens, any sort of fault that might put power on the metal chassis gets shunted to ground, protecting you and, hopefully, popping your home's circuit breaker, or at least blowing the fuse in the faulty device. When the plug has only two blades, though, there's no way to ground the chassis. So, with only two wires, why does it matter which connection is used for what?

Well, which would you rather have connected to the chassis, hot or neutral? I'll vote for neutral! That way, if you make contact with it and also touch anything grounded—which is *lots* of stuff in your home—you won't be turned into a fried appetizer. And, since neutral and ground are connected back at your home's electrical panel, it's almost as good as a real chassis ground.

On a polarized plug, the wider blade is neutral (white wire) and the narrower one is hot (black wire). Never defeat the polarization or replace the polarized plug with an unpolarized one. It could be the last thing you do.

MOVs

Metal oxide varistors, or MOVs, are used to absorb transients that come down the power line. These can arise from lightning strikes or utility oopsies that put a sudden surge into your equipment. The purpose of the MOV is to absorb the overage, shorting it from one side of the line to the other, or to ground in a three-wire setup (Figure 7-3).

Typically placed directly across the incoming AC power, an MOV exhibits high resistance under normal circumstances. So, very little current passes through it—it's essentially invisible. When the voltage suddenly rises much higher, as it does during a spike or a surge, the resistance instantly drops and the spike passes from one side of the MOV to the other, keeping it from reaching the delicate circuitry of your device. This nonlinear change in resistance is similar to what a zener diode does, except that it's in both directions, and an MOV can absorb a lot more current. (We'll cover zener diodes shortly.)

FIGURE 7-3 MOV.

MOVs have many characteristics, but the main ones are the varistor voltage, which tells you what voltage makes the resistance go way down, and the energy absorption in joules. Most MOVs are intended to be placed across the AC power line, so the varistor voltage is set above that. A joule is how much heat gets dissipated when 1 amp of current flows through 1 ohm of resistance for 1 second. While voltage spikes are very short in duration, on the order of microseconds, they can provide a couple of thousand amperes, particularly when lightning is involved, so those joules add up.

A typical MOV made for American-standard power line use might have an operating voltage rating of 130 volts AC, with a varistor voltage of 200 volts. The energy absorption may be around 70 joules, with peak current of 6,500 amps. MOVs for 220-volt service will have higher voltage ratings.

Symbol

Markings

MOVs are marked with manufacturers' part numbers, not their characteristics. To get the specs, search online for the specific part number on the MOV and look at its data sheet.

Uses

There's only one common use for these things—to absorb power line voltage spikes in an attempt to keep them from causing damage.

What Kills Them

MOVs are designed for quick overvoltage conditions in the microsecond range, which are a typical result of nearby lightning strikes. These components can't survive anything lasting much longer than that. Because some spikes may persist for more than a few microseconds, the MOV can stop working. Your device will still function but it won't be protected anymore, and you have no way to know it.

Repeated strikes can degrade the MOV, lowering the varistor voltage until it falls into the range of the properly operating power line, resulting in current flow at normal line voltage. This or the more catastrophic failure from a big zap can cause the MOV to burst into flames, which kinda defeats the purpose! Yes, it happens, especially in cheap, poorly designed power strips. To avoid fires, some MOVs incorporate thermal fuses that open before the part gets hot enough to burn up. You may lose the protection and your device, but at least you won't lose your house. Whether you'll find such a protected MOV in a particular electronic product is anybody's guess. You'll have to look up the part number to find out. If the part isn't thermally protected, it relies on the product's fuse to blow when the MOV shorts . . . hopefully before the MOV catches fire.

If an MOV fails by shorting, it'll look burned or melted. This does not indicate a circuit failure. It just means there was a serious power surge and the MOV absorbed it, or that many smaller surges have occurred, causing the degradation and unwanted current flow described earlier.

Out-of-Circuit Testing

Test MOVs with your DMM's ohms scale. The part should show some resistance, well over 100 ohms, and possibly a lot more. If it reads 0 ohms or close to it, the MOV is shorted. If the resistance reading is infinite, the part is open and no longer providing protection. Either way, it needs to be replaced.

By the way, if the product's AC power line fuse is blown, be sure to check for a shorted MOV! If you find one, you can remove it and replace the fuse to see if that was all that was wrong. If the device comes to life, replace the MOV, and you're all done.

Safety Capacitors

These are much like the capacitors we'll discuss in the next section, except that they're rated to withstand the voltages caused by spikes and surges without self-destructing, and they have some special characteristics intended to make them safe for use when directly connected to the AC power line. Most are ceramic, while some are made from plastic film. They look a lot like MOVs and standard ceramic capacitors, but a little thicker. Many are blue or orange (Figure 7-4).

There are three classes of safety capacitors. X-class capacitors are intended for placement across the AC line. A shorted X-class capacitor will blow a fuse or a circuit breaker, but it can't hurt you, as long as it doesn't catch fire, because neither end is connected to anything you could touch. Y-class capacitors are intended to be placed between the AC line and ground. A short in one of these could put AC line voltage on the chassis of a device, exposing you to a serious shock. So, these parts have to meet more rigorous standards to make that extremely unlikely. X/Y-class parts can be used in either configuration and feature the higher standards of a Y-class capacitor.

Symbol

FIGURE 7-4 Safety capacitor.

Markings

Safety capacitors are marked with their capacitance values as well as their class, X, Y or X/Y, and a number after it. Most of these parts intended for household 120-volt AC power line use are rated X2 or Y2. The number tells you the maximum continuous and peak voltages the part can withstand. You'll also see logos of the various organizations that have certified the capacitor as safe.

Uses

Safety capacitors are placed across the incoming AC power line to filter out EMI, or electromagnetic interference. (When it's at radio frequencies, it's called RFI.) They can help keep the interference from getting into your device, but more typically they keep it from getting *out*. Most modern products incorporate switching power supplies, as we discussed earlier, and those darned things generate lots of electrical noise that can interfere with radios, TVs and other items employing RF (radio-frequency) signals. The antenna radiating those signals is the AC power cord! So, to meet regulations prohibiting equipment from generating radio interference, products that produce it will include safety capacitors.

How much a capacitor opposes the passage of AC current depends on the part's capacitance and the rate of change of the current. For a given value of capacitance, the faster the rate of change, the more current will get passed through the capacitor. For 50- or 60-Hz (Hertz, or cycles per second) household power, the rate of change is pretty slow. Using a small capacitance value means that very little will pass through it, so no significant power is passed at normal line frequencies.

With radio energy, its faster rate of change (it could be alternating millions of times per second) results in lots more getting through the capacitor, shorting the offending energy to the other side of the power line or to ground, significantly reducing what gets radiated by the power cord to nearby devices.

What Kills Them

These capacitors are built to withstand fairly high voltages, but they can break down from a big spike or go bad from age. X-class parts are designed to short when they fail, so that the product's fuse will blow. Because a shorted Y-class capacitor could put hazardous voltage on the chassis or metal knobs of the device, these parts are made to fail by becoming open circuits. This renders the capacitor useless, but at least you won't get killed. Any part can short under extreme conditions, though, so Y-class capacitors are built super-tough and have to pass rigorous tests to be certified as such.

Older gear may contain a brand of safety capacitors called Rifa. These parts are infamous for exploding without provocation! If you see a burned-up Rifa cap across your device's AC line, it doesn't indicate that any fault occurred. Rifa parts can be replaced with modern safety capacitors.

Out-of-Circuit Testing

Safety capacitors can be tested like any other capacitors, as described in the next section. While it is possible for a safety capacitor to measure as good with a capacitance meter but break down under actual operating conditions, it's not likely. And if the part reads as leaky or shorted, it is!

SCRs, TRIACs and Thyristors

SCRs (silicon controlled rectifiers) and TRIACs (triode for alternating current) are devices that turn on when a voltage at their control terminals, called the *gate*, is above a certain level. The difference between them is that SCRs are diodes, conducting in only one direction when activated, while TRIACs conduct in both directions. Thyristors are very similar to SCRs, but with slightly different internal construction. They do pretty much the same thing (Figure 7-5).

FIGURE 7-5   SCR and TRIAC.

Symbol

Anode ▷|— Cathode

Gate

SCR

Gate

MT1   MT2

TRIAC

Markings

Sometimes looking like power transistors, these components will be marked with manufacturers' part numbers. Look those up online to get the characteristics.

Uses

While TRIACs are found in lamp dimmers, SCRs and TRIACs are widely used as automatic fuse blowers called *crowbars* in power supplies, stopping the supplies' operation when the DC output voltage goes too high due to a failure in the circuitry. The aim is to protect whatever is being powered from overvoltage destruction.

What Kills Them

When tripped, SCRs and TRIACs handle a lot of current for a short time. The big current surge can result in a blown part if the fuse doesn't stop the current fast enough.

Out-of-Circuit Testing

Set your DMM to read resistance (ohms). On an SCR or a thyristor, connect the positive (red) lead to the anode of the part and the negative lead to the cathode. You should see an open circuit, which reads as infinite resistance. Now, touch the gate lead to where the red lead connects to the anode. The resistance should drop to a low reading. Disconnect the gate lead, and the resistance should stay low, indicating that the SCR has latched in the "on" state. If any of this doesn't happen as described, the part is bad.

The one exception is a GTO thyristor, or gate-turn-off type. These turn back off when the gate is disconnected, rather than staying latched on.

A TRIAC can be tested the same way, except that it doesn't matter which of the two leads, labeled "MT1" and "MT2," you connect to which lead of the DMM. You still need to connect the gate lead to the positive lead of the meter, though, to make the TRIAC turn on.

Some SCRs and TRIACs require gate voltages higher than what your DMM offers, so they won't turn on with this test. You can find out how high a voltage you need by checking the part's spec sheet online. To test such a part, you'll need to supply it with a high enough voltage to exceed the turn-on voltage and put a load, such as a light bulb, in series with one of the terminals other than the gate. Please don't try this by connecting it directly to the AC power line! It's just too dangerous. Use a power supply or an isolation transformer, and observe all precautions to avoid contacting the applied voltage. I really like it when my readers live to read the rest of the book!

BMS for Lithium Batteries

BMS, or battery management systems, are not individual components. They are entire circuit boards, although they may be very small. While they might be part of the main circuit board of a device, most are separate boards incorporated into flat lithium-polymer batteries of the kind used in phones, tablets and laptop computers. You'll also find them in Bluetooth speakers and other small, lithium-powered gadgetry. Large LiFePO4 batteries, the ones that look like car batteries, feature internal BMS as well.

When they're part of an internal, soft battery that has no shell, they will be placed under the tape at the top, where the wires exit. Peeling back the tape (carefully, and not with metal implements!) will expose the BMS (Figure 7-6).

Symbol

Not being individual parts, BMS have no symbols. On a schematic diagram (see Chapter 8), the BMS will be shown as its individual components if it's part of the main circuitry. If it's integral with the battery, it probably won't be shown at all.

FIGURE 7-6 Lithium battery with BMS board.

Markings

There ain't none! What happens in the battery stays in the battery, and the BMS is considered part of it. Even worse, the inclusion of a BMS isn't noted in any standardized way on the battery itself. Some batteries have a BMS, some don't.

Uses

The purpose of a BMS is to prevent overcharging, because putting too much charge into a lithium battery can be catastrophic. The battery might swell, burst or even explode violently, destroying the product and possibly starting a very hot, hard-to-extinguish fire. Early laptops and pocketable devices like vape pens were notorious for nasty fires. The problem has been so common that airlines implemented regulations regarding the transporting of lithium-powered items and loose batteries.

Rigorous use of BMS has greatly reduced the risk of lithium fires. While fairly complex, BMS boards are so cheap today that nearly all batteries have 'em. However, 18650 cells, the common type that look like overgrown AAs and are used in many products, do not; they're intended for use with lithium chargers that have their own BMS on board. When the cells are formed into premade packs, though, the BMS will usually be part of the pack.

What Kills Them

BMS boards are pretty reliable, but they can fail if the output of the battery pack gets shorted, resulting in a huge current flow. They also fail randomly now and then, like any complex circuit.

The ones used in "smart batteries" of the sort attached to laptops may have their operating software in memory that's maintained by the battery itself. If the battery is allowed to drain all the way to zero, that software is lost, and the BMS will never work again even if you open up the battery and apply charge directly to the cells—which is not recommended, by the way. Without that software, the BMS won't let the laptop try to charge the battery, either, so the pack is ruined. It's for the best, because lithium cells that have gone to zero often swell if charge current is applied.

Out-of-Circuit Testing

If your battery has a BMS and won't charge, and there's no voltage at the battery terminals, either the BMS or the cell itself is shot, and you'll need to replace the battery. If the BMS is part of the product, instead of being integrated into the battery, check for charging voltage at the battery terminals while the charger is connected. If it's there, the battery may be dead but the BMS is probably working. If there's no voltage being applied to the cells, the BMS could be bad. However, some BMS will check for a minimum voltage on the cell (or cells, if it's a multi-cell pack). If they're at zero, the BMS will not try to charge them. The only way to know for sure is to replace the cells.

Capacitors

Capacitors consist of two plates separated by an insulator, or *dielectric layer*. A charge builds up on the plates when voltage is applied, which can then be discharged back into the circuit. The unit describing how much energy can be stored in a capacitor is the *farad*, named after famed electrical experimenter Michael Faraday. Capacitors come in many types, including ceramic, electrolytic, tantalum, polystyrene (plastic) and trimmer (variable) (Figure 7-7).

Symbols

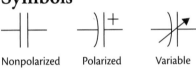

Nonpolarized   Polarized   Variable

Markings

A farad is a huge amount of capacitance rarely encountered in practical circuits. Typical capacitances found in devices you'll work on range from microfarads (μF, or millionths of a farad) down to picofarads (pF, or trillionths of a farad). Most capacitors are marked in a straightforward manner, with the numerical value followed by one of those two

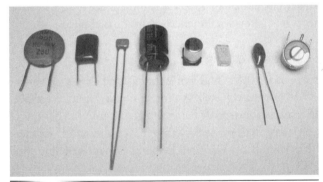

FIGURE 7-7 Ceramic disc, plastic, monolithic, electrolytic surface-mount electrolytics tantalum and trimmer capacitors.

designations. Leading zeros to the left of the decimal point, such as 0.01 μF, are not shown, so we won't use them here either. Some European gear has caps marked in "nF" (nanofarads, or billionths of a farad). Thus a cap marked "1 nF" = .001 μF. That convention never caught on in the United States, but you may run into nanofarad markings in European equipment sold here.

Some capacitors are marked with three numbers, with no indication of μF, nF or pF. With these, the last number is a multiplier indicating how many zeros you need to tack on, starting from picofarads. For instance, a cap marked "101" is 100 pF because there is one zero after the two numbers indicating the value. A marking of "102," then, is 1,000 pF, or .001 μF. And "103" is 10,000 pF or .01 μF, and so on. Here's a handy list:

- XX0 = less than 100 pF. The two X's are the value in picofarads. Sometimes there's no zero on the end. A tiny cap marked "27" is 27 pF. One marked "270" is also 27 pF.
- XX1 = value × 10 pF
- XX2 = .00XX μF
- XX3 = .0XX μF
- XX4 = .XX μF
- XX5 = X.X μF
- XX6 = XX μF

Any value greater than these will be marked directly in microfarads, as in "2,000 μF."

This scheme is absolutely, always how it's done . . . except when it isn't! Some caps are marked with three numbers that really are all the value in picofarads, with no multiplier. I've seen caps marked "470" that were 47 pF, as they should be, and a few with the same marking that were 470 pF. Beyond guessing from the part's size or looking it up on the device's schematic, the only way to tell for sure is with a capacitance meter. The ambiguity arises only when the final number is a zero, however. Nobody makes 471- or 273-pF capacitors, so you can be sure those last numbers are multipliers.

On ceramic disc capacitors, you may also see a marking like "N750." This specifies the temperature coefficient, or how much the capacitance drifts with temperature and in which direction. Keep an eye out for "NP0," which means no drift in either direction. NP0 caps are used in *time constants* and tuned circuits so they won't change frequency as the unit warms up. Should you ever need to replace an NP0 cap, be sure to use the same type.

While many capacitors can accept voltage applied in either direction, polarized types are constructed such that the voltage has to be connected with positive to + and negative to − or the capacitor will self-destruct. There's no real benefit to polarization; it's just a necessary consequence of the chemistry required for stuffing a lot of capacitance into a small package. To avoid calamity, such parts are marked for their polarity. With can-style electrolytics, which contain a liquid *electrolyte* that stores the energy, the marking is a long arrow or black line, and it indicates the *negative* lead. Some very old electrolytics may show a + sign instead, indicating the positive lead, but they haven't been made that way for many years. Look for them in antique radios and such.

Tantalum electrolytics, which look like little dipped candy drops with wires or tan rectangles (second and third from the right in Figure 7-7), denote the positive lead with a + sign or sometimes a red or silver line or dot.

Surface-mount electrolytics of the can type are marked with a black line or semicircle, usually on the top, indicating the negative lead. Flat plastic electrolytic caps (third from the right in Figure 7-7) have one end painted silver or white, and this denotes the *positive* terminal. Don't confuse it with the negative-indicating line on can-style caps. Some small surface-mount caps of well under 1 μF have no markings at all, making it impossible to determine their values without a capacitance meter. They are neither polarized nor electrolytic.

If you find a can-type capacitor with no polarity marking, look for "NP," which indicates a non-polarized electrolytic cap. These are uncommon, but sometimes you run across them in audio gear where AC audio signals will be passing through them. You must replace NP caps with the same type because the polarity reversals in the signals will wreck polarized parts.

Uses

Different styles of capacitors cover various ranges and are used for different purposes. Here are the common ones:

- *Ceramic.* These cover the very small values, from a few pF up to around .1 μF, and are used in resonant radio circuits (those that tune to specific frequencies) and bypass applications, where high-frequency noise and stray signals are deliberately shunted to ground.
- *Plastic.* These start at around .001 μF (1,000 pF) and may go as high as .47 μF or so. They are used for bypass and coupling (passing signals to succeeding circuit stages) and are sometimes found in time constants (circuits that take a precise time to charge and discharge) because of their excellent stability over a wide range of temperatures.
- *Electrolytic.* These start at around .47 μF and cover the highest ranges, on up to tens of thousands of μF. Smaller values are used for coupling signals, and larger ones for filtering noise in power supplies.
- *Tantalum.* These range from .1 μF to around 47 μF and are a special type of electrolytic capacitor with lower impedance (resistance to signals) at high frequencies. They are used in filtering and bypass applications when high frequencies are present.
- *Trimmer.* These range from the low picofarads to around 200 pF and are used as frequency adjustments for tuned (resonant) circuits and oscillators. You'll find trimmer capacitors mostly in radio and TV gear.
- *Supercapacitor.* This is a special type of electrolytic capacitor. It crams a huge amount of capacitance, as much as half a farad (500,000 μF), into a tiny package. Supercapacitors are used for temporary memory storage in products like shortwave radios, preserving

the frequencies you've saved while you change the batteries. They were common once, but today's prevalence of non-volatile memory has largely obsoleted them.

The tradeoff to putting so much capacitance into such a small space is that the capacitor can't take in or return energy very fast. High ESR (internal resistance) is a natural consequence of their construction. So, they're not used for normal purposes of filtering or passing signals. But, a supercapacitor holds enough energy to keep a memory chip alive for minutes, which makes these parts perfect for their intended use. I've seen them in radios and early computer products but haven't run into one in anything newer.

A huge type of supercapacitor is used in car audio installations where tremendous amplifier power is required. Thanks to their size, these caps don't have the speed limitation for delivering current inherent in the little ones, so they can deliver tons of juice on audio peaks, permitting that car next to you at the red light to blow your ears off with booming bass from 15 feet away. They can dump so much current so fast that some automotive jump starters use them instead of batteries, charging the caps from what's left in a near-dead car battery over several minutes and then supplying the quick dump of hundreds of amps required to turn over the engine for a few seconds. I have one of these jump starters, and it really works!

What Kills Them

Different styles of capacitors fail for different reasons, depending on how they're used and to what conditions they're subjected. Generally, application of a voltage above the cap's ratings can punch holes in the dielectric layer, heat can dry or crack them, and some wear out with age.

Ceramic and Plastic   These types very rarely fail. In all my years of tech work, I've found two bad ceramics and one bad plastic cap! It just doesn't happen. If these caps appear unharmed, they are almost certainly okay.

Electrolytic   These caps are the most failure-prone components of all. Part of their charge-storing layer is liquid, and it can dry out, short out, swell, and even burst the capacitor's seals and leak out. Heat, voltage and age all contribute to their demise. The constant charging and discharging as they smooth ripple currents in normal operation gradually wears the caps out. Today's predominance of switching power supplies, with their fast pulse action, has accelerated electrolytic capacitor failure. And application of even a little reverse-polarity current will trash polarized electrolytics in a hurry. A power supply's leaky rectifier (changer of AC into DC) will permit some reverse-polarity voltage to hit the filter caps connected to it, resulting in ruined caps. If you replace them without changing the rectifier, the new parts will fail very quickly.

Capacitor failure modes include shorts, opens, loss of capacitance from age or drying out, electrical leakage (essentially a partial short) and high ESR. If you see even a slight

FIGURE 7-8   Bulging 3,300-µF electrolytic capacitor with greatly reduced capacitance.

bulge in the top of an electrolytic cap, or anywhere on it, for that matter, it's bad and must be replaced. Don't even bother to check it; just put in a new one. Look at the bowed top of the capacitor in Figure 7-8. Keep in mind that an electrolytic cap also can exhibit high ESR or decreased capacitance with no physical signs.

Many surface-mount electrolytics made in the 1990s leaked, thanks to a defective electrolyte formula. If the solder pads look yellow or you see any goo around the cap, the part has leaked and must be replaced. Sometimes the yellow pads are the only clue. You may also notice a fishy smell.

Tantalum   These parts use a solid electrolyte in the dielectric layer that is quite thin. Consequently, they are prone to shorts from even momentary voltage spikes exceeding their voltage ratings and punching holes in the layer. And they are even less tolerant of reverse current than standard electrolytics. Tantalum caps short more often than they wear out. Almost all of the bad ones I've seen have been shorted. Sometimes they short out from age, even when nothing has abused them. That's especially true of the old candy-drop type. The newer surface-mount tantalum caps are hardier.

Trimmer   Trimmer capacitors, or *trimcaps*, use a plastic or ceramic insulating layer that is very reliable. They rely on a mechanical connection, though, between the rotating element and the base, making them prone to failure from corrosion (especially oxidation) over a period of many years. The capacitance may go to zero, or it might get flaky, changing with temperature or vibration. Sometimes, rotating the adjustment through its range a few times can clear it up, but doing so causes loss of the initial setting, requiring readjustment afterward. Never spray the plastic variety with cleaner spray, as it may damage the dielectric layer and destroy the component. I've tried it, and it never worked anyway. The best fix for a bad trimcap is replacement.

Out-of-Circuit Testing

To test a capacitor after removal from a circuit, use your capacitance meter or ESR meter. If you don't have those, you can use the ohms scale on your DMM or VOM to check for shorts, at least. (But you *did* buy an ESR meter, didn't you?) Small-value caps will appear open on an ohmmeter whether they are or not; they charge up too quickly for you to see the resistance rising.

For electrolytics over 1 μF or so, you can do a quick test with an analog VOM, watching for the initial needle swing and slow release back toward infinity. Make sure the cap is discharged before you try to test it. With a polarized cap, connect the test leads + to + and − to −. Most polarized caps will survive a reversed test, but tantalum types may not; even momentary reversed voltage can send them to capacitor heaven.

Crystals and Resonators

Quartz crystals and ceramic resonators are made from slices of quartz or slabs of ceramic material with electrodes plated on the sides. They exploit the *piezoelectric effect*, in which some materials move when subjected to a voltage and also generate a voltage when mechanically flexed. Crystals are always found encased in metal (although there can be a plastic jacket, as shown second from the left in Figure 7-9), while ceramic resonators are usually in yellow or orange plastic or in a dipped ceramic package. Crystals have two leads, and resonators may have two or three (Figure 7-9).

FIGURE 7-9   Quartz crystals and ceramic resonators.

Symbols

Crystals and resonators use the same symbol unless the resonator features internal capacitors and has three leads, like the symbol on the right.

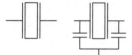

Markings

Crystals and resonators are marked with their frequencies. For crystals, assume that the number means megahertz (MHz). On resonators, assume hundreds of kilohertz (kHz), although some may be in megahertz as well. Crystals can sport other numbers indicating the type of cut used (how the material is sliced, which affects how it behaves), which is quite a complicated topic. It's not a concern in most service work, though. Either the crystal oscillates or it doesn't.

Uses

Quartz crystals are used as frequency-determining elements in oscillators (generators of fixed-frequency signals) in all sorts of gear. Digital devices are driven by *clocks* that step them through their paces. Most require pretty accurate timing and use crystals to provide it. Ceramic resonators are used the same way, but in applications requiring less stability and accuracy, and usually at lower frequencies. You are more likely to find a quartz crystal in the clock oscillator running a digital device like a laptop, DVD player or LCD TV, with a ceramic resonator lurking in a remote control or some radio circuit stages.

What Kills Them

Crystals and resonators are mechanical. They actually move on a microscopic level, vibrating at their resonant frequency. They are also made of crystalline material, so they're somewhat brittle. Heat and vibration can crack them, as can a drop to the floor. Quartz crystals, especially, can develop tiny internal fractures and just quit on their own, with no apparent cause.

Some flaws don't stop them outright; they become finicky and unpredictable, as I mentioned under "Silent Shortwave" in Chapter 4. Touching their terminals may cause them to stop or start oscillating. Also, crystals drift in frequency a little as they age, sometimes drifting past the point at which the circuit will operate properly.

They can also become sensitive to temperature. I had a digital thermometer that stuck on the outside of my dining room window and ran on one AA cell. Though exposed to the elements, it was well sealed and ran for 5 years with nothing more than the occasional battery change. One day, the display was blank. Figuring the cell had died, I took it inside, only to find that the voltage was over 1.4 volts. It was fine. The solder on the battery terminals looked a little funky, so I resoldered them to the board, and the thermometer came to life. Back it went on the window.

A week later, after a cold snap, it was dead again. I took it inside, verified that the battery was still good, and sat there scratching my head. What was going on here? I tried resoldering the contacts once more and changing the battery. The thermometer started working, so I put it back in service. Another cold snap and it went blank. I was done soldering contacts. Something else was causing this.

The unit's microprocessor was clocked by a tiny cylindrical crystal of the sort used for digital watches, kitchen timers and other low-frequency, non-critical gadgets. All of those crystals run at 32.768 kHz. I wondered if the crystal might be getting flaky and temperature-sensitive. There wasn't much else I could try with this thing, so I decided to replace it. I was hunting in my parts box for an old digital watch to cannibalize when I discovered that I had a bag of brand-new crystals of this type. I have no idea where I got it, and had never needed one. I soldered a new crystal into the thermometer, and it has worked perfectly ever since.

Out-of-Circuit Testing

Without a crystal checker, which is simply an oscillator with an indicator light, there's no way to tell whether a crystal works without scoping its signal in an operating circuit. Even a crystal checker may lie to you, indicating a good crystal that still won't start in the circuit for which it's intended.

Crystal Clock Oscillators

Crystal clock oscillators are complete clocking circuits in a small metal box with four or six pins, and are very common in today's digital products. Figure 7-10 shows a really small one. (You can see a much larger unit in Figure 10-8.) On four-pin parts, two pins are for power and ground, a third for output, and the fourth, called *output enable*, activates or inhibits the output. Six-pin versions sport two complementary outputs (one goes high while the other goes low as the signal switches back and forth at its rated frequency), output enable, power, ground and one unconnected pin. In this case, the line over one of the OUT pins indicates that it flip-flops opposite the other OUT pin.

FIGURE 7-10 12-MHz crystal clock oscillator.

Symbols

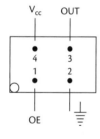
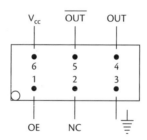

Markings

The frequency will be marked on the case. You may also see a manufacturer's part number.

Uses

More and more, crystal clock oscillators are replacing separate crystals, especially in products using multiple frequencies for various tasks. It's just cheaper and easier to use 'em. Because the oscillator is in the can, no extra circuitry is required, so cost and required space are reduced.

What Kills Them

These oscillators contain quartz crystals, so they're subject to the same mechanical issues inherent to separate crystals. Because the cans also include a complete circuit, they're vulnerable to heat and overvoltage failures as well. For the most part, though, crystal clock oscillators are very reliable.

I ran into one case of a truly odd crystal clock oscillator. This one produced seemingly normal output, but the frequency jumped around randomly every few seconds, making a mess of the product's behavior. A new oscillator fixed it.

Out-of-Circuit Testing

Applying power at the correct voltage, usually 5 volts but sometimes 3.3 volts, and ground to the appropriate pins should produce an output waveform at the frequency shown on the can, within whatever frequency tolerance is specified for the oscillator by the manufacturer. Most of these oscillators produce square waves, but some are designed to generate sine waves. The peak-to-peak voltage of the waves should be fairly close to the DC voltage powering the oscillator. Unless no power is getting to the oscillator, it's easier to test one in-circuit than out. If there's power, the waveform should be there. (If there is no power getting to it, the oscillator isn't the problem anyway!) The output enable pin must not be low or no output will appear. It's fine for it to be left unconnected, and in many circuits, it is, because the oscillator needs to run all the time in most applications.

Diodes, Rectifiers and LEDs

Diodes are one-way valves. Current can flow from their cathodes (–) to their anodes (+) but not the other way. They are made from silicon slabs doped (deliberately contaminated) with materials that cause the one-way current flow, with two dissimilar slabs meeting at a junction point between them. Large diodes used in power supplies are called *rectifiers*, but they do the same thing. Two of them in one package, sharing one common terminal (for a total of three leads), are called a *double diode* or *double rectifier*. Four of them arranged in a diamond-like configuration are called a *bridge rectifier*, whether they are separate parts or integrated into one package (Figure 7-11).

FIGURE 7-11   **Diode, rectifier and bridge rectifiers.**

Symbols

Markings

Single diodes and rectifiers have a band at one end indicating the cathode. They are marked by part number, which does not indicate their operating parameters like maximum forward and reverse operating voltages and maximum current. To get those, you must look up the part number in a book or online.

Bridge rectifiers may be marked for peak forward voltage and current, but they often have part number markings or none at all. Usually, you'll see "~" at the AC inputs and "+" and "−" at the DC outputs.

You may see a marking like "200 PIV," for peak inverse voltage. This is the maximum voltage the diode can withstand in the reverse (nonconducting) direction before breaking down and allowing the current to pass the wrong way. Exceeding the PIV rating usually destroys the diode.

Uses

Diodes and rectifiers are common parts in pretty much every electronic product. Small-signal diodes rectify signals for detection of the information they carry, as in a radio, and direct control voltages to particular places that turn various parts of the circuitry on and off. They may also be used as *biasing* elements, providing a specific current to the inputs of transistors and other amplifying elements, keeping them slightly turned on so that they can conduct over the required portion of the input signal's waveform.

There are specialized categories of small-signal diodes, including gallium arsenide (GaAs), Schottky and germaniums. They all act like standard diodes but have different characteristics, such as lower voltage drop or extra-high speed so they can handle higher-frequency signals.

Rectifiers convert incoming AC power to DC by blocking or redirecting the current when it switches direction. Bridge rectifiers are commonly used to change AC line current to DC by directing opposite sides of the AC waveform to the appropriate + and − output terminals. This is called *full-wave rectification*. In *half-wave rectification*, the reverse-polarity side of the waveform is simply blocked with one rectifier.

LEDs, used to make light for just about everything from control panels to laptops, monitors and TVs these days, are special diodes, but they're still diodes, passing current only in one direction, just like their non-illuminating cousins. Now and then, you'll see an LED on a circuit board, where nobody can see its light during normal product operation.

It's there to show some state of circuit operation for a technician to observe, and it can be part of the circuit's function, rather than just an indicator. Depending on the design, it may light to denote normal circuit operation, or it might suggest a specific fault.

What Kills Them

Diodes and rectifiers can fail from applied voltage that exceeds their limits, but the most common cause is too much current and the heat it produces. Often, a short elsewhere in the circuit has pulled excessive current through them. Sometimes they just fail with age, too. Failure modes include opens, shorts and leakage. Opens are the most common.

Out-of-Circuit Testing

Most DMMs include a diode test function that shows the voltage drop across the part, indicating whether it is passing current. Be sure to test the diode in both directions. A standard silicon diode or rectifier should show a drop of around 0.6 to 0.8 volts in one direction and appear open in the other. A germanium diode, sometimes found in small-signal stages of radios, or a Schottky diode should drop only around 0.2 to 0.3 volts instead of 0.6 to 0.8. An open reading will show the applied test voltage, the same as when the meter's leads are unconnected to anything. Even a small voltage drop in the reverse direction suggests a leaky diode that should be replaced. Diodes with excessive voltage drop in their forward direction should be replaced as well.

With an analog VOM, you can use the high resistance ranges to get a good idea of a diode's or rectifier's condition, but small reverse leakage currents are hard to detect. With either type of meter, you're most likely to see a total failure, either open or shorted, if the part is bad. Leaky diodes are rare, but not so rare that you shouldn't keep the possibility in mind.

To test a bridge rectifier, hook your meter's black lead to the + output of the bridge. That might seem counterintuitive, but it's the cathode, where positive will *end up*, so it's the more negative part of the circuit compared to the anode side of the diode string.

Now, connect the red lead to one of the bridge's AC inputs. You should see indication of current flow. If you're using the ohms scale, you'll see some resistance instead of an open circuit. If you're using the diode test function, it should indicate 0.6 to 0.8 volts. You should see the same reading with the red lead connected to the other AC input. Connect the red lead to the – output of the bridge, and do the same tests with the AC inputs by connecting the black lead to each one. You should see the same results. Testing with the leads connected across the AC inputs should show an open circuit in both directions.

Finally, reverse the leads, and do these tests again. You should see an open circuit on all of them. Anything else indicates a leaky rectifier.

LEDs can be hard to test with an ohmmeter or even the diode test function on a DMM. The voltage drop across an LED is much greater than that across a standard diode, and it varies as well, depending on the type of LED, so a meter might not have enough voltage to make it conduct. If you can't get a reading on your meter, your best bet is to power the LED with your bench power supply. If you can see one, the flat side of the case goes to negative. Start out with the power supply at 0 volts and turn it up *slowly*. If the LED is good, it will start to glow with 1.5 to 3 volts across it. The trick here is to avoid burning out the LED by raising the supply voltage high enough to push excessive current through it. If it doesn't light, reverse the leads and try again, being sure to turn the supply down to zero once more. Reverse polarity on an LED will not hurt it; like any diode, it'll block reverse current. No light in either direction means the LED is bad.

Inductors and Transformers

Inductors, or coils, generate a magnetic field when current passes through them. When the current through the inductor stops or reverses direction, the field collapses and creates a current in the wire, opposing the changes. The effect is to store some of the energy and put it back into the circuit. This is called *inductance*.

Coils may be wound on nonferrous cores having no effect on the magnetic field, or they may be wound on iron cores that play a significant role in the inductor's behavior. Two inductors in close proximity, often wound on the same iron core, can be used to convert one voltage and current to another by creating a magnetic field in one coil and using it to generate a current in the other coil, which may have a different number of turns of wire, thus producing a different voltage. This arrangement is called a *transformer*. In Figure 7-12, inductors are in front, and transformers are in back.

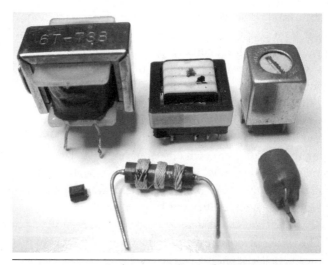

FIGURE 7-12   Inductors and transformers.

Symbols

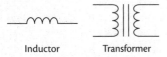

Inductor   Transformer

Markings

Many coils are not marked. Some small ones are encapsulated in plastic or ceramic and marked in microhenries (µH, or millionths of a henry) or millihenries (mH, or thousands of a henry). Others use a color code like that used on resistors and capacitors. In this case, though, the colors denote inductance, starting from microhenries instead of picofarads. There are no picohenry coils; even a short length of wire has more inductance than that.

Uses

Inductors and transformers are used to convert AC voltages, to couple signals from one stage to another (especially in radio equipment) and to filter, or smooth, voltage variations. They are also essential parts of many resonant circuits, working together with capacitors to establish time constants (circuits that take a specific time to change states).

    Their inherent opposition to rapid voltage change makes coils useful for isolating radio-frequency (RF) signals within circuit stages while allowing DC to pass. When used in this manner, coils are called *chokes* because they choke off the signal. Look for them where DC power feeds from the power supply into RF stages. Chokes also can be found in power supplies, especially of the switching variety, where they perform the same function on the rapid pulses used in such designs, keeping those pulses from appearing in the DC output.

What Kills Them

In low-power circuits, coils are highly reliable. Failure is pretty much always due to abuse by other components. It's very rare for small-signal inductors to fail on their own, since little current passes through them during normal operation. In circuits where significant supply current is available, a shorted semiconductor can pull enough current through the inductor feeding it to blow the coil. If you find an open inductor, assume something shorted and killed it.

    The windings in power transformers sometimes open when too much current overheats them and melts the wire. Often, the primary (AC line) side will burn out, even though the excessive current draw is on the other side of the transformer, somewhere in the circuitry. Also, some transformers feature *fusible links* inside. These act like fuses, melting at a specific current and opening the circuit, stopping the power. Unlike normal fuses, though, you can't change these! If a transformer's fusible link has blown, a short someplace in the product has pulled way too much current through the transformer. The primary winding will measure as an open circuit. In some cases, you can cut the paper insulation over the windings and repair the link, but it doesn't always work, and you must take care not to cut into the windings. Just be sure to repair the short before you fix or replace the transformer, or it will blow again as soon as you apply power.

Windings also may short to the iron core. The insulation of the windings in high-voltage transformers can break down and arc over to other windings, create a short between windings, or arc and short to the core. That was fairly common in backlight inverters for older, fluorescent-lit LCD panels and *flyback transformers* in CRT TVs and computer monitors. It still happens now and then in switching power supplies, but the voltages are lower, so arcing is less likely.

I ran into an odd case that typified my generalization about too much current being the cause of inductor failure, except this time it wasn't due to a bad part. This one was an amateur (ham) UHF/VHF FM radio transceiver. The transmitter had died, and I traced the fault to a low-level amplifier stage in the chain leading to the RF power output transistor. The inductor supplying power to the stage was open, but the transistor pulling current through it was fine. It turned out that, like the "Goodbye Darkness" projector case in Chapter 4, this was an example of bad design. The transmitter ran the inductor too close to its current rating, heating and fatiguing the tiny wire inside until it finally opened. After I found it, I went online and discovered that this model radio was known for this problem! I put in a bigger inductor of the same inductance but higher current capability, and it has worked great for several years now.

Out-of-Circuit Testing

Use your DMM to test for continuity from one end of a coil or section of a transformer to its other end. You can also use the ohms scale to check for shorts from windings to the core and between unconnected windings in transformers. Any resistance at all between sections intended to be isolated, or from a winding to the core, indicates a fault. You should see an open circuit (infinite resistance) on everything except across the windings themselves.

It's very hard to tell whether windings are shorted to each other in the same coil. Using the ohms scale, the difference can be so slight that it's undetectable. If you have an inductance meter and know the correct inductance, it will give you some idea. With unmarked coils and just about all transformers, though, you won't know what the inductance should be without access to a schematic or an identical part for comparison. In some circuits, such as LCD backlighting inverters, there can be two identical transformers, so you may be in luck.

Back in the days of CRT TV servicing, techs employed a *ring tester* to diagnose the high-voltage transformers used to generate the tremendous voltage applied to the picture tube's anode. This test instrument applied short pulses of energy to the transformer's high-voltage winding, with the tester's signal output connected to an oscilloscope.

When such pulsed energy is applied to a coil, the energy zips back and forth, getting a little smaller each time due to resistive loss in the wire. The result is electrical ringing analogous to the mechanical ringing in a bell as it slowly dies away. When turns in the coil are shorted to each other, the ringing dies out faster than it would with a good coil, resulting in a shortened waveform on the scope.

Ring testing works on audio and power transformers, too, although the waveform differences between good and bad coils aren't as pronounced. If you work a lot with transformers, a ring tester might come in handy. You're unlikely to find one anymore, but they're easy to build and diagrams are available online.

Integrated Circuits

ICs, or chips, come in thousands of flavors. In today's advanced products, ICs do most of the work, with transistors and other components supporting their operation. Digital chips are at the hearts of computers, DVD players, digital cameras, phones, tablets, TVs, you name it. In audio/visual gear, they work side by side with analog ICs handling radio, audio and video signals (Figure 7-13).

ICs integrate anywhere from dozens to millions of transistors on a small square of silicon, with microscopic structures printed using photographic techniques. A failure of even a single transistor will render the chip defective. It's pretty amazing that they ever work at all! Obviously, there's no way to test the individual structures; all you can do is verify whether the chip is performing its intended function properly.

FIGURE 7-13 Small- and large-scale ICs in a DVD player.

There are some common, off-the-shelf chips used in many products, but custom ICs specific to a model or product category dominate modern gadgets' innards. Each specialized chip can include more product-specific functions, so it takes fewer of them to make a device. The fewer parts and interconnections, the more reliable an item is likely to be. And that gadget can be smaller and cheaper to build. Phones and tablets are made mostly from chips designed for such products.

Large-scale integrated chips (LSIs) can have up to a few hundred pins spaced so closely that you can't even put a probe on one without shorting it to its neighbor. Unless you have a rework station, it's difficult to unsolder or resolder these parts. There is a trick for touching up suspected bad joints on some of them with a soldering iron, but it takes practice and doesn't always work. We'll look at it in Chapter 12.

Luckily, ICs are very reliable. Most of them handle small signals, so they don't dissipate a lot of power and get hot enough to fail. There are some exceptions to this, however—notably, central processing units (CPUs) and video graphics chips (GPUs) in computers. Some graphics cards have fans over the chips, as do CPUs. You know the thing gets mighty warm when it needs its own fan! These parts get so hot because they have millions of transistors switching from millions to a few billion times per second. That many microscopic heat generators add up to a serious temperature rise.

Other hot-running chips include audio power output modules in stereo and multichannel audio receivers, some types of voltage regulators, motor controllers and anything else that pumps real power. These parts usually are mounted on heatsinks, and they fail as often as power transistors.

Symbols

Chips are denoted on schematics by the number of pins and their general shape. Internal structures are not shown. Some simple chips, like logic gates and op amps, may include a diagram of their general internal functions but not the actual schematic of transistors on the chip. The data sheets for simple chips like op-amps and voltage regulators do sometimes show that information, though, and it can be useful in understanding how the chip is supposed to behave. LSIs have too much in them to show on a data sheet, however. Even if it could be fitted on a page, deducing the operation from millions of transistors would be impossible anyway.

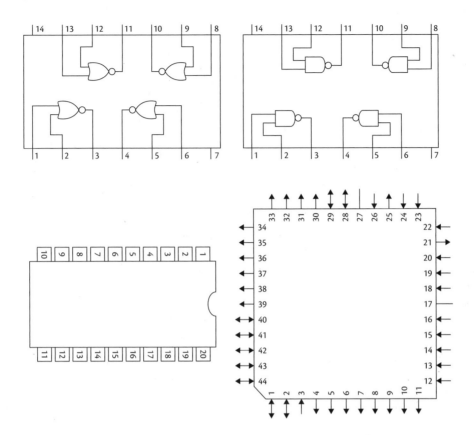

Markings

The parts are marked with part numbers that can mean many things. Small-scale, industry-standard chips are produced in *families* sharing some part number commonality. For instance, older CMOS logic gate chips are called "4000 Series" and have part numbers like 4011B and 4518. They all begin with 4. Those are still in use, and you may run into them. In particular, the 4066B analog switch chip, used to route audio and other low-frequency signals, shows up often and is prone to failure. LSIs often have proprietary part numbers, and they certainly will if they're custom parts made for a specific type of product.

The pins on an IC are numbered going counterclockwise around the chip in a ring. Pin 1 will have a dimple or spot next to it on the chip's plastic casing, and it will be in a corner (see Figure 7-13).

Uses

ICs are used for just about everything: audio amps, data processors, logic gates, oscillators, signal processing and any other function you can think of.

What Kills Them

ICs are highly reliable, but heat is a major danger, especially when it's internally generated. Voltage spikes can destroy some chip families, but more modern varieties are fairly voltage-tolerant. Still, a static charge may deliver a voltage too high for any chip to withstand. A short in another area of the circuit that pulls too much current from a chip's output can blow it. If the power supply feeding the chip fails but signals are still fed to it, the chip can *latch*, causing a permanent internal short. Finally, some ICs, especially LSIs, have internal features so small that atomic forces may gradually eat through the microscopic wires and connections, creating holes that wreck the chip. Essentially, the part dies of old age. This problem has been researched by chip makers for many years, and today's ICs hold up rather well. Now and then, a chip still dies without apparent provocation.

Flash RAM, much like what's in a thumb drive, is used to store user preferences in many common products. In applications where it gets written to a lot, it wears out and finally stops accepting new data. The device can still read what's stored on it but cannot store anything new.

Out-of-Circuit Testing

There's no way to put a meter on a chip to see if it's good. In fact, doing so may damage the chip if the meter's test voltage happens to be high enough and touches the wrong pins. Simple logic gates can be plugged into a test board and hooked up to test their functions, but the exercise is more trouble than it's worth. For the most part, chips must be tested in-circuit by observing their actions with your oscilloscope.

Op Amps

The op amp, or operational amplifier, is a common type of chip. It's an analog general-purpose amplifier configurable to do many different jobs, such as buffering (providing added current to a signal), oscillating and signal filtering, by connecting resistors and capacitors to its inputs and outputs. Op amps are commonly found in analog circuit stages like tone controls on audio amplifiers. The variety of circuits that can be made from op amps is staggering! One chip may contain several op amps. Dual and quad op amps are used in many products (Figure 7-14).

FIGURE 7-14 Op amp.

Symbol

Markings

The chips are marked with part numbers, and there are some, like LM358, you'll see often.

Uses

You'll find op amps in audio circuits, radio circuits, motor controllers and power supplies. Any place requiring an amplifier is a prime candidate for an op amp. Most op amps are for small-signal applications, but some power op amps are capable of driving significant current into heavy loads such as motors, mechanical actuators and speakers. Expect those op amps to be mounted on heatsinks.

What Kills Them

Heat and overcurrent are the primary culprits.

Out-of-Circuit Testing

As with logic gates, op amps have to be in some kind of circuit for you to test them, so they're best checked in the product as you operate it by scoping the input and output signals and verifying that power to the chip is present.

Resistors

Resistors offer opposition to current, dissipating the opposed power as heat. The opposition limits how much current can flow, as described by *Ohm's law*. Limiting current is a vital function in any circuit, so every electronic product has resistors. They come in many shapes and sizes, but those with leads are easily identifiable by their color bands indicating the resistance value in ohms. Some large resistors are marked numerically, as are some of the tiny surface-mount parts (Figure 7-15).

At one time, resistors were made of a granulated carbon compound and were called "carbon composition." Nowadays, the basic type of resistor, found in virtually everything, is made from a carbon film. These resistors run the full range of values, from less than 1 ohm (Ω) to 10 megohms (MΩ, or millions of ohms) or more. Most you'll see will be over 10 Ω and under 1 MΩ. The tolerance for standard carbon resistors is ± 5 percent. That is, the measured value should be no more than 5 percent higher or lower than the stated resistance.

Some applications require the use of wire-wound resistors. These look a lot like carbon parts, but you can see the coil of wire under the paint on the body. Wire-wound resistors

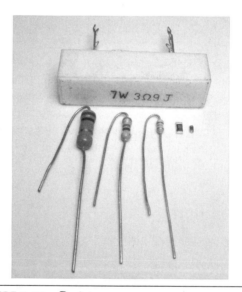

FIGURE 7-15 Resistors.

can dissipate more heat than can carbon types. They also can be manufactured to very tight tolerances, but they have some inductance because they are coils, and thus are ill-suited to high-frequency circuits where the inductance might matter. The two types should not be interchanged.

In addition to the resistance value, the power-dissipating capability of a resistor, measured in watts, is an important factor. Because power is dissipated as heat, knowing how much heat the part can take before disintegrating is vital. Standard carbon resistors with leads are rated at ¼ watt, with some slightly larger ones able to dissipate ½ watt. Very tiny ones are rated at ⅛ watt, while surface-mount versions typically vary from ⅛ watt down to ¹/₃₂ watt.

Other resistor formulations include metal film and metal oxide. Metal film types introduce a bit less noise into the circuit and are used in areas where that's a significant issue. Metal film and metal oxide parts have tighter tolerances, in the range of 1 or 2 percent of stated value. Some circuits require such tight tolerances for proper operation.

Single-inline-package (SIP) resistor packs are used when multiple data lines in a digital system need a *pull-up* or *pull-down* resistor on each line to make the line go high or low so that the chip providing data has to pull only one way to change the data from 0 to 1 or vice versa. This is a common scheme and would require a lot of individual resistors. Using SIPs reduces the parts count and required space on the circuit board (Figure 7-16).

Most SIPs you'll encounter have a common ground at pin 1, with the same resistance from it to each of the other pins. There are other schemes, including series resistors and sets of isolated resistors, but you're not likely to run into those. SIPs are used in low-current signal applications, so they stay cool and are pretty reliable. Not all resistor packs look like SIPs. Tiny ones with surface-mount packages look somewhat like other chips, except that they're noticeably narrower.

FIGURE 7-16   SIP resistor pack.

Symbols

Resistor   SIP

Markings

The use of color coding dates back to the vacuum tube days, when resistors got so hot that printed numbers would evaporate. While it's not used for surface-mount resistors, color coding is still the norm for through-hole resistors (those with leads). To read the color code, first you must determine which end of the resistor is the start and which is the far end. Look for a gold or silver band; that's the tolerance marking, indicating the far end, and there's usually a little extra space between that band and the others. The first digit will be at the end farthest from the tolerance band.

Each number is represented by a color. The scheme is as follows:

Black	0	Green	5
Brown	1	Blue	6
Red	2	Violet	7
Orange	3	Gray	8
Yellow	4	White	9

The tolerance bands at the far end are as follows:

Brown	±1 percent
Red	±2 percent
Gold	±5 percent
Silver	±10 percent

To determine a resistor's value, read the first two bands as numbers. The third band is a multiplier indicating how many zeros you need to tack onto the numerical value. So, for instance, red–red–brown would be 2, 2, and one 0, or 220 Ω. Yellow–violet–orange would be 4, 7, and three 0s, or 47,000 Ω (a.k.a. 47 kΩ). Be careful not to confuse a black third band with a zero; it means no zeros.

This scheme works great until the resistance value gets below 10 Ω. Try it—there's no way to mark such a value. For resistors this low, the third band is set to gold, meaning that there's a decimal point between the first two bands. So, green–blue–gold–gold would be a

5.6-Ω resistor with a ±5 percent tolerance. You won't see too many resistors under 10 Ω, but you might run across one in an audio output stage or a power supply.

Some resistors have four bands plus a tolerance band. These are higher-precision parts with tighter tolerances and must be read slightly differently. With these, read the first *three* bands as numbers and the fourth band as the multiplier. So, a 47-kΩ resistor would read yellow–violet–black–red.

How can you tell what type of resistor you have? Sometimes you can't. If the resistor has a gold or silver tolerance band, assume it's a standard carbon part, unless it's in a very low-noise circuit like an audio preamp, in which case metal film might be a more appropriate replacement. If it has a red or brown tolerance band, indicating higher precision, it might be a metal film or metal oxide component.

Many surface-mount resistors are marked numerically using the same idea. The last number is a multiplier. If you see an "R" between the numbers, that's a decimal point. You'll see that only on resistors of rather low value. For instance, "4R7" means 4.7 Ω. If you see a number with a zero at the end, don't confuse that to mean a number; it indicates *no* zeros. Thus, "220" means 22 Ω, not 220 Ω, and a 220-Ω resistor would be marked "221." Also, a letter *after* the Ω symbol on any type of numerically marked resistor is not part of the numerical value, even if it's a "K." So, "47 ΩK" is still 47 Ω, not 47 kΩ. Used this way, the "K" denotes 10 percent tolerance. Pretty crazy, huh?

Some tiny SMD (surface-mount device) resistors are too small for numerical value markings. Instead, they sport a two-digit number that must be cross-referenced from a list. The number itself has no direct relation to the value. This marking scheme has several permutations, depending on the size of the part. At the end of this chapter, you'll find a table of SMD sizes and designations. It's unlikely that you'll ever need to replace one of these tiny resistors, because they carry very little current and rarely fail, but sometimes other circumstances can require you to do so. I had to replace one not long ago because it snagged on a plastic screw post as I was removing the circuit board from the product's case. The grain-of-salt-sized component snapped off and promptly disappeared.

SIP markings indicate the number of pins, type of construction, electrical configuration, resistance value and tolerance. Unfortunately, the marking scheme is not standardized, so it varies among manufacturers. The resistance is the rightmost three-digit number and follows the usual arrangement of a two-digit value and a multiplier.

Uses

Resistors are found in just about every circuit. They limit the current that can pass through other parts. For instance, transistors amplify a signal by using it as a control for a larger current provided by the power supply, somewhat like the handle on a spigot controls a large flow of water. A resistor between the supply and the transistor sets how much current the transistor has to control. Without the resistor, the transistor would have to handle all of the supply's current and would self-destruct.

What Kills Them

Resistors rarely fail on their own. Heat caused by passing too much current burns them out, sometimes literally. Carbon resistors can go up in flames or become a charred lump when a short in some other part pulls a lot of current through them. It's not uncommon to see one with a burn mark obscuring the color bands.

Out-of-Circuit Testing

Use your DMM's ohms scale to see if the resistance is within the specified tolerance range. Most resistors do better than their tolerances, but expect a small difference between what's on the code and what you measure.

When the resistor is charred beyond your ability to read the color code, it can be a real problem unless you have the unit's schematic diagram. Luckily, resistance values follow a standard pattern because there would be no point in producing resistors whose values fell within another resistor's tolerance range. So, you may be able to infer a burned-out band's value from others that can still be seen. If you ever need it, you can look up the list of standard values online. Keep in mind that higher-precision resistors, indentifiable by their extra color bands, offer more standard values because the tighter tolerance means each value covers a narrower range.

Potentiometers

A potentiometer, or *pot*, is a variable resistor. A small one used for internal circuit adjustments is called a *trimpot*. Pots and trimpots may have two, three or (rarely) four leads, but most have three. The outer two are connected to the substrate on which the resistive element is formed, with one lead at each end of the resistor. The center lead goes to the wiper, a movable metal contact whose point touches the resistive element, selecting a resistance value that rises relative to one outer lead while falling relative to the other as you turn the knob (Figure 7-17).

A two-wire pot has no connection to one end of the resistive element but is otherwise the same. Some two-wire pots connect the free end of the element to the wiper, which slightly affects the resistance curve as you turn it, but it doesn't matter a whole lot. Two-wire pots are sometimes called *rheostats*. Four-wire pots, used mostly in stereo receivers to provide the "loudness" function that increases bass at low volume levels, are like three-wire pots but with an extra tap partway up the resistive element.

The amount of resistance change you get per degree of rotation is even from end to end on *linear taper* pots. On *log taper* pots, also called *audio taper*, a logarithmic resistance curve is used so that audio loudness, which is perceived by our ears on a log curve, will seem to increase or decrease at a constant rate as the control is varied. Trimpots, whose primary use is for set-and-forget internal adjustments on circuit boards, are always linear taper.

FIGURE 7-17   Potentiometers and trimpots.

How much power the pot can dissipate depends on its size. It's not marked on the part. In most applications, only small signals are applied to it, so dissipation is not much of an issue. Some pots have metal shafts, while others use plastic. Plastic provides insulation in applications like power supplies, where you might come in contact with a dangerous voltage. Also, some pots have switches built into them, and those may be rotary, operating at one end of the wiper's travel, or push-pull. Pots with switches will have separate connections at the back for the switch.

Most trimpots rotate through less than 360 degrees, just like pots. Those used in applications requiring high precision may be multi-turn, with a threaded gear or rod inside providing the reduction ratio. Multi-turn trimpots use a small screw for adjustment, usually off to one side of the body of the component. Be wary of turning the screw, because there's no visual indicator to help you set it back where it was.

In stereo receivers, pots can be ganged together onto one shaft so that turning the shaft will affect both channels together. Each pot is internally isolated from the other, though you may find one end of all of them tied to ground at the terminals.

Symbols

3-wire   2-wire

Markings

Pots are marked numerically with their resistance values using the same scheme as resistors. A "B" in the code indicates a linear pot, while a "C" means a log pot. An "A," though, can mean either, because the codes have changed over the years. Some pots have "LIN" or "LOG" printed on them. Most don't, though.

Uses

Pots are used to adjust operating parameters for analog signals and power supply voltages. Once, they were the primary method of setting just about everything. In this digital age, volume, treble, bass, brightness, contrast and such are more often selected from a menu or adjusted with up/down buttons. Trimpots on circuit boards are also less common, but you'll still find some, especially in power supplies, including switchers.

What Kills Them

Most failures are caused by wear or dirt where the wiper makes contact with the resistive element. The symptom is scratchiness in audio, flashes on the screen in video or an inability to set the pot to specific spots without the wiper losing contact with the element. Contact cleaner spray will usually clear it up, but if the element is too worn, replacement is the only option.

Occasionally, the resistive element cracks, resulting in some weird symptoms because the wiper is still connected to one end but not the other. Plus, which end is connected reverses as you turn the control over the broken spot. Audio may blast through part of the control's range and disappear below the break point.

Out-of-Circuit Testing

Test the outer two leads on a three-lead pot with your DMM's ohms scale. It should read something close to the printed value. If the element is cracked, it'll read open.

To check for integrity of the wiper's contact, connect the meter between one end of the pot and the wiper. Slowly turn the pot through its range, observing the change in resistance. This is one test better done with an old-fashioned analog VOM; the meter needle will swing back and forth with the position of the wiper, and it's easy to interpret. A two-lead pot should read something close to its stated resistance at one end of its control range and 0 Ω at the other.

To verify whether a pot is linear taper or log taper, measure the resistance at one third of the rotation and again at two thirds. See if those values are about one third and two thirds of the total resistance from end to end. If it's a log taper part, they won't even be close.

Switches

Switches permit or interrupt the passage of current. There are many, many kinds of switches, and they're used in just about everything. Toggle switches, slide switches, rotary switches, leaf switches, pushbuttons, internal switches on jacks . . . there's practically no end to the varieties (Figure 7-18). Many newer products do not have "hard" power switches that actually disconnect power to the unit. Instead, a low-current switch signals the microprocessor, which then shuts down power using semiconductors to interrupt the flow.

Switches can have any number of contacts for switching multiple, unconnected circuits at the same time. Rotary and slide switches, especially, may have many sets, while toggle switches rarely have more than two or three. Each set is called a *pole*. Some contacts may be *normally open* (NO), meaning that they are not connected with the switch in the "off" position, and some may be *normally closed* (NC), meaning the opposite. Each direction is called a *throw*. Thus, a double-pole, double-throw (DPDT) switch would have two separate sets of contacts, each with three elements: the NO side, the movable contact and the NC side. The dotted line in the symbol indicates mechanical linkage between poles but not electrical connection.

Symbols

The dotted line indicates mechanical linkage, not electrical connection.

FIGURE 7-18   Switches.

Markings

If marked at all, switches may show their maximum voltage and current ratings.

Uses

Expect to find switches everywhere. From pushbuttons on front panels and remote controls to tiny slide switches on circuit boards, they handle power and information input in essentially all products. Leaf switches, with bendable, springy metal arms, sense the position of mechanisms like laser optical heads in disc players, informing the microprocessor of the state of moving parts. Switches inside jacks sense when accessories are plugged in, altering system behavior to accommodate them. This is how your phone knows when a headset is plugged in and puts up that little headphone icon on the screen.

What Kills Them

Age, oxidation and contact pitting from arcing usually do switches in. If the switch's construction permits any access, try spraying some contact cleaner inside, and then work the switch a bunch of times.

Out-of-Circuit Testing

Test switches with your DMM's continuity or lowest ohms scale. The contacts should read nearly 0 Ω when closed and infinity when open. There's an exception, though: Some pushbuttons, such as the kind on remotes, laptop keyboards and tiny products like digital cameras, use a carbon-impregnated plastic or rubber contact to make the connection. You can identify them by their soft feel when pressed; they don't click. These switches are intended only for signaling, not for handling significant current, and they may have a few tens of ohms of resistance when in the "on" state. While such a reading would indicate a bad toggle switch, it's fine with these little guys.

Relays

Relays are switches controlled by putting current into a coil. When the coil is energized, the resulting magnetic field pulls a metal plate toward it, pressing the attached switch contacts against opposing contacts. In this way, a small current can control a much larger one, just as a transistor does in *switch mode* when it is turned fully on and off by signals at its input terminal.

Relays may have any number of contacts for switching multiple, unconnected circuits at the same time. As with switches, each set is called a *pole*, and some contacts may be *normally open* (NO), meaning that they are not touching until the relay is energized, and some may be *normally closed* (NC), meaning the opposite. Each direction is called a *throw*. Thus, a double-pole, double-throw (DPDT) relay would have two switches, each with three contacts: the NO side, the movable contact and the NC side (Figure 7-19).

Most relays return to their "off" state when power is removed from the coil. An unusual type called a *latching relay* has two coils. Its mechanical construction keeps it in the "on" state after one coil is no longer energized. To turn it off, power has to be applied to the other coil, pulling the switch in the opposite direction. The advantage of such an arrangement is that no power is required to hold the relay in the "on" state, reducing battery consumption in small devices.

Another style of latching relay is really just a normal type where one set of NO contacts is used to keep the coil energized after the rest of the circuit has turned it on by feeding power supply voltage through those contacts and back to the coil. This type requires power to stay energized, so it does not conserve battery power.

Symbols

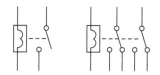

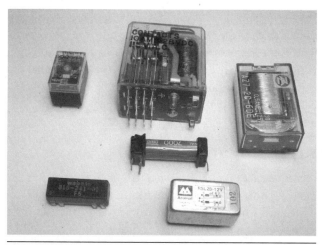

FIGURE 7-19   Relays.

Markings

Relays may be marked schematically for their internal construction—that is, where the coil connections are and what kinds of switches are inside. You may find a voltage rating; this specifies the coil voltage, not the maximum voltage permitted on the switch contacts. A resistance rating is also for the coil. If you find one, you can calculate the current, called the *pull-in current*, that the coil needs to actuate the switch with Ohm's law by dividing the power supply voltage by the coil resistance. Knowing the resistance rating also helps you when testing the coil with your DMM.

Some relays include an internal diode across the coil to prevent the reverse voltage its inductance generates when power is removed from the coil from feeding back into the circuitry and damaging it. If the relay shows polarity markings (+ and −), it has a diode.

The maximum current the switch contacts can handle usually isn't shown, but it might be. If the markings read "12 VDC, 3A," this indicates a 12-volt coil intended to be driven with DC power, with switch contacts capable of switching 3 amps. No relay coil requires 3 amps to pull in! Typical coil current is a few hundred milliamps.

There's variation on this, though. Some relays show two sets of markings, and both are for the contacts. If you see one that says something like "12V 3A" and under it "120V 1A," this means the contacts can handle 3 amps at 12 volts, degrading to 1 amp at 120 volts. Neither marking is for the coil. To help alleviate the confusion, such relays often have a separate marking for the coil voltage.

Some relays are made specifically for AC coil operation, too, and are marked as such. Those will not have a diode.

Uses

Relays were once used widely to switch large currents with smaller ones. These days, semiconductors usually do that job, but some applications still employ relays. Power supply delay circuits, which prevent power from reaching the circuitry for a few moments after the supply is turned on, often have relays. Speaker protection circuits in high-power audio amplifiers use them too, because the high-current audio signal is not altered or impeded at all by a relay, but it would be by passing through a semiconductor. Most relays make an audible click when they switch, giving their presence away.

What Kills Them

Relay troubles usually involve the switch contacts. Corrosion from age and oxidation, as well as pitting from arcing when large currents are switched, cause resistance or flaky contact. Once in a while, a relay will become sticky, not wanting to open back up after power is removed from the coil. This condition can be caused by arcing in the contacts welding them to each other and by weakening of the spring used to pull the plate away from the coil's iron core. Occasionally, a relay will stick because the pull-in plate gets magnetized, so it stays stuck to the metal core around which the coil is wound after the coil current is turned off. Demagnetizing it with a tape head demagnetizer or a magnet of opposite polarity can restore proper operation. Do *not* do this to a latching relay of the "no power needed to stay on" type! Those relays contain permanent magnets to hold the switch in place and will be ruined if you demagnetize them.

If the relay has a removable cover, you may be able to pop it off and clean or unstick the contacts. Be careful when prying off the covers of miniature relays. Those covers can be quite tight, requiring using an X-Acto knife to pry up the plastic. It's easy to destroy the coil of such a relay if the knife slips inside and slices across the hair-thin windings. I've done it. See *curse words* in the Glossary.

Often, just pulling a piece of paper soaked in contact cleaner between the contacts will spiff them up and restore proper operation. If that's not enough, very light wiping with fine sandpaper may remove the outer layer of gunk. Silver polish works too, but make sure to get it all off when you're done. Be careful not to bend the contacts, and don't sand off the plating; it's vital for the contacts' long-term survival. Whichever method you use, wipe the contacts with cleaner-soaked paper to remove residue before you put the relay's cover back on.

I've seen a few relays with slightly bent contacts that didn't press hard enough on their mates to make reliable contact when the relay was energized. It happens with relays that spend most of their time pulled in, such as in power supplies and the speaker protection circuits of audio receivers. Over time, the movable contacts' metal fatigues and loses some of its springiness. Even after a good cleaning, those relays just wouldn't make a reliable connection until I bent the movable contacts a little to straighten them out. If you do this, make sure not to bend them so far that they touch their mating NO (normally open) contacts when the relay is not energized. Also be careful that they still touch the NC (normally closed) contacts when the relay is off and don't touch them when the relay is pulled in.

Two other causes of relay problems aren't actually the relay itself! With a larger relay, solder joints where it meets the board can crack and fail for two reasons. First, they're big and are often not well soldered during assembly because they absorb too much heat for a good joint to form. Second, movement of the relay contacts causes a little sonic shock wave—that click you hear—that can break the solder joints over many years.

Finally, if the relay is in a socket, oxidation can degrade the connections to it. If you pull the relay and see dull or discolored metal on the contacts, clean 'em up before putting it back in. Especially in old equipment, this is a frequent problem.

Out-of-Circuit Testing

Use your DMM's continuity or lowest ohms scale. Check for coil continuity. If there's a reverse-polarity protection diode, be sure to check in both directions. The coil shouldn't read 0 Ω; there's enough wire there for dozens to a few hundred ohms. If it reads very near zero on a relay that has a diode, suspect a shorted diode, especially if the symptom suggests that the coil doesn't want to pull in the switch or the transistor energizing the relay runs hot or burns out. If the diode is on the board, outside the relay, try pulling up one leg and checking it again for a short. Also, verify that there is no continuity between the coil and its metal core. If there is any, the coil has a short, and you need a new relay.

A blown transistor can also be caused by an *open* diode across the relay coil. The purpose of the diode is to absorb the backward current spike that occurs when power to the coil is removed and its magnetic field collapses. If the diode opens, the spike gets applied to the transistor and can destroy it. Any time you find a blown relay driver transistor, suspect a shorted coil or an open or shorted protection diode. Most of the time, the diode is at fault.

Check the relay's NC contacts with the meter. They should read 0 Ω or very close to it. Use your bench power supply to energize the coil. If it has a diode, be certain to get the polarity correct, with + to the diode's cathode, not its anode. (Remember, the diode is supposed to be wired backward so it won't conduct when power is applied.) With no diode, polarity doesn't matter. Once you hear the click, check the NO contacts. They should also read 0 Ω or very close to it. When you disconnect the power supply, the contacts should return to their original state. The NO contacts should open, and the NC contacts should close. If there are multiple poles, check each one.

Transistors

Along with ICs, transistors are the active elements that do most of the work in circuits, amplifying, processing and generating signals; switching currents; and providing the oomph needed to drive speakers, headphones, motors and lamps (Figure 7-20).

FIGURE 7-20 Transistors.

Transistors act like potentiometers, but instead of your hand turning the shaft, a signal does. The current fed by the power supply through the pot can be much greater than that required to turn the shaft, providing *gain*, or amplification, as the wiggling signal molds a bigger version of itself from the power supply's steady DC.

There are thousands of subtypes of transistors, but most fall into three categories: bipolar, also called BJT (bipolar junction transistor), JFET (junction field-effect transistor) and MOSFET (metal-oxide–semiconductor field-effect transistor). Bipolar transistors are the standard types used in products since the 1950s. They come in two polarities, NPN and PNP, and consist of three elements joined by two junctions. The three elements are the *base, emitter* and *collector*, each with its own lead. Current passing between the base and the emitter permits a much larger current to pass between the collector and the emitter, with one of them being fed from the power supply. The greater the voltage difference between the base and emitter, the more current will pass, and a proportionally higher current can pass between collector and emitter. In most configurations, the signal is applied to the base, causing a bigger version of the signal to be formed from the flow between the collector and emitter. The ratio of collector-to-emitter current to base-to-emitter current is the transistor's *current gain* and is inherent in the component's design. For example, if it takes 2 mA (milliamps) of current from base to emitter to cause 100 mA to pass from collector to emitter, the transistor has a current gain of 50.

In an NPN transistor, the base must be positive with respect to the emitter for collector-to-emitter current to pass. In a PNP transistor, the base must be negative. So, the two types are of opposite polarity and cannot be interchanged. Most transistors used today are NPN, but you will find circuits with some PNP parts.

JFETs work on a similar principle. They label their three elements differently. Instead of the base, the controlling terminal is called the *gate*. The emitter is called the *source*, and the collector is the *drain*. Instead of a current passing from gate to source, application of a voltage to the gate goes nowhere but results in an electric field that controls a channel in the transistor, permitting current to pass between drain and source. This gives the JFET a very high input impedance, which is another way of saying that it does not take much signal current to turn it on.

MOSFETs are similar to JFETs, and they use the same terminal names, but their internal construction is a bit different. They have even higher input impedance and some other desirable characteristics that have resulted in their pretty much dominating FET applications. You may see JFETs in older gear, but you're much more likely to see nothing but MOSFETs in newer equipment.

Like bipolars, FETs come in two polarities, P-channel and N-channel, corresponding to PNP and NPN bipolar transistors. They also come in *enhancement-mode* and *depletion-mode* types, specifying what happens when the gate voltage is zero. An enhancement-mode FET will be turned off with no voltage at the gate; like a bipolar transistor, it requires a *bias* voltage to turn it on. A depletion-mode FET will be turned on with zero gate voltage. The only way to turn it off is to apply a voltage of opposite polarity to the one that will increase current flow. Essentially, it's like a spigot that stays open part way until you actively open it even more or turn it all the way off.

Luckily, you don't need to worry too much about these arcane details. If a FET is bad, you'll look up its part number and replace it with a compatible type. Still, knowing the basics of how these parts work is essential for understanding how to troubleshoot circuits using them . . . which are pretty much all circuits made since the early 1990s.

As you can see, transistors have many parameters, so it's not surprising that there are thousands of subtypes with different gains, power-dissipation capabilities, frequency limits and so on. Some are similar enough that they can be interchanged in many circuits, but most are not. To replace one part number with another, you need a transistor substitution book or an online cross-reference guide.

Symbols

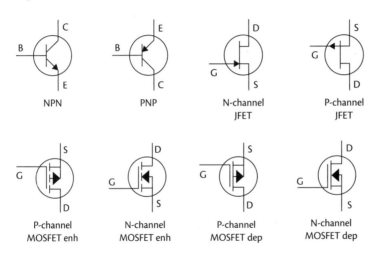

NPN   PNP   N-channel JFET   P-channel JFET

P-channel MOSFET enh   N-channel MOSFET enh   P-channel MOSFET dep   N-channel MOSFET dep

Markings

Transistors are marked by part number, called a *type number*. There are thousands of these numbers! Some numbers indicate whether a bipolar part is NPN or PNP. If the number starts with "2SA" or "2SB," it's PNP. If it starts with "2SC" or "2SD," it's NPN. Sometimes the "2S" will be left off, and there are plenty of type numbers that don't follow this scheme at all. A "3N" indicates a FET, but some of them have numbers starting with "2N," just like bipolar parts. "2SJ" is a P-channel FET, and "2SK" is an N-channel FET. Other number schemes are used as well.

Some transistors have *house numbers*, which are proprietary numbering schemes used by different manufacturers to mean different things. There is no way to ascertain what the industry-standard number would be for such a part. Tiny surface-mount transistors often have no numbers at all, and the only way to find out what they are is to obtain the schematic diagram. Sometimes, you can infer the polarity from the voltages applied to the terminals, but doing so takes a fairly deep understanding of circuit configurations, and it's still hard to guess whether the part is bipolar or a FET. However, if the gate terminal in an amplifier stage has no DC bias applied to it through resistors, it may very well be a depletion-mode FET. These kinds of no-bias amplifier configurations are commonly found in small-signal stages in radio receivers.

The arrangement of the leads varies with transistor type. Small Japanese parts with leads are usually laid out emitter–collector–base (ECB), left to right, with the flat side up, while American parts are often emitter–base–collector (EBC). Metal-encased power transistors have only two leads and use the metal casing as the collector. Plastic power transistors are usually base–collector–emitter (BCE), with the metal tab, if there is one, also being C.

Uses

In discrete (non-IC) stages, transistors are the active elements doing the work, with passive parts like resistors and capacitors supporting their operation. You'll see this kind of construction in radio and TV receivers, audio amplifiers and some sections of many other products. In stages where an IC is at the center of the action, transistors frequently do the interfacing between the IC and other parts of the circuitry, especially areas requiring more current than the IC can supply. MOSFETs are used as switches, permitting the microprocessor to turn power on and off to various parts of the circuitry. Some very sensitive MOSFETs are used to amplify and detect radio and TV signals. Many audio power amplifier output stages are made from bipolar transistors. It's hard to find any function that transistors *don't* do. After electrolytic capacitors and bad connections, transistors will be the focus of much of your repair work.

What Kills Them

Transistors are not especially fragile, but they work hard in many circuits and fail more often than most components. Overheating due to excessive current will burn them out, as will too high a voltage. Generally, circuits are designed to avoid applying voltage above their transistors' ratings, but it can happen due to a fault, especially in power supplies. MOSFETs are particularly prone to shorts from static electricity. Sometimes the internal structure of a transistor develops a tiny flaw, and the thing self-destructs with no apparent cause. In fact, many random product failures occur for precisely this reason. You change the part and the unit works again, and you never find any reason for the dead transistor.

Out-of-Circuit Testing

If you have a transistor tester, use it! Nothing's easier than hooking up the leads and getting the test result. If you don't have one, you can use your DMM's diode test or, lacking that, the ohms scale to check for shorts. If you get near-zero ohms between any two leads, check in the other direction. If it's still near zero, the part is shorted.

Checking for opens is a bit more complicated because some combinations of terminals *should* appear open, depending on to what the control terminal is connected. Connecting the base of a bipolar transistor to its collector should result in its turning on, showing measurable resistance between the emitter and collector in one direction. Connecting the base to the emitter should turn it off.

Similarly, connecting the gate of a FET to its drain should turn it on, and connecting it to the source should turn it off. However, some FETs are symmetrical and will turn on with the gate connected to either of the other terminals as long as the polarity of the applied voltage is what the gate needs. And the whole situation is complicated by the enhancement/depletion-mode issue because a depletion-mode FET will stay on. Depletion-mode FETs are not easy to test with an ohmmeter!

Here's a quick and dirty way to check power MOSFETs, the type used as choppers in switching power supplies and output stages in other applications, like pulse-width-modulated (PWM) digital audio amplifiers and motor controllers, where the transistor operates as a switch rather than as a variable-resistance element. These are enhancement-mode FETs, and this test will work only with them; it won't work with the depletion-mode FETs commonly used in stages that handle small signals.

Take a 9-volt battery and a 1-kΩ resistor and connect them as shown in Figure 7-21. Set your DMM to read voltage, and connect it across the drain and source of the FET.

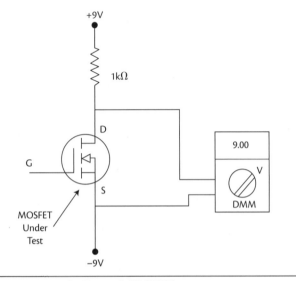

FIGURE 7-21   Power MOSFET test setup.

Leave the gate terminal unconnected. Touch two of the fingers of one hand between the gate and drain terminals momentarily. The voltage shown on the meter should drop quite a bit and stay at the lower value after you release your fingers. This indicates that the gate of the FET is positively charged and is turning on the drain-to-source path, as it should. Now, touch your fingers between the gate and source terminals and let go. The voltage should rise back up to the battery's voltage and stay there because you've discharged the gate. If you get these behaviors, the part is good.

Most of the power MOSFETs you'll encounter are N-channel, and their pinout, left to right, is usually gate–drain–source (GDS). For a P-channel part, you can do the test the same way. Just reverse the battery polarity and expect to see a minus sign on your DMM's display.

Voltage Regulators

Voltage regulators take incoming DC and hold their output voltage to a specific value as the incoming voltage fluctuates or the load varies with circuit operation. Yesteryear's simple products often had no voltage regulators, but practically everything made today does.

Linear regulators act like automatic variable resistors, passing the incoming current through a transistor called the *series pass transistor*. They monitor their output voltage and set the base current to keep the output voltage at some specific value. When the load changes and the voltage increases or decreases, the regulator detects that and adjusts the base current to compensate, altering the transistor's resistance and permitting more or less current to go through it. The power lost in the resistance of the transistor is wasted and dissipated as heat.

Switching regulators chop the incoming current into fast pulses whose widths can be varied. The pulses are then applied to a capacitor, which charges up, converting the pulses back to smooth DC. By monitoring the output voltage, the regulator detects changes and alters the pulse width. The wider each pulse, the more current can charge the capacitor, raising the output voltage. Narrower pulses lower it. While more complicated, this approach, called *pulse-width modulation* (PWM), supplies only the current required to keep the output voltage constant, without wasting the excess as heat. Thus, it is much more efficient.

PWM regulation is an inherent feature of switching power supplies, and some products use switching regulators internally as well. They are complicated, though, and also generate a fair amount of RF (radio-frequency) noise, so they aren't suitable for all uses.

Linear regulation, while wasteful, is still very common in low-current applications because the amount of power wasted is trivial. The linear approach generates no noise and is a lot simpler and cheaper, making it attractive.

While both types of regulators once took a bunch of components to implement, they can be had in chip form today, requiring just a few external parts to support their functions. The three-terminal linear voltage regulator, available in both fixed- and variable-voltage varieties, is the most common type you'll find. It looks like a transistor but is really an IC, with one terminal for input, one for ground and one for output (Figure 7-22). Hang

FIGURE 7-22 Linear voltage regulators.

a capacitor or two on it and it's ready to rock. Small, surface-mount regulators may look like typical flat ICs and have extra leads, some of which are used to inhibit or enable the output voltage, and some of which parallel other leads or do nothing. Many products have several regulators supplying separate sections of their circuitry.

Symbol

Markings

Voltage regulators are marked by part number, like transistors. Some standard markings tell you the voltage, which is handy. In particular, the widely used 7800/7900 series of linear regulators offers useful marking information. All regulators starting with "78" are positive regulators with negative grounds. A "7805" is a 5-volt regulator, a "7812" is 12 volts and so on. All parts starting with "79" are negative regulators with positive grounds. They use the same voltage numbering scheme. An "L" or "M" in the middle of the number, such as "78L05," indicates low or medium power-handling capability compared to parts without the letters. Other regulators don't necessarily indicate their voltages in the part number, and you will have to look them up.

Uses

Voltage regulators provide stable voltage to entire devices or sections of them. In some applications, a regulator may feed a single stage or area of the circuitry that requires a different voltage than does the rest of the circuit. The output of a linear regulator is always lower than the input, but some switching regulators can boost the voltage and regulate it at the same time. This is especially useful in battery-operated devices employing only one or two 1.5-volt cells or a single 3.7-volt lithium cell but requiring a higher voltage.

What Kills Them

Pulling too much current through a regulator can overheat and destroy it. This is especially true with linear regulators. Most of those include overcurrent protection and will shut down if they try to feed a short circuit, but running them only slightly under their full-power ratings, or employing inadequate heat sinking, can get them hot enough to fail over time. Voltage spikes can cause internal shorts, and random chip failures occur too. Switching regulators are prone to blowing their chopper transistors, just like switching power supplies.

Out-of-Circuit Testing

You can use the ohms scale of your DMM or VOM to check for shorts, but that's about it. To evaluate a regulator properly, it needs to be in a circuit, receiving power.

Zener Diodes

Zener diodes are special diodes used in voltage regulators. In the forward direction, they conduct like normal diodes. In the reverse direction, they also act like regular diodes, blocking current. But, when the reverse voltage rises above a predetermined value set in manufacture, the zener breaks down in a nondestructive manner and conducts. This results in a constant voltage drop across the part, making it useful as a voltage reference. Zeners dissipate power as heat, so they are rated in watts for how much they can take before overheating and burning out. Zener diodes look much like other diodes, but many have somewhat beefier cases and thicker leads to increase heat-dissipation capability (Figure 7-23).

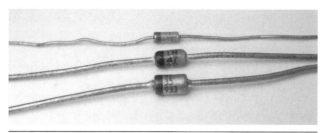

FIGURE 7-23   Zener diodes.

Symbol

Markings

The band on a zener's case indicates the cathode, as with any diode. Because zeners are used for their reverse breakdown action, though, expect them to seem to face the wrong way in a circuit, with the band connected to positive. In fact, that's one way to help determine whether a diode you see on a circuit board is in fact a zener and not just a normal diode. Except for reverse-polarity protection diodes at a DC power jack or across relay coils, diodes that aren't zeners will pretty much always have their cathode bands connected to a part of the circuit more negative than where their anodes are connected. Otherwise, no current would flow through them.

Zeners are marked with part numbers, when they are marked at all. If the number begins with "1N47" and is followed by two more digits, that's a zener. Some have numbers like "5.1" or "9," which would seem to suggest their breakdown voltages. This is *not* always the case! For instance, a "1N4733" is a 5-volt zener, not a 33-volt one. Look up the part numbers to determine a zener's characteristics.

Uses

Zener diodes provide a stable voltage reference in many circuits, especially power supplies and regulators. Both linear and switching regulators may use them for reference. Zeners can function as linear regulators by themselves when only a small current supply is required. A resistor will be used to limit the current going to the zener, and the regulated voltage will appear at the cathode, while the anode goes to ground (in normal, negative-ground circuits).

What Kills Them

Putting too much current through a zener will exceed its wattage rating and overheat it, destroying the part. Over time, even zeners in proper service may fail from the cumulative effects of heating. It's quite common in older products. In general, zeners are troublesome parts, so always check them if voltage regulation is out of whack.

Out-of-Circuit Testing

If you have a multi-component tester that can evaluate zeners, now's the time to use it. If not, test zeners for basic diode operation using the diode test or ohms scales on your DMM. Zeners can short, but most fail open, or at least they appear to do so. In fact, they may short and pass so much current that they melt inside, quickly opening.

If the zener tests bad as a normal diode, it is bad. If it tests good, it may have lost its breakdown ability and still be bad, though. There's an easy way to tell using your bench power supply. For this to work, the supply must be able to deliver a voltage at least a few volts higher than the zener's expected breakdown voltage.

Take a 1-kΩ resistor (brown–black–red) and put it in series with the zener's cathode, using clip leads. Connect your bench power supply with its + terminal to the other end of the resistor and its – terminal to the anode of the zener. Set your DMM to read DC volts, and hook it across the zener. Turn the bench supply as low as it will go, and then switch it on. Increase the supply's voltage while watching the DMM. As the indicated voltage rises, it should hit the zener's breakdown point, and the DMM's reading should stop rising, even though you continue to crank up the power supply (Figure 7-24).

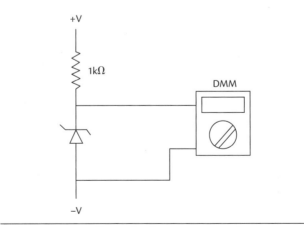

FIGURE 7-24   **Zener diode test setup.**

If the voltage keeps going up past the zener's breakdown voltage, the part is bad. If it stops very near the rating, it's fine; standard zeners are not high-precision devices and may be off by a few fractions of a volt. Also, the exact stopping point depends somewhat on the current going through the zener. An open zener won't stop at all.

This test is also handy for characterizing unmarked zeners as long as they are good. When you encounter a dead unmarked zener, determining what its breakdown voltage was supposed to be can be a real problem unless you can find a schematic diagram of the product. Sometimes you can infer the breakdown voltage from other circuit clues. We'll get into that in Chapter 11.

There are many other kinds of less frequently used components. If you run across one that doesn't fit into any of the categories we've discussed, look up its part number to find out what the part does. An online search will usually turn up a data sheet describing the component in great detail.

The Small Things in Life

For a few decades now, through-hole components with leads soldered to the back side of a circuit board have been gradually phased out, with only big items still featuring leads— things like power transistors, transformers, larger voltage regulators and capacitors, and resistors that can dissipate several watts. Pretty much all the small-signal-processing circuitry these days is SMD. As discussed earlier, these parts have no leads. Instead, they have solder pads that affix to the same side of the board the part is on. This change in component packaging permits much denser boards and lower cost of assembly. The parts are placed by robots, stuck in position with glue, and then the entire board is soldered at once.

At first, SMD parts were not really tiny, so handling and soldering them by hand were feasible, if a bit harder than with through-hole components. Our insatiable demand for smaller, more complex gadgetry has forced the size down, down, down to the point that some parts are barely visible! This is not an exaggeration; I've had to fire up my video microscope to determine whether I was looking at a resistor or a speck of dust. How anybody manufactures these parts is hard to imagine. They are supplied on reels that look a bit like old-fashioned movie film, but each "frame" is actually a little plastic bubble in which resides the minuscule component, ready to be extracted and situated on the board by a computer-controlled *pick and place* machine.

Most electrolytic and tantalum SMD capacitors are still large enough to have markings, and there are standards defining them. The capacitance value and voltage rating are the prime considerations, so those values are most likely printed on the part, using the multiplier marking scheme discussed in Chapter 6. If you see a capacitor marked "105 016," that's 1 μF at 16 volts. You may also see codes like "A" and "M." The "A" indicates the case size and the "M" the tolerance of the capacitance value (how far from the stated value the part can measure while still being good). Tables of these markings and their meanings are plentiful on the web.

However, much of the rest of today's SMD componentry is too small to be marked. This presents a real problem when you're trying to fix something without a schematic—a problem compounded by the modern manufacturing precept that we are consumers and should never open any of our products, so there's no need to publish any diagrams.

Luckily, those ultra-tiny resistors and capacitors usually aren't the problem. Most repairs today concern power supplies, electrolytic capacitors, solder joints and the odd transistor or diode. But the need to deal with super-small components does come up sometimes. For instance, a shorted transistor, itself not absurdly tiny, might burn up a barely visible resistor rated to handle $\frac{1}{20}$ of a watt by pulling 10 times more than that through it.

To get around the marking issue, the various sizes of SMD parts have been codified based on their dimensions, and you can measure the part with a pair of calipers and determine the code for the required replacement. There are both imperial and metric codes, sometimes using the same code for different-sized parts, due to the differences in the size units of inches and millimeters. In each case the code is a description of the imperial or metric dimensions. As an example, here's a table of SMD sizes and codes for resistors.

Imperial Code	Metric Code	Power Rating	Size in Inches	Size in Millimeters
01005	0402	$\frac{1}{30}$ watt	0.01 × 0.005	0.4 × 0.2
0201	0603	$\frac{1}{20}$ watt	0.024 × 0.012	0.6 × 0.3
0402	1005	$\frac{1}{16}$ watt	0.04 × 0.02	1 × 0.5
0603	1608	$\frac{1}{10}$ watt	0.06 × 0.03	1.6 × 0.8
0805	2012	$\frac{1}{8}$ watt	0.08 × 0.05	2 × 1.2
1206	3216	$\frac{1}{4}$ watt	0.12 × 0.06	3.2 × 1.6
1210	3225	$\frac{1}{2}$ watt	0.12 × 0.1	3.2 × 2.5
2020	5025	$\frac{3}{4}$ watt	0.2 × 0.1	5 × 2.5
2512	6332	1 watt	0.25 × 0.12	6.3 × 3.2

From 2512 to 0805, handling and soldering these parts isn't too rough. The 0201 parts are something else, though! Keeping one in place for resoldering is quite a challenge. And check out the 01005 size. It's literally the size of two grains of salt next to each other. There are many more codes, depending on the type of component, so keep that benchtop computer booted up, and make good use of it!

Chapter 8

Road Maps and Street Signs: Diagrams

Today's products can contain hundreds or even thousands of components. Even after you've considered a unit's failure history and pondered a preliminary diagnosis, there can still be lots to examine and test. How the heck do you find your way around what looks like a city on Mars?

There are three types of road maps to help you navigate the innards of an electronic device:

- *Block diagram.* This lays out the device by the function of each section and its basic interconnections. It does not show individual components or specific connection points. It's the most general, conceptual view, analogous to a map showing cities and route numbers for major highways between them but not street-level detail (Figure 8-1).
- *Schematic diagram.* This shows all the components and interconnections but does not indicate their purposes by specifying sections or overall structure. This is the street-level map (Figure 8-2).
- *Pictorial diagram.* This uses drawings of the parts and shows their interconnections. The pictorials included in service manuals are really layout diagrams detailing the placement of components as they exist on circuit boards and chassis. This is the drawing of landmarks and where to find them (Figure 8-3).

All three diagrams work together to guide you to your destination. The block diagram helps you grasp the signal flow and interactions between circuit sections so you can see how they are supposed to work with each other. The schematic shows you individual components and stages so you can zero in on specific components you may want to scope or pull for testing. The pictorial helps you find the darned things!

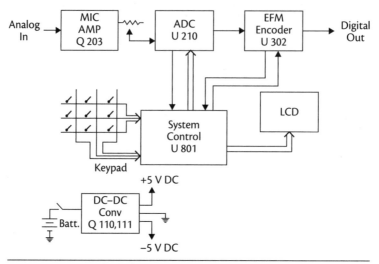

FIGURE 8-1 Block diagram.

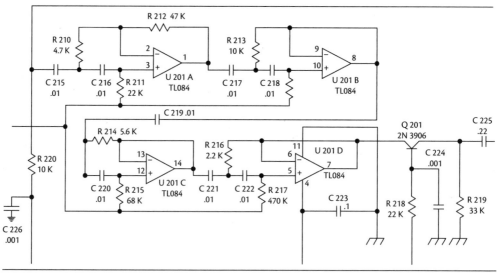

FIGURE 8-2 Section of a schematic diagram.

If you can obtain only one style of diagram, get the schematic because it offers the detail necessary for troubleshooting at the level of individual components. In years gone by, most products included a schematic, printed inside the case or in the instruction booklet. That became impractical as gadgets got more complex; there were just too many parts to fit the diagram in such a small space. For a while, manufacturers supplied fold-out schematics with their instruction booklets, but they finally began omitting the sheets because products no longer had incandescent lamps, snap-in fuses or any other parts the user could change.

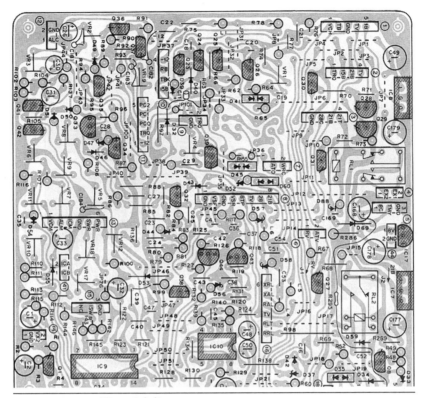

FIGURE 8-3 Pictorial diagram.

After all, only a service tech could really make use of diagrams anyway, so why spend the money to print them by the hundreds of thousands? Instead, service manuals were made available to repair shops, and the end user, soon to be called the *consumer*, was left high and dry, no longer privy to the products' insides. "No user-serviceable parts inside. Refer service to qualified personnel" replaced the diagrams. Is it any wonder they call today's gadgetry *consumer* electronics?

Service manuals were cheap and plentiful, and shops kept huge rows of filing cabinets bursting with them. In addition to manuals generated by the products' makers, the Howard W. Sams company produced its own comprehensive line of Photofact schematics for just about everything out there. If you couldn't get a schematic from Zenith, you could get one from Sams easily enough for a buck or two.

As products got still more complex, manuals grew from a few pages to a few hundred, with large, fold-out schematics and detailed color pictorials. Producing these big books became quite expensive, so their prices skyrocketed. Shops continued to buy them—they had little choice—but no consumer would spend more for a manual than the product cost in the first place! Companies gradually reduced and finally abandoned the infrastructure for selling manuals to the public, and today's age of "use it, wear it out and toss it" was in full swing. Many manufacturers will no longer sell schematics or service manuals to consumers, thanks in part to fear of potential lawsuits by injured tinkerers. Some companies won't even sell manuals to service shops unless they're factory-authorized warranty service providers. And, believe it or not, some refuse to provide diagrams even to those facilities! Secretive computer makers, in particular, only let authorized servicers swap boards; the techs work on their machines for years without ever seeing a schematic of one.

Where does this leave you? Forget contacting the major manufacturers; most won't sell you a service manual no matter how you plead. There are online sources, though, continuing to provide this vital information for many products. Howard W. Sams continues as Sams Technical Publishing at www.samswebsite.com. Its Photofact and Quickfact manuals aren't $1.50 anymore. As of this writing, most of them cost $20 and up. Still, for a tough case that has you going around in circles, it may well be worth the investment. Numerous other sites offer service manuals and diagrams, often for free. Doing an internet search may turn something up, and it's always worth a try. These days, it's where I head as soon as a broken gadget hits my workbench.

Hooked on Tronics

Reading a schematic is a bit like reading music: Learning to name the notes is just the beginning. To really understand what's going on, you need to recognize the larger harmonic and rhythmic structures and how individual notes fit into and connect them. Identifying components on a schematic is a good start, but seeing how they form stages and sections, and how those work with each other, is vital to being able to find the ones that aren't properly performing their functions. The best techs have a good grasp of circuit fundamentals, but there's no need to be an engineer or a math whiz. It's far more useful to be familiar with the overall structure and with how basic circuit elements like transistors work.

In Chapter 7, we reviewed the component symbols for the most common parts. Along with those, schematics include symbols for other items. Here are some you're likely to encounter:

AC voltage   The presence of AC voltage is indicated by a sine wave.

Antenna   The antenna symbol represents any type of antenna, even if it doesn't resemble the symbol.

Battery   In addition to the main batteries powering portable devices, small backup batteries may be found on circuit boards, either soldered or in holders. When wires connect batteries or battery holders, it is standard to use a red wire for positive and a black wire for negative.

Conductors, joined   This is where two wires or circuit traces meet and connect.

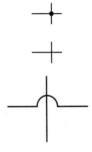

Conductors, not joined   This is where two wires or circuit traces cross on the schematic (but not necessarily physically in the device) without connecting. Older schematics use the three-dimensional–looking loop shown on the right. Newer diagrams don't, because today's products are so densely packed with interconnections that the loops become unwieldy and take up too much room on the schematic. Remember, if there's no dot where they cross, they are *not* connected! When a conductor meets another one without crossing, though, it is connected whether or not there's a dot.

Conductors, merged   This shorthand description of multiple wires is widely used in digital gear, particularly when parallel data lines all go to the same chip. Instead of showing a separate conductor for each line, only one is shown, making complex schematics a little less cluttered and a bit easier to read. Typically, there are letters or numbers at the origin of each wire and corresponding designations where they fan out to connect to whatever is at the other end, so you can follow what goes where.

DC voltage   The lines represent what DC looks like on an oscilloscope, with the dotted line indicating ground.

Ground   There are four types. *Earth* (at left in the illustration) indicates a connection to the AC line's ground lug. *Chassis* (at right) means the connection goes to the unit's metal chassis or, lacking one, a common point on the circuit board. The earth symbol is often used in place of the chassis ground symbol—you'll see it in battery-operated gear that never gets connected to the AC line—but not the other way around.

Analog and digital ground symbols are used in devices having separate ground points for their analog and digital sections. This arrangement helps keep electrical noise generated by the digital circuitry from intruding into sensitive analog circuits. Many disc players have separate analog and digital grounds. So do DVRs. In some products, the grounds meet, and only the length of a circuit board trace separates them. In others, analog and digital grounds remain separate, and connecting them externally will cause malfunction or undesired noises in the output signal.

Jack There are many styles of jacks, so jack symbols are somewhat pictorial. Also, some jacks have internal switches that sense when a plug is inserted, and those will be shown too.

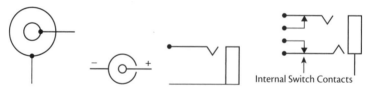

Speaker The speaker symbol looks like a classic loudspeaker, but it can be used to indicate headphones as well. Sometimes, a drawing of a headset will be shown instead.

Call Numbers

Each component will have a part number unique to that product's schematic. Some techs refer to it as the *call number*. This number is unrelated to the part number printed on the component, the one by which you can look up the part and learn its electrical characteristics. Instead, the call number is derived from the parts list found in the service manual.

Each call number begins with a letter specifying the type of component, followed by a few numbers. The first number tells you in what section the component resides, and the others are unique identifiers. Designations like R201, L17 and Q158 are call numbers.

The section number is arbitrary and varies from product to product, but parts with the same first number will live in the same neighborhood. For example, R201, Q213 and C205 will all be in the same area, but R461 and Q52 won't. And C206 is probably right next to C205, or at least not very far away. If the circuit board is labeled, it'll show those numbers next to each component.

A labeled board makes finding parts *so* much easier! But you get what you get. In a lot of today's products, the parts density is too high to make room for call numbers, and in others, the manufacturer just doesn't bother with them. I consider it an indication of whether the maker considered the product worth repairing or it's just a throwaway—not that I won't try to fix it anyway!

Every part normally has its own call number, but multisection integrated circuits (ICs) such as op amps may be shown with each section as a separate device, even though they're really in one package. In Figure 8-2, the TL084 chip has four sections, each labeled A, B, C or D. On another schematic, the same component could be drawn as one rectangle resembling the shape of the real thing, with all four sections inside. In that case, each section's triangular op-amp symbol might be drawn inside the little box, but you can't count on that. Presenting the part as separate sections helps keep the signal flow clearer and is the preferred method.

Although each schematic has unique call numbers, the letters denoting component types are somewhat standardized. Here are the letters in common use:

Component	Designator	Component	Designator
Capacitors	C	Relays	RL
Connectors	J or CN	Speakers	SP
Crystals and resonators	X or Y	Switches	S or SW
Diodes	D	Transformers	T
Fuses	F	Transistors	Q
Coils (inductors), but not transformers	L	Test points (places to put your probes)	TP
IC chips	IC, U or Q	Voltage regulators	IC, U or Q
Resistors	R	Zener diodes	Z, ZD or D
Potentiometers	R or VR		

Because "Q" is the designator for transistors, it gets used for just about anything made from them, even if they are microscopic structures in an IC chip. Newer schematics, though, are more likely to differentiate, calling ICs "U," transistors "Q" and voltage regulators "IC."

Even if you don't have a parts list, call numbers are handy because they help you identify mystery components. Especially in this age of ultra-tiny surface-mount parts, some look so similar that it's hard to guess what they are. If you see a call number starting with an "R," you know the part is a resistor. An "L" tells you it's an inductor, and so on. Should you run into a part with a designator not covered in this book, a quick trip to the internet will turn up its meaning.

Good, Not Bad and Miserable Schematics

Not all schematics are alike. There are good ones, even great ones. There are average ones. And there are the dreadful diagrams that can lead you down rabbit holes with really devious bunnies inside, just waiting to hop you along to the Avenue of Frustration.

The Good

A good schematic is laid out logically, showing most stages with signal flow from left to right, with enough space between the stages to make the organization clear. It includes call numbers and part numbers, with resistor and capacitor values specified. A really good one may have arrows or thickened lines indicating signal flow through and between stages. A truly great diagram even has voltage readings and—it doesn't get better than this—

snapshots or drawings of scope waveforms at various test points! If you're lucky enough to work with such a schematic, it'll greatly speed up your hunt. Touch a probe, compare what you see to the diagram, and either it looks the same or it doesn't. In real life, it's rarely as simple as that, but having those guideposts is a wonderful help.

The Not Bad

A merely okay diagram is clear, with a reasonable sense of organization. It has call numbers but probably no part numbers or component values like ohms or microfarads. Forget about signal flow arrows or waveforms. It's no GPS, but it's a serviceable road map. Everything is accurate, and nothing is left out. Which brings us to the dark side, that malevolent maw of misleading misery, the incorrect schematic.

The Miserable

A really bad schematic may have reversed diode polarity, wiring errors, incorrect connection indications on conductors crossing each other, or omission of some parts, any of which can confuse the living heck out of you and send you off in the wrong direction. Switches show no indication of what they do in what position, and the drawing might not even be clear enough that you can read parts of it. Even worse, sometimes manufacturers make major changes to their designs and never update the manual, so your diagram may not even be close to what's on the board. Still, even a miserable schematic can be better than none, as long as you remember not to trust everything you're seeing. Occasional errors crop up even in the best diagrams, of course, but it's rare to find a truly rotten schematic from a major manufacturer. I've seen some doozies from off-brand companies, though.

Once Upon a Time . . .

To get started reading schematics, consider the organization of a book. It begins with letters that form words, which make sentences. Those are grouped into paragraphs and finally into chapters. Each paragraph links with the others to tell a story, and the chapters present a progression driving toward the finish. Some conditions are introduced at the start of the book and resolved at the end.

Electronic devices are organized much the same way. Components work together to form *stages*, each one feeding others, resulting in a signal flow proceeding from some starting point, such as a microphone, antenna or DVD, to some ending point, perhaps a display screen or a speaker. Each stage performs a function contributing to the overall processing. A group of stages involved in a particular part of the device's operation constitutes a *section* dedicated to a specific purpose.

With occasional exceptions, a schematic's signal flow in each stage proceeds from left to right. Signal flow *between* stages normally goes left to right as well. So, the most sensitive stages handling the weakest signals are usually on the left side of the page. If you're looking for an antenna or microphone input, look for it there. In a power supply, the AC line connection is probably on the left side, too, as it's considered the supply's input. Very complex schematics sometimes violate these conventions, simply because they run out of room on the page.

Look for power supply sections at the bottom of the page. Output sections, LCD screens and speakers should be on the right. Processing stages, such as the IF (intermediate-frequency) stages of a radio receiver or the microprocessor in a disc player, will be in the middle.

When reading a schematic, keep your eye on the story and its central characters. Not all players are equally important. The plot is driven by the active elements like transistors and ICs, since they do most of the work. Crystals and resonators generate signals, so they're crucial characters without which the story never gets moving. After those, look for coupling elements such as transformers, capacitors and resistors linking one stage to the next. They move the plot forward because signals will be flowing through them on the way to subsequent stages. The other resistors, capacitors and coils set the voltages, currents and various conditions the big shots need to do their jobs. Those subplots are necessary to the overall story but not central to its theme. Try not to let them distract you from the primary action.

When following the signal flow, keep in mind that the base connections of transistors are pretty much always inputs. There are some grounded-base amplifiers that use one of the other terminals for input, but they're not common. Even then, the base is never an output! So, you can look for the base (or gate, in the case of a field-effect transistor) and figure that it's probably the input. Once you get the hang of recognizing the input and output points of each stage, the overall organization will become much easier to follow. Let's look at the schematics of a few circuits and how to interpret them.

Amplifier Stage

For our first adventure, we'll look at a single stage. Let's examine every part in it, what it does and what would happen if it malfunctioned.

This one is an inverting amplifier, typical of what you might find in just about any product (Figure 8-4). *Inverting* means that the output signal rises as the input signal falls, and vice versa, producing a replica that's upside down. Sometimes inversion is necessary to the circuit's operation, while other times it's just an irrelevant consequence of getting voltage gain from the stage. Either way, you'll see lots of inverting amplifier stages.

The stage has four major points where things go in and out. At the top, power is applied through the transformer, reaching the collector of the transistor. Signal input is on the left and goes to the base of the transistor, Q1, via C1. The output is at Q1's collector, which is why TP1, the test point, is there. Finally, R2, R3 and C2 go to ground. Ground

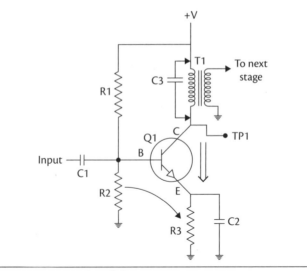

FIGURE 8-4 Amplifier stage.

is as important an input-output point as the others; without ground, there's no place for current to go, and you've got a paperweight.

To be a sharp troubleshooter, you need to understand how the circuitry is supposed to function. So, let's see how this thing works, starting at the top. Power goes through the winding of the transformer and reaches the collector of the transistor, Q1. It's the *active element*, so it's the central character. When the transistor is turned off, no current passes through the transformer because the lower end of its winding sees no connection to ground. When the transistor is turned on, the path between collector and emitter connects, effectively grounding the transformer's winding and completing the circuit, pulling current through the winding.

It's not an all-or-nothing proposition, though. Remember, transistors act like variable resistors (potentiometers), except that signals, rather than your fingers, turn them up and down. Here, the signal is applied to Q1's base through C1. C1 couples only the AC component of the incoming signal to Q1, preventing any DC in the signal from reaching the transistor, and also preventing any DC *from* Q1 getting back into the previous stage. This means the signal as it appears on the right side of C1 tries to swing both positive and negative.

Where R1 and R2 meet establishes a voltage somewhere between the power supply's value and ground, to bias the transistor's base, or put a little DC on it, keeping it turned on through whatever portion of the incoming signal's waveform the amplifier is intended to amplify. The bias voltage depends on the ratio of R1 to R2. Let's say the power supply, V+, is 10 volts. If R1 is 100 kΩ (kilohms, or 100,000 ohms) and R2 is 20 kΩ, that's a ratio of 5:1. So, the voltage where they meet will be a fifth of 10 volts, or 2 volts, and the audio signal will be centered on that, with 2 volts as the midpoint of the signal, keeping its swings in both directions well above zero.

Well, that would be true if there were no transistor! In practice, some of the current arriving via R1 goes from the base of Q1 through its emitter and then through R3 to ground. So, the bias voltage will be lower because the path through Q1 and R3 is essentially in parallel with R2, which also goes to ground, lowering its apparent resistance and altering the ratio.

The higher the bias voltage, the more current will go through the transistor, turning it on harder (lowering its resistance from collector to emitter), thus pulling more current through the transformer. Biasing is a critical function, as discussed in the "Stereoless Receiver" example in Chapter 4, and should not be ignored while troubleshooting. Lots of audio amplifier faults can be traced to bias problems.

As with many (but not all) amplifiers, we want the whole waveform to appear at the output of this one, so the transistor has to be biased with enough positive current that when the incoming signal goes negative, the transistor's base never gets below about 0.6 volts, which is the cutoff point for a standard bipolar silicon transistor. Otherwise, the amplifier will cut off the bottom of the incoming waveform, a result called *clipping*, as discussed in Chapter 5 and shown in Figure 5-4. Clipping can be useful in certain applications, but it's extremely undesirable in an audio amplifier.

The transistor passes current from collector to emitter in proportion to how much current passes from base to emitter, as shown by the arrows. The ratio of the two currents determines how much the transistor can amplify a signal. It is an inherent part of the component's design and is given in its specifications, but it can be manipulated up to its maximum value by the surrounding components. As the incoming signal wiggles up and down, the base current varies with it, causing the transistor to pull a proportionally larger current through the transformer on its way to R3 and finally ground. This forming of the power supply's DC into an enlarged replica of the signal is called *gain* and is the essence of amplification. It's the foundation of all modern electronics, analog or digital.

Ah, R3 and C2. What're they there for? R3 limits the total current through the circuit; without it, the transistor would attempt to pull the power supply's entire current capability to ground, dragging down the supply and probably blowing the transistor or the transformer. R3 limits the total base current as well because it's in series with that path too. Thus, it sets a limit to how much signal current the previous stage has to supply.

C2 is a little trickier to explain. Transistors, having adjacent regions of semiconductor material in them that are not at the same voltage at the same time, also behave like capacitors. They store some charge and take a little time to discharge. The presence of R3 slows this down because the discharge has to reach ground through its resistance. This forms a *time constant*, which is a fancy way of saying that there's an upper limit to how fast the transistor can get rid of its charge. The bigger the value of R3, the longer the discharge process takes, and the slower the transistor can react to incoming signal changes. When the frequency of the incoming signal is faster than the time constant, the transistor's residual charge fills in as base current drops, acting like any filter capacitor and smoothing out the waveform. As a result, the transistor can't react quickly enough to respond and amplify the signal. Thus, the upper speed limit, or *frequency response*, of the amplifier drops off.

C2 allows the rapidly changing parts of the signal to reach ground with less resistance, discharging the transistor faster. The apparent resistance of a capacitor drops with increasing frequency because it never gets the chance to charge fully and oppose the incoming current. So, C2 compensates for the transistor's capacitance, giving it a lower-resistance path to ground with increasing signal frequency. The result is to even out the amplifier's frequency response by restoring the lost high-frequency response without also increasing the low-frequency response.

T1, the transformer, plays a crucial part in the amplifier's operation. As current is pulled through its *primary* coil, it generates a magnetic field that impinges on, or cuts across, its *secondary* coil, the one on the right. As the current in the primary gets stronger and weaker in step with the signal, the changing magnetic field generates a current in the secondary that makes its voltage rise and fall in step. This changing voltage couples the new, amplified signal to the next stage. There are other ways to couple a signal, without a transformer, but using one has advantages in some kinds of circuits, especially those employing the transformer as a *tuned circuit* resonating at a specific frequency. *Superheterodyne* radio receivers use lots of tuned amplifiers of this sort to pick out the selected signal from all the others hitting the antenna. A capacitor across the transformer, shown in Figure 8-4 by the optional C3, is a dead giveaway that an amplifier is tuned. C3 could be on the other side of the transformer, too, and still have the same effect in tuning it to a desired frequency.

Thar She Blows

Now that we've explored how the amplifier stage is supposed to work, let's see what the effects of malfunctioning parts would be. Again, starting at the top, how would a bad transformer affect the performance of the circuit?

If the transformer were open, no current would pass through its primary, so the collector of the transistor would read 0 volts. Transformers in small-signal circuits don't pass much current, so an open winding is unlikely, but it can happen. By the way, to estimate how much total current the winding might have to handle, just divide the power supply voltage by the value of R3, à la Ohm's law. This pretends that the transistor and the transformer's winding have no resistance, so the real value will be a bit less, but at least you'll know the approximate upper limit.

If the transformer had a short, its resistance would decrease, and it'd be harder for the transistor to pull its lower winding connection toward ground, so the voltage at the transistor's collector would be close to that of the power supply, with little signal variation. Depending on the total current through the circuit (limited by R3, as described earlier), the transistor might get hot or even be blown.

If R1 were open, there'd be no bias current going to the transistor's base, so the transistor would be turned off. Its collector would be at the same voltage as the power supply, and its emitter would be at ground potential, 0 volts. The signal itself might have enough current to turn the transistor on a little bit during the positive half of its waveform, resulting in a weak, distorted mess of downward signal excursions appearing at its collector. (Remember, the amplifier inverts.)

Shorted resistors are pretty much unheard of, but, for the sake of this thought experiment, we'll consider what would happen if one did short. If R1 shorted, the base would be biased to the power supply's full voltage. The transistor would be fully turned on no matter what the incoming signal did. The collector would read somewhere close to 0 volts, and the transistor might be hot, as might be R3.

If R2 were open, the bias would be too high, with much the same result. The signal's influence might result in some output, though only on the negative-going half of its waveform. Those negative incoming peaks would cause a rise at the output, thanks to the inversion. As before, you'd get a weak, distorted mess, but the collector's DC level would be low, not high. If R2 were shorted, the base would be pulled down to 0 volts, resulting in the same no-bias conditions you'd get if R1 were open, except that the signal could not produce any output at all because it would be shorted to ground as well.

If C1 were open, no signal would get to the transistor, but the DC voltages on Q1 would look just fine. If C1 were shorted, the base bias would be influenced by the previous stage, resulting in unpredictable behavior. If the bias were pulled toward ground, Q1's collector would rise, clipping off the top of the output signal if it got too high. If the bias went high because of voltage being fed in from the previous stage, the transistor would turn on too hard, the collector would be low, and the signal would be cut off toward the bottom.

A shorted or leaky input coupling capacitor is something you might actually run into, particularly when it's an electrolytic or tantalum cap. With a leaky one, it can be maddening to try to deduce why the stage behaves so oddly. If you disconnect the input cap and the DC voltages on the transistor change, the cap is letting some DC pass through it. It's leaky. Rip that puppy out of there and put in a new one!

Now to the heart of the stage: the transistor. This is the component most likely to cause trouble. Transistors can fail in numerous ways. They can open from emitter to collector, usually as a result of overcurrent. They can also short that way, and often do. A short or an open can occur from base to emitter as well. I've seen transistors with all three leads shorted together like one big piece of wire!

Transistors can be leaky, too, allowing current to pass when it should be cut off, or even to move backward through their junctions. A very small amount of reverse leakage is normal, actually, but it's not enough to affect the circuit's operation. A leaky transistor allows much more reverse current, and that'll produce all kinds of unpredictable effects. The little monsters can also become thermal, changing their gain and leakage characteristics as they warm up.

An open between any two junctions will result in no output. TP1 will be at the power supply voltage because the transistor will not pass any current toward ground. Even if the collector-emitter junction is fine, an open base junction will prevent the transistor from being turned on.

A shorted collector-emitter junction, which is quite common, will appear as if the transistor were turned on all the time, all the way. The collector voltage will be at or very near zero. A shorted base-emitter junction, also common, will pull the base bias and incoming signal down toward ground through R3, and the transistor will not turn on. So, the collector voltage will not pull down and will be at the power supply voltage. A short from collector to base will feed current into the base and turn the transistor on, having the opposite effect.

A quick and dirty way to hunt for transistor shorts is to check the DC levels on all three leads while the product is operating. If any two are exactly the same, the part may very well be shorted. If they're even a little bit different, a short is far less likely.

If a transistor has leakage, it can act strangely. If collector current flows into the base, for instance, it can overbias the part, as described earlier. Depending on how much current leaks, the transistor may still work to some degree. If a stage acts wonky but everything seems to measure okay, and especially if the behavior changes with temperature, leakage is likely. Most bad transistors I've run into have been shorted or open, though, not leaky.

Switching Power Supply

Switching supplies range from moderately complex to ridiculously so. Rarely will you find one that looks especially simple. If you think I'm kidding, crack open a modern AC adapter. Even those inexpensive little wall warts are stuffed with chips, diodes, transistors and regulators, along with a smallish transformer and the usual plethora of resistors and capacitors.

Switchers producing only one output voltage are less dense, with a straightforward arrangement for regulating it. More elaborate circuits can have multiple output voltages, overcurrent sensing and other fail-safe protection measures adding to their parts counts.

At their core, though, switchers are not that complicated. Their basic operation is pretty much the same, regardless of the frills. So, let's strip away the doodads and look at what makes these omnipresent beasts purr.

Take a look at Figure 8-5 for a simplified schematic of the sort of power supply you're likely to encounter in many modern products, from LCD monitors to big-screen TVs and audio gear. How do we know it's a power supply? The presence of the AC line input at the far left is a dead giveaway. Is it a linear supply or a switcher? Notice that the AC line goes directly to a bridge rectifier and then to a transistor, before reaching the transformer. That's

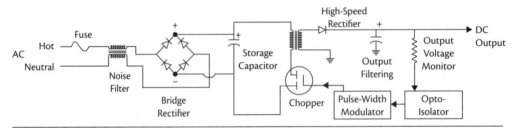

FIGURE 8-5   Switching power supply.

the classic switcher design: the AC gets changed to DC, stored in a big electrolytic capacitor and then chopped by a transistor that feeds the transformer. So, this definitely is a switcher. In a linear supply, the AC would go straight to the transformer; rectification, filtering and regulation would be done on the other side.

That big transistor below the transformer is the *chopper*. It switches on and off at a high frequency, pulling current through the transformer to generate a pulse of magnetism each time it turns on. The chopper is connected to the AC line, along with everything else on that side of the transformer.

The transistor is driven by pulses from the pulse-width modulator. That circuit (usually a chip) keeps an eye on the supply's output voltage and adjusts the *duty cycle*, or on/off ratio, of the pulses it feeds to the transistor's gate. The wider the pulses, the more energy will flow across the transformer, and the more power will fill up the output capacitor, keeping the voltage from sagging when the supply's load (the current taken by the circuit being powered by the supply) increases. If the load decreases and the output voltage starts to rise, the chip notices it and narrows the chopper's pulses, bringing the output back down to its correct value. This regulation effect keeps the output voltage steady as the circuit being powered varies its demand for current.

Between the output stage and the pulse-width modulator is an opto-isolator, which is nothing more than an LED and a light-sensitive transistor in one case. It's there to pass information about the output voltage back to the chip without forming an electrical connection between the two. Lack of a connection keeps the output side of the supply (everything to the right of the transformer) isolated from the AC line and thus safe. Pretty simple story, isn't it?

Pop Goes the Switcher

Let's look at what would happen if the major components failed. Starting with input from the AC line, our first major component is the bridge rectifier. It could fail in several ways, but the most common problem is an open circuit in one of the four diodes. That'll result in half the AC waveform not getting transferred to the output of the bridge. With no capacitor to smooth things over, it would look like Figure 8-6.

Because a capacitor is storing the charge, you'll see a much lower-than-normal voltage at its positive terminal, along with a droop where the missing segment should be replenishing it (Figure 8-7). The chopper may still run in this condition, but it probably won't.

One of the diodes could short, resulting in reverse-polarity voltage getting where it shouldn't. In this case, expect the supply's fuse to be blown.

The most common failure in a switching supply is a bad chopper transistor. It operates at high voltages and takes a lot of stress. If it's shorted, the fuse will be popped, and the supply will be dead as a doornail. If the transistor is open, the supply will still appear dead, of course, but the fuse will be good, and the big capacitor will have a full charge of a few hundred volts on it. That is, unless the transistor shorted first and then opened because of lots of current through the short. If it did, the fuse probably will be blown.

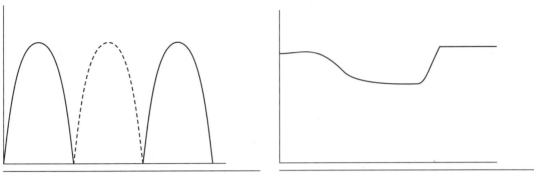

FIGURE 8-6 Missing rectified AC segment.   **FIGURE 8-7** Droop due to missing segment.

If the PWM (pulse-width modulator) IC is dead, there will be no pulses at the base or gate of the chopper. The IC could appear bad, though, for other reasons. First, it needs some voltage to run, even before the chopper starts, so there could be a bad diode, zener or small cap in its standby voltage supply. Also, if the output voltage of the entire switching supply goes abnormally high, the chip will sense it and shut down. Typically, though, it will try to restart every second or so, resulting in a chirping noise from the transformer. Most switchers will also do that if the load they're driving pulls too much current, dragging the voltage down below what the supply can replenish.

Some supplies include a *crowbar* protection circuit. Should the DC output voltage rise too much, the crowbar trips and blows the supply's fuse. See more about crowbars in Chapter 15.

Figure 8-8 shows the schematic of a real, fairly simple switcher. Just after the fuse in the upper left corner is an MOV, or *metal oxide varistor*. As discussed under "Safety Components" in Chapter 7, MOVs are surge suppressors that absorb power line spikes. The parts look like thick ceramic disc capacitors. In the event of a voltage spike on the AC line, the MOV conducts it across to the other side of the line, protecting the circuitry. A longer surge or a shorted MOV will blow the fuse. Notice the opto-isolator shown as a four-pin box at the bottom. The bridge rectifier is shown as a diamond, with a diode symbol in it. Schematics are like that—there can be lots of variations in the symbols, especially with European gear. On the right are the output components, and on the left is the AC input section. Energy flows across the transformer magnetically, and information regarding the output voltage comes back to the pulse-width modulator (the big eight-pin chip on the left) through the opto-isolator. This power supply provides two output voltages, both positive with respect to ground. How do you know they're positive? The diodes coming off the right side of the transformer both have their anodes going to the transformer, and the positive leads of the electrolytic capacitors go to the output terminals, with the negative leads going to circuit ground, the 0-volts terminal.

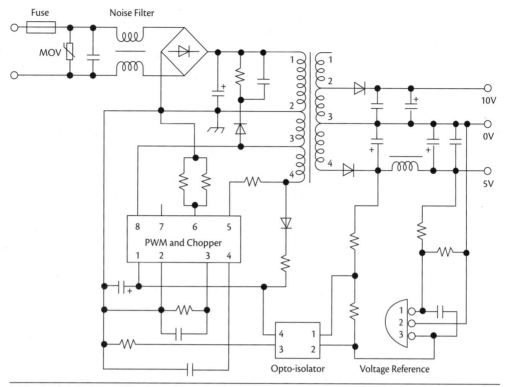

FIGURE 8-8   Simple switching power supply.

Although current actually flows from negative to positive, an easy way to remember diode polarity is to imagine positive flowing in the direction of the symbol's arrow. Whichever way the arrow points, the positive voltage will appear to go. This type of convenient, imaginary flow opposite to physical reality is called *conventional current*.

Push-Pull Audio Amplifier

Let's try another example. Figure 8-9 shows a slightly simplified channel of a typical analog audio amplifier. It might seem like everything is digital these days, but a lot of stereo and multichannel amplifiers are still analog. The design shown has the somewhat humorous but also descriptive name *push-pull* because it splits the incoming audio waveform into two halves, separately processing the positive and negative portions of the signal. One half of the amplifier pushes the speaker cone outward, and the other half pulls it back in. It's how most modern analog audio amplifiers are built.

The example shown in Figure 8-9 is a true bipolar circuit, powered by positive and negative voltages with respect to ground, shown by +V and −V. The only components

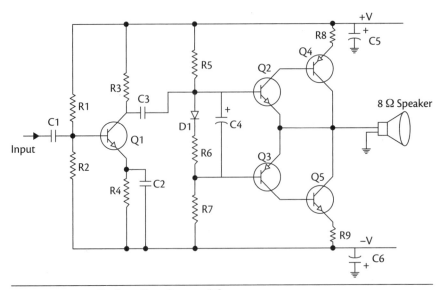

FIGURE 8-9   Push-pull audio amplifier.

connected to ground are the filter capacitors, C5 and C6, on the power supply *rails*, and the speaker. This is called a *complementary* amplifier.

Some similar amplifiers, called *quasi-complementary*, use only a single power supply polarity. They still split the waveform, but the voltage never goes negative because there's no negative power supply. This works fine, but it means that there will be a DC offset at the output, set to half the supply voltage so the signal can swing equally up or down before hitting the limits of the *rail* (power supply voltage feeding the amplifier) or ground. Such a unipolar design will have an output capacitor to block that DC offset from reaching the speaker and keeping its cone pushed halfway out (not to mention wasting a lot of power and heating up the speaker's voice coil).

Working from left to right, as usual, we see the input stage, which amplifies the incoming signal enough to drive the next stage, consisting of Q2 and Q3. The opposite polarities (NPN and PNP) of those transistors mean that opposite halves of the signal will turn them on. The signal is coupled to them by C3, which blocks any DC component from Q1 from influencing their biasing. The bias network of R5, D1, R6 and R7 keeps the transistors turned on just slightly, so there's no dead spot to cause *crossover distortion* when the input signal's waveform is less than ±0.6 volts or so. C4 couples the signal to both transistors. Their outputs drive Q4 and Q5, which provide enough current gain to move a speaker cone. In a real amplifier, a little bit of the output would be fed back through a few resistors and capacitors to the input stage in a *negative feedback loop* to correct for distortion introduced by the imperfect nature of the amplifying elements (transistors) by reintroducing the same distortion upside down, canceling it out. We're omitting those parts here to keep things clear.

Sounds Like a Problem

Let's look at how malfunctions in each stage would affect the amplifier's behavior. If Q1 or its surrounding components broke down, no signal (or perhaps a very distorted signal) would emerge from C3. Because C3 blocks DC, the badly skewed voltages at the input stage would not affect the rest of the circuit, so further transistors would not be damaged by being turned on too hard and pulling too much current.

In a *direct-coupled* circuit, though, there'd be no C3. Should the input stage of a direct-coupled amplifier malfunction and generate a lot of DC offset, as may happen with a bad transistor, that offset could wreak havoc on the rest of the transistors, possibly blowing all of them. I've seen it happen in high-end stereo receivers, which are usually direct-coupled because that type of circuitry sounds especially good, thanks to the lack of capacitors in the signal path.

If D1 opened, the top half of the amp would be turned on very hard by the bias provided by R5. Q2's base would go very positive, turning it all the way on. This would pull the base of Q4, the PNP transistor, down toward ground through the speaker, turning that transistor all the way on as well. Both transistors would get quite hot and might be destroyed.

Meanwhile, R7 would pull the base of Q3 very negative, which would turn that PNP transistor all the way on as well. This would turn on Q5, too, and funerals for Q3 and Q5 would likely be in order.

Normally, a push-pull amplifier has one half on at a time, with the other half conducting only slightly until the signal's polarity flips, reversing the process. Power supply current passes through the output transistors, one at a time, through the speaker and then to ground. In this case, both halves would be turned on at the same time, effectively shorting the +V and −V lines through the output transistors. Yikes! You can imagine the results. Smoke, burned emitter resistors (R8 and R9), blown transistors, an unholy mess . . . all from one bad diode.

If D1 shorted, though, the results would be different. The bias would become unstable and the amp's DC offset would swing around with the signal and distort it badly, but the transistors would probably survive because the bias wouldn't be so far out of whack that it'd turn them all the way on.

If C4 opened, the top half of the amp would work, but the bottom would get no signal, so severe distortion would occur, with only one half of the waveform present at the output. If C4 shorted or got leaky, the result would be similar to what you'd see if D1 shorted: The bias would get wonky and the amp would distort, but parts probably wouldn't be damaged.

An open in Q2 or Q3 would turn off the corresponding half of the amplifier, with loss of one half of the signal waveform. A short in one of those transistors would be a much more serious matter.

Q2 and Q3 are referred to as *driver transistors* because they drive (feed signal to) the output transistors, Q4 and Q5. At this point, the circuit is direct-coupled, so a short in a driver will turn its output transistor fully on, probably blowing it. At the very least, there will be a lot of DC at the speaker terminal, and the speaker's voice coil also might be blown by all the current passing through it.

Mega Maps

Highly complex schematics can be tough to follow, with all kinds of confusing signal and power lines running every which way. Especially with such schematics, a block diagram can be incredibly helpful. Devices with multiple boards may have many connectors and cables shuttling signals back and forth. The connectors are valuable focal points. If you're not sure what goes where, trace the schematic back from a connector to see if you're in the right place. Still not sure? Grab the block diagram and see if that area goes where you think it does.

Another great place to find a signpost on a big schematic is an input or output point. Jack and speaker symbols stand out because there are so few of them on any given diagram. Trace the lines back to find what's feeding them.

Transformers also jump off the page. Whichever symbol you choose, you can train yourself to scan a schematic and find it, disregarding the others. It's a lot harder to do that with resistors and capacitors, of course, because there are so many of them.

Give It a Try

If you have any schematics, now's a good time to pull them out and practice your reading. If you don't have any, go on the internet and search for some. Radios, TVs and disc players all make for good reading material. See what you can identify. Try to find the following stages and sections.

Radios

Until recently, virtually all radio receivers, including those in TVs, were of the same basic superheterodyne design (Figure 8-10). A radio-frequency (RF) amplifier increased the signals coming from the antenna. Then, those signals were mixed with a *local oscillator* that converted them to a different frequency from the one at which they were received so that they could be amplified without interfering with the signal coming in. As you tuned the radio, the local oscillator would track the tuned frequency, keeping the frequency difference between them constant. After the mixer, this resulted in a copy of the incoming signal at whatever that difference frequency was. The new, unvarying frequency, called the *intermediate frequency* (IF), was passed through a series of tuned amplifiers to make the signals bigger while rejecting those on nearby frequencies—a phenomenon known as *selectivity*—and finally *demodulated* to extract the audio, data or video originally encoded

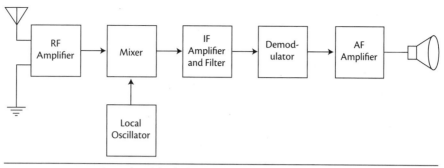

FIGURE 8-10   Superhet block diagram.

on the radio signal. Though an analog process, the superhet technique could be used to receive both digital and analog information.

In just a few years, things have changed. Many new receivers are based on a technique called *software-defined radio* (SDR). There are two flavors. In some SDR receivers, there are no tuned amplifiers, no mixers and no demodulators (Figure 8-11). Incoming signals are amplified a little bit and then applied to an *analog-to-digital converter* (ADC) and a *digital signal processor* (DSP). The incoming signals are *digitized* (converted to the 1s and 0s of digital data), and the DSP performs all the necessary functions of selectivity and demodulation in software by applying mathematical transformations to the data. If the receiver is intended for analog output, a *digital-to-analog converter* (DAC) follows the DSP stage. If the receiver is purposed for digital output, a format encoder packages the data into whatever stream, such as USB or HDMI, is required. Many SDRs combine the ADC, DSP, DAC and format encoder into one LSI chip. You might see several chips of lower density instead, depending on the design.

In other SDRs, conventional superhet techniques are used to generate an IF signal. From there on, a DSP chip takes over. SDR is a powerful technique that permits receiver

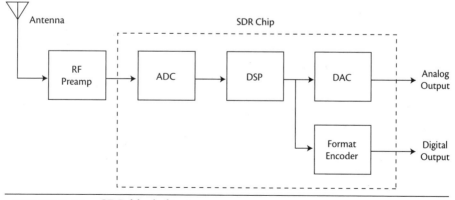

FIGURE 8-11   SDR block diagram.

characteristics to be tailored to the signal being received, with no changes in hardware. It also enables a drastic reduction in the components required and needs no adjustments. Entire radios can be built on a chip. Bluetooth and Wi-Fi are both implemented with SDR. So is your cell phone, and so are those inexpensive family radio service (FRS) UHF walkie talkies. Let's have a look at both superhets and SDRs.

Superhet: Mixing It Up

Although the signal processing in superhets is analog, pretty much all of them made in the last few decades use digitally controlled oscillators for tuning, either with phase-locked loops (PLLs) or direct digital synthesis (DDS), as we discussed in Chapter 5. Regardless of the tuning style, superhets are especially distinctive on paper. Look for the *front end*, which is the first section accepting input from the antenna. In an old AM pocket or table radio, you'll see tuned circuits with variable capacitors. Many of these simple receivers used the front-end transistor as the RF amplifier, local oscillator and mixer, all in one (Figure 8-12)!

PLL and DDS sets don't have variable capacitors for tuning. Instead, look for a big chip and a bunch of surrounding parts forming a *frequency synthesizer*. How do you find the synthesizer? It'll be connected to an LCD and probably to a keypad, too. If it's a PLL, you'll see one or more varactor diodes, as discussed in Chapter 5.

Any superhet with a frequency synthesizer—which is most of them these days—will have a separate front end and mixer. Follow the signal through the mixer, where it gets combined with the local oscillator generated by the frequency synthesizer. The mixer may be a chip, or it may be four diodes in a ring configuration that looks a bit like a bridge rectifier, except that the diodes are facing different ways. Some mixers use a dual-gate MOSFET transistor, with the incoming signal fed to one gate and the local oscillator fed to the other. The mixer's output goes to the IF stages. Those are the tuned amplifiers. Most feature transformer coupling between stages, but some do it via crystals or ceramic resonators. Most modern radios, analog or digital, convert the frequency twice (and sometimes even more), so you should see a fixed-frequency oscillator feeding another mixer stage, followed by more IF stages. To find the fixed-frequency oscillator, look for a quartz crystal.

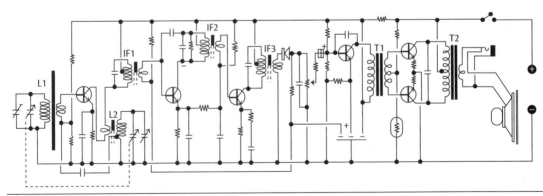

FIGURE 8-12 Classic superhet AM radio schematic.

From the IF section, the signal gets *demodulated*, or detected. This is the process of extracting the information—audio, video or data—that was originally impressed on the signal at the transmitter. AM detection may involve nothing more than a diode and a capacitor. FM is a bit more complex, and data can involve all sorts of decoding circuitry. Data detection used to require a lot of chips, but these days a small microprocessor or a DSP chip can do the work. DSPs get used for enhancing voice signals and reducing noise, too, especially in modern communications receivers and transceivers, both superhet and SDR.

With SDR taking over, why are we going into such detail about superhets? There are zillions of 'em out there! The signal path is fairly complex, and a lot can go wrong with them, so it pays to be familiar with the overall scheme if you want to fix radios or TVs more than a couple of years old. Heck, even if they stopped making superhets tomorrow, it'd take at least a decade for them to fall out of common use. While the receivers in today's TVs and all cell phones, WiFi and Bluetooth devices are SDR, you're likely to find superhets in your home theater system's receiver and your car stereo.

SDR: By the Numbers

While the operation of an SDR receiver is quite complex, most of the action occurs in software, so you'll never see it or have anything to fix. As I mentioned earlier, many SDR receivers used in consumer devices have an RF preamplifier to get the incoming signal strong enough to be digitized, and then a big DSP chip with some supporting components. That's it! SDRs made for satellite reception and other frequency bands too high for direct digitizing are more like superhets, with a front-end RF amp, a digitally tuned local oscillator and a mixer. There might even be some IF stages. Then, the preselected band of down-converted frequencies goes to the DSP, where the fancy math is applied, and out comes the audio or data. These are really hybrids, part superhet, part SDR, and will look more like superhets on paper, except for that big chip on the end.

Digitizing chips keep getting faster, and pretty soon no frequencies in practical use will be too high for them. Some advantages of superhet design are still hard to achieve with straight SDR, though, so local oscillators, mixers and IF stages may continue to be incorporated into new products for quite a while.

I haven't included an SDR schematic here because there are so many varieties that there's no "typical" SDR. If your receiver has nothing more than a transistor or two and maybe a few coils feeding a big chip, it's a safe bet that you're looking at an SDR.

Disc Players

Despite the low prices of disc players, getting the data off an optical disc is not a simple task. It involves three *servo* systems working together to find the microscopic tracks, follow them as they pass by, and keep constant the rate at which their data bits are read out. The laser head must properly track the absurdly tiny groove, even though normal eccentricities in the geometry of the discs are many, many times the size of the grooves themselves. This is a three-dimensional problem, with the wobbling distance between the head's lens and the disc surface requiring a dynamic focusing servo to keep the beam size at the point it meets the track small enough to grab just one bit of data at a time.

Unlike analog records, CDs and DVDs use *constant linear velocity* (CLV) to pack the data in with maximum space efficiency. With CLV, the speed at which the laser head sees the track go by is constant, regardless of whether the distance around the track is short, as it is near the center of the disc, or long, as it is near the outer edge. So, the disc must spin faster at the start and gradually slow down as it plays, reaching minimum rotational speed as the head plays the longest tracks, nearest the outside. Accomplishing this automatic speed control requires yet another servo to keep the disc spinning at exactly the required rate.

DVD and Blu-Ray players are complex enough that their diagrams are composed of many pages and sections. I'd love to include a player's schematic here, but there's no way to fit it on the page! Download one and take a look through the sections. You'll see pages for the power supply, LCD and system control, analog output stages, digital signal processing and servos. You might see separate boards for pushbuttons, too. Can you find the laser head? On some diagrams, it'll be called the *pickup*. Notice that it has more than one photodetector (light-sensitive transistor). Three are used to keep the laser beam centered on the track. Look also for the servo-driven *voice coils* used to float the lens and make it dance in step with disc wobbles.

Trace back from the head and see if you can locate the head *preamp*, which boosts the weak signals from the photodetectors. It should be a chip of medium density and will have a few test points at or very near some of its pins.

Trace back from the servo coils and see if you can find the focus servo section and the tracking servo, too. In newer machines, both might be together in one chip.

Find the sled motor, which moves the head across the disc as it plays, and the circuit driving it. Most motor driver circuits have transistors between the chips controlling the motion and the motor because the chips can't supply enough current to run the motor directly. Since the sled motor has to run in either direction, look for what's called an *H-bridge configuration* of the driving transistors, in which the connection to them is in the middle, with each wire going to the motor coming from where two transistors meet. It looks like the letter H, hence the name. Neither motor wire goes to ground, so the controlling circuitry can flip the polarity to the motor at will, reversing its direction. Many H bridges are implemented on a single chip, but some are still made from separate transistors (Figure 8-13).

Now, search for the disc motor, which spins the disc. Its driver circuitry will look similar to the sled motor's circuit, but not identical. It'll probably have a transistor as well, but it has to spin only one way, so no H bridge is required.

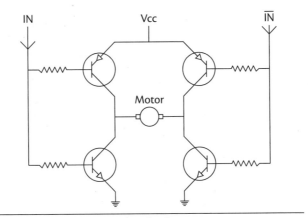

FIGURE 8-13   Typical H bridge.

Keep Reading

You'll find lots of schematics and block diagrams on the internet for every type of product. You can zoom in on your computer and examine the various sections in great detail. Get a few diagrams and practice reading them, focusing on signal flow and organization. Try to deduce which components are generating or passing signals and which are support systems for the central players, even if the overall complexity makes following the entire signal path baffling. Remember, when you work on these machines, you're not going to be chasing data streams through LSI chips. Look for coupling components, filter and bypass capacitors, power supply sections, digital control systems, voltage regulators, oscillators and so on, so you'll know how to find and service the stuff that usually fails and can be fixed. After a while, reading a schematic will be as familiar as reading a book. You'll be able to take one look and recognize the sections and stages in just about any product.

But I Ain't Got One!

As important and useful as a schematic is, you will find yourself working on many devices without one, simply because you can't get it or it costs more than you want to pay. Without the road map, how do you find your way around?

It's a lot tougher, but if you keep the overall circuit functions in mind, you can find the major sections and determine whether they're working properly. When you get to a suspicious stage, try drawing your own mini-schematic by tracing the connections of the active element and its surrounding components on the board. Sometimes, seeing it in front of you will illuminate the concept of the design and lead you to good troubleshooting ideas.

Once you locate the malfunctioning section, your understanding of stages and signal flow will help lead you to the problem. At least, that's the way we want things to go.

Sometimes it does, sometimes it doesn't. In very complex devices, it's easy to get lost and go around in circles when you're flying blind, without a diagram. Luckily, most malfunctions are due to power supply problems, bad connections and faulty output stages, which are relatively easy to track down, even without a schematic. When complex signal-processing stages don't work right, you may not be able to determine why, especially in modern devices with hundreds of parts the size of grains of salt. Heck, some of those things can be unfathomable even *with* a schematic! In those situations, you can learn a lot by tracing back from input and output jacks and poking a scope probe on what look like input and output points of successive stages. You just might find the spot where the signal disappears and zero in on the bad component.

Your Wish Is Not My Command

Let's take a case and work through it without a diagram. This will be a nice little LCD TV that refuses to respond to its remote control. It functions fine with the front-panel buttons, but the remote does nothing.

Is the remote working? How can we tell? If only we could see infrared (IR) light! Grab your cellphone. Most of those can see infrared. Even though they have filters to block it, some IR light gets through. Bring up the camera app, point the remote at the lens and hit one of the buttons while looking at the display screen. If you see a flashing light, the remote is working. Could it work but not be sending the right codes? Yes, theoretically, but I've almost never seen it happen, except in the case of a universal remote set up to operate the wrong device. That's user error, not a repair problem. If the remote's IR LED lights up when you press a button and stops when you let go, you can assume the remote is working properly.

In this case, it works. So, why can't the TV see it? Something in its remote receiver circuitry is out, and we're going to hunt that problem down. Naturally, this newer product has no available schematic, so we're on our own.

The first thing to find is the photodetector that picks up the remote's signal. Most products use a prefab remote receiver module containing a photodetector and a preamp. The module is usually in a little metal box (Figure 8-14), though in small, battery-operated gadgets it's more likely to be a bare plastic part. It sits just behind the front panel of the unit so that the photodetector can see out through the panel's plastic bezel. Yup, there it is. It's mounted to the circuit board and connected by three lines. Let's stick our scope probe on them and see what's there.

First things first: We have to connect the scope's ground lead to circuit ground and then power up the TV. To begin, let's set the scope for DC coupling, 1 volt per division, and put the trace at the bottom of the screen, since we expect positive voltages in this negative-ground TV. (We did verify that the negative power supply line went to circuit ground, right? Of course we did.) Hmm . . . one of the module's three lines seems to have nothing on it at all. Why would a connection have no signal? Could finding the fault be

FIGURE 8-14   Remote control receiver module.

this easy? Nah! The line goes to a big area of copper on the board. It's ground! Okay, one down. The second line shows a steady 5 volts that doesn't vary when the remote is pointed at it and a button is pressed. This one must be the power supply voltage running the module. The third line has what looks like a little noise on it. When a remote button is pushed, this line becomes an irregular pattern of pulses at just under 5 volts peak to peak. Aha! That's the code the remote is flashing. We've found the module's output line.

The good news: We've proved the receiver module is okay. If it weren't, the output line wouldn't do a darned thing when the remote sent its optical signal. It's showing the transmitted code, so it's working. The bad news: We still have no idea why the TV won't respond to it.

So, we follow the output line and see where it goes. It appears to terminate at a diode that feeds a transistor. The diode's purpose isn't obvious, but there it is, so we'd better check it. We scope it to see if the signal from the module is getting there. It is. At the other end of the diode, the signal is smaller, but it's still there. Because the signal is smaller, we can figure that the diode is not shorted. And, because it's there at all, the diode is not open. It could be leaky, but most likely it's fine. Time to focus on the transistor.

One of the transistor's three leads, presumably the base or gate, is connected to the diode. Sure enough, the remote's signal is there. The other two leads are at 0 volts, though! No DC, and neither shows any activity when the remote is activated. No wonder the TV never sees the remote's signal. It's a dead end. Blown transistor! At least, that's what a novice would think. In truth, we don't have enough information to draw any reasonable conclusion yet, so we look further.

One lead goes to circuit ground, so we wouldn't expect anything there. The other should have some voltage, though, right? Tracing from that one, we come to a couple of resistors, and it isn't clear where they go. Because the transistor needs power to function, one of those resistors must go to a power supply feed point. And, since the transistor's output signal has to feed some other part of the circuit that knows what to do with it, the other resistor must be coupling it to another stage. But which one does what?

Putting the scope probe on the opposite end of each resistor, we find that neither has anything on it there either. To use the technical term, *deadus doornailus*. If the transistor were shorted, it would pull its output line to ground, but not the other end of the resistor feeding power to it. That's the whole point of having a resistor there: to limit the current and avoid pulling the whole supply down when the transistor turns on and connects the line to ground. So, the resistor's far end, fed from the power supply, should still have voltage on it no matter what the transistor does. But the voltage isn't there. This suggests the problem lies elsewhere, and the transistor is probably not the culprit.

Tracing each resistor, we see that one goes directly to a huge chip with lots of leads and a crystal next to it. The other goes someplace far away we can't easily find. The one going to the chip is most likely the transistor's output, feeding the remote's signal to the microprocessor for decoding, because that big chip is clearly not part of the power supply. The other resistor, then, has to lead to the missing voltage. So, we follow it as it snakes along. Eventually, we find it leading to a little jumper wire, the other end of which connects to a fairly large land with lots of other parts and traces going to the same place. This should be the power supply feed point. Poke ye olde probe, and there's five lovely volts. Yee hah, it works! Oh, wait a minute, why isn't it getting to the resistor?

We turn the set off and watch the voltage on the feed point die away to zero. Out comes the DMM, set to the ohms scale. Hmm, there seems to be no connection from the feed point to the resistor. A check of our tracing confirms that we haven't made a mistake. Nope, we're in the right place. We flip the board and take a good look at that jumper wire's solder joints. One of them looks cracked! We touch it up with the iron and recheck the line for continuity. Now there's a connection. You can guess the rest. We power up the set, and it works, remote and all!

This is a *very* typical repair case. And, as you've seen, it can be solved without a diagram, by understanding basic circuit function and applying a little logic. Had there been voltage at the power supply side of the resistor but not at the transistor it was feeding, yanking the transistor and checking it for shorts would have been the next step, and it's a pretty safe bet the part would indeed have been shorted from collector to emitter, pulling the applied voltage to ground. The lack of voltage at the resistor's far end was the critical clue in acquitting the transistor, tracking down the real culprit and solving the case. Verdict: time served, and free to go!

And Sometimes You Lose

This one was a 5-amp charger for my 50-amp-hour LiFePO4 lithium battery. I bought the charger new, and it died during the first use. The seller was playing games to avoid reimbursing me for it, so I decided to try and fix it, since it was too cheap to warrant shipping back, and I suspected I'd never see my refund anyway.

No schematic, of course. I opened it up and found the typical switching power supply configuration. The fuse, bridge rectifier and 450-volt capacitor checked out fine. Not surprisingly, the chopper transistor, a MOSFET, was blown. I figured that must be the trouble, so I dug through my parts supply and found the only suitable replacement in my lab. The characteristics on the spec sheet were similar, but the new transistor had a higher voltage breakdown rating than the original, which was great. I always recommend replacing chopper transistors only with specified substitutes, but this was all I had, so in it went.

I plugged in the charger, and the new transistor literally exploded! Yet, the fuse was still good, proving the old adage that transistors are there to protect fuses. Surmising that the PWM chip must be bad and keeping the transistor turned on continuously instead of pulsing it, thus passing my wall plug's entire 20-amp electric service through the poor thing, I gave up and tossed what was left into my parts board bin. It would've cost me more to order another high-voltage transistor plus the apparently bad chip than to just buy a new charger. And who knows, maybe the real problem was elsewhere, and the new parts would have gone up in smoke as fast as that transistor did. You can't save 'em all!

Chapter 9

Entering Without Breaking: Getting Inside

Once upon a time, getting at the circuitry of a consumer product was a piece of cake. Remove a few screws, pop off the back, and there it was. Providing access to the inner workings was a tradition begun in the vacuum tube days, when the unit's owner needed to get inside on a frequent basis to change tubes, lamps and fuses. After semiconductor technology replaced the troublesome tubes with considerably more reliable transistors, there was still the expectation that the buyer might have a legitimate need to reach the circuit board. Early solid-state products even put the transistors in sockets! As semiconductors got sturdier, the sockets went away, which was good because their flaky connections caused more failures than did the transistors themselves. There were still lamps and fuses to contend with, though, and access was expected and easy.

That was before today's age of complex, ultraminiaturized circuits, complete lack of user-serviceable parts (thanks mostly to the LEDs that replaced lamps) and lawsuits. These days, no manufacturer wants you anywhere near the stuff under the hood.

Consequently, many equipment cases are deliberately sealed, or at least made pretty tough to open. Most AC adapters are ultrasonically welded together and have to be cracked apart; even the expensive ones are considered nonrepairable items by their makers. Many cellphones and tablets are glued together with the intention that they will never be taken apart.

Lots of laptop computers, video projectors, disc players and TVs sport hidden snaps, so they won't pop open even after you remove the screws (Figure 9-1). And anyone who's ever tried to take apart a certain American computer company's sleekly designed products knows the meaning of frustration; they're clamped together seamlessly and tightly in a clearly deliberate attempt to keep you out.

So, how do you open up these crazy things? Often, others have suffered before you and have posted step-by-step disassembly instructions on their websites, with clear photos of the whole process. If you can't fathom how to get something apart, it pays to do a web search on "disassemble _____ ," with the blank filled in by the name or model number of your

FIGURE 9-1 Hidden plastic snap.

product. Some companies selling parts post these helpful tutorials, facilitating your getting to the point of having a reason to order their wares.

If you can't find disassembly instructions, you'll have to wing it. With or without help, this is the stage at which you have the most opportunity to wreck your device! Without x-ray vision, you can't know if that little screwdriver you're using to pry the case halves apart or unhook the hidden snaps you're not even sure are there might be ripping a tiny component off the circuit board or causing some other drastic damage. You also have no idea whether a ribbon cable may join the two halves and be torn when they suddenly separate. And nothing is more frustrating than thinking you're about to repair something and destroying it instead, before you ever get a chance to look for the original problem.

Despite all this gloom and doom, you can get into nearly any product successfully and safely if you're careful and take your time. Disassembly is not a trivial process; expect it to take a significant portion of the total repair time, at least with the smallest, most complex devices. There are some tricks to the endeavor, and we're going to explore them now. But first, some rules:

- *Rule number one. Always* disconnect power before taking something apart. This is true with battery-operated products as well as AC-powered ones. Even if there's no danger to you, you have a much higher chance of damaging the device if power is connected when things come apart, whether the unit is turned on or not. Pull the plug. Yank the batteries.

- *Rule number two.* Remove everything you can before going for the screws. Battery covers, recording media (e.g., discs and memory cards), rubber covers over jacks, lanyards—if it comes off, take it off. There's no need to pull knobs from a front panel if you don't anticipate having to remove the panel, but that's about it. Everything else should go. And don't be too surprised if you wind up having to take off those knobs later on.
- *Rule number three.* Never force anything. If the case won't come apart or some corner seems stuck, there's a reason. Perhaps snaps are hiding on the inside of the plastic. Screws are sometimes hidden under labels and rubber feet. Run your fingernail over labels, feeling for the indentation where a screw head might be lurking. If you find one, peel back the label just enough to get to the hole. Peel back the rubber foot near the stubborn area; just because the other feet aren't hiding screws doesn't mean this one isn't. You'll probably have to glue the foot back on later, but it's a necessary consequence of peeling the original adhesive.
- *Rule number four.* Don't let frustration drive you to make a destructive mistake. Even the calmest tech can get riled up when a recalcitrant patient tries that 'ol patience badly enough. The most common errors are to start moving too fast, to force something or, in extreme cases, to smack the casing, hoping it'll loosen up. Bad idea! Okay, I admit it: I once threw a really nice, rather expensive pocket stereo cassette player against the wall after a maddening, futile hour of trying to take it apart, but I don't recommend the exercise. It felt good, but, needless to say, that repair job was over before it began. Plus, for weeks I was stepping on pieces of that poor little thing I'd murdered. I couldn't help but think of Poe's "The Tell-Tale Heart" every time something went crunch under my foot. At least it was my own player, and not something I'd have to replace for a customer or lose a job over.

The order of disassembly matters, especially for damage prevention. Obviously, you have to get inside first, so that's where we'll start. Place the unit in whatever orientation is required for access to the screws or snaps holding the thing together. Usually, this means face down. You don't want to mar the face or scratch the display screen, so find something soft on which you can lay the device. A few sheets of notebook paper will work well, but don't use just one, because a flake of solder or other grit on your workbench can poke through it and still cause damage if you slide the product around.

Other good options are cardboard and expendable towels. Avoid pillows; you don't want the repair item to shift position while you work on it. The only times I've ever used pillows were when I was servicing the undersides of turntables or reel-to-reel tape recorders. The conformity and support of a pillow really help prevent damage to the arm and other protrusions, and such big machines are heavy enough not to move around too much.

Removing Outside Screws

Some companies stamp little arrows by the screws that must come out, but that's not an industry standard (Figure 9-2). If you do see arrows, remove only the screws that have them. If there are no arrows, start with the screws near corners to see if that frees up the case. No? Then you'll have to remove them all and hope for the best.

As you unscrew a screw, observe how it comes out. It should rise, indicating that it's screwed into a fixed object, not a nut. If it seems very loose as soon as you start to turn it counterclockwise, screw it back in again and see if it tightens down without slippage. If not, then there probably *is* a nut on the other side, and you don't want to unscrew it! Nuts are almost never used on the screws holding case halves together. How would the manufacturer tighten such a screw without having access to the nut? Once in a great while, you do find a nut or a little metal bracket on the inside that's glued into a plastic shelf so it'll stay put after the case is assembled. It's rare, though, and found mostly on older gear.

Resist the temptation to save time by loosening multiple screws and then turning the unit over! That's a guaranteed way to lose some screws as they bounce away into never-never land. If turning the unit over is the only way to get a screw all the way out, do so one by one. It's slower but a lot safer. When screws are recessed, they may not want to come out even after you unscrew them fully. A magnetized screwdriver helps, but sometimes the friction against the sides of the channel is too great, and the little buggers refuse to yield. If you have a stuck screw like this, leave it where it is, and then push it out from the screw's tip after you get the device open. Just keep an eye on the screw until then, in case it decides to let loose and head for the hills.

Take a good look at each screw after it's out. Pay careful attention to the length, comparing it to the last one. Quite often, otherwise identical-looking screws are of different

FIGURE 9-2 Arrow indicating screw removal.

lengths, and putting one that's too long in the wrong hole when you reassemble the device can make it poke into something, causing a short or other serious damage.

As you take out each screw, count aloud how many you've removed. Remember those little cups I suggested you collect way back in Chapter 2? Here's where you will use them. Put the screws into one of the cups, and make a small written note of how many there are. Tear off the note and put it in the cup with the screws. If they weren't all the same length, draw a little diagram of which went where, and put that in the cup too. You'll see why in Chapter 13. When you're done removing screws, put an empty cup into the one you just used, covering the loose screws.

Separating Snaps

Popping apart hidden snaps is almost an art form in itself. First, be absolutely sure you've removed all the necessary screws. Take a look at the bottom of the unit to see if there might be slots into which you can put a screwdriver to pop the snaps. Slots are common on AC-operated devices whose bottoms face shelving, but not on pocket toys. If you do find slots, shine a flashlight into one to see if you can deduce what needs to be pressed in which direction to unhook the snap. Pop one open while pulling the case halves apart with your free hand. To prevent accidental snap reinsertion, keep holding the halves apart while you do the next snap on that side of the case. Once you have a couple of snaps open, you won't need to continue the forced separation—the rest should open easily.

If you find no slots, pick one side of the case and press on the seam around the edge, looking for inward bending of the plastic. Move slowly and feel for slight movements indicating where hidden snaps may lie. When the plastic gives a little, press harder and attempt to pull the seam up. If it won't budge despite your best efforts, move to another part of the case and try again. After you get one snap undone, the rest will release a lot easier.

If no amount of effort will release the snaps, you might be tempted to slip a small screwdriver into the seam to pry it open. This is a last-ditch procedure, but it usually works. It's almost certain to break snaps, though, and cause some visible damage on the outside of the case.

Even with the best technique, you'll break a few of those darned snaps while you get the hang of it, and occasionally even after you're an expert. Luckily, it's not the end of the world. Often, the loss of a snap or two has little or no effect on a product's integrity, but sometimes the reassembled case can feel loose or have a gap along the desnappified seam. Just save the broken pieces in case you need to melt them back on later. Remelted snaps are never very strong, but they can be better than nothing.

Removing Ribbons

Ribbon cables have replaced wires in small products. They offer much higher density with a lot less mess, and we couldn't have today's complex pocket devices without them. The ribbons are delicate, though, and removing them from their sockets requires care.

Some ribbon connectors have latches that press the ribbon's bare conductor fingers against the socket's pins. Others have no latches and rely on the thickness of the ribbon to make a firm connection. Either style may have a stiff reinforcement tab at the end of the ribbon, but the latchless style always does.

Before pulling out a ribbon, take a Sharpie marker and put a mark on both the ribbon and the socket so you'll know how to orient the cable during reassembly. Use a Sharpie; other markers may rub off. If other nearby ribbons are similar enough that you could possibly confuse which goes where, use a unique mark on each one.

Now, examine the socket closely. If you see small tabs at either end, it has a latch. Even without tabs, it might have a flip-up latch. Pulling a ribbon from a latch-type connector without opening the latch first can easily tear the ribbon. You don't want that! Ribbon cables are custom made for each product, and you aren't going to find a replacement unless you can dig up a dead unit for parts. A torn ribbon usually means a ruined device.

If there is no latch, grasp the ribbon at the reinforced tab and pull steadily. Don't jerk or you'll almost certainly destroy it. Pull gently at first and then harder if the ribbon doesn't move. With some of the larger cables, you might be surprised at how much force it takes to get them out of their sockets. If you need to pull hard, hold the socket down on the board with your other hand to prevent ripping it from its solder joints. It's rare for a ribbon to need that kind of force, but I've seen it a few times. Whatever you do, don't pull on the unreinforced part of the ribbon; it won't withstand the stress.

If there's a latch, open it first. There are two basic styles: slide-out and flip-up. Have another look at Figures 3-2 and 3-3 so you'll know 'em when you see 'em. Slide-out latches have little tabs at the ends of the socket. A fingernail or the end of a flat-blade screwdriver will pull them open. It's best to open them at the same time to keep the sliding part from getting crooked or breaking off, but gently opening them a little at a time, going back and forth between the sides, usually works fine.

Flip-up latches open easily with a fingernail. You can use a screwdriver, but be careful not to scrape and damage the ribbon cable. Especially with a very small connector, open the flip-up slowly and carefully, as they break quite easily. It helps to pull them from near their ends, lifting both ends at the same time, rather than from the middle. If you break one, keeping enough pressure on the ribbon to make a proper connection with the socket is next to impossible.

Pulling Wire Connectors

Larger items may have a mixture of ribbon cables and good old wire assemblies. Circuit board–mounted connectors for wiring rarely have latches. If they do, the latch will be large and obvious, something you squeeze with your thumb while pulling on the connector. Most wire connectors simply pull straight up. Reorientation is not an issue with these, but marking is still advisable if other nearby connectors could cause confusion. We've all been taught through the years never to pull on the wires when removing plugs, but that's what you have to do with these because the plug fits entirely into the socket, leaving nothing else to grab. Grasp the wires and pull steadily without jerking, and the connector should pop out. Sometimes it helps to rock the wires left and right a little bit while pulling. Don't overdo this; it puts a lot of stress on the wires at either end of the connector, and redoing the connection after you tear one out is a royal pain in the aft region.

Before you pull, though, be certain there actually *is* a connector! Groups of wires sometimes terminate in what look like connectors, but the wires go right through the plastic and are soldered directly to the board. Obviously, you don't want to pull on those.

Layers and Photos

Unlike the simpler products of yesteryear, modern gadgetry is often built in layers. Perhaps the topmost layer is a display. Under that lies a metal shield. Beneath that is a circuit board, and there's another board under that one as well. Behind all of it is the battery compartment and a little board for connectors. Ribbon connectors join the layers, with several on each side of the main boards, and you can't get to the lower layers without removing the upper ones first. Sounds like a huge product, doesn't it? Perhaps a home theater receiver or a laptop computer? Hah! I just described a typical digital camera!

To reach the innermost spaces, the layers have to be stripped away in precisely the reverse order of their original assembly. And, naturally, the problem you're chasing is at the bottom layer, right? Ol' Murphy knew what he was doing.

Each layer is held by screws, and they're probably different sizes from those holding the next layer down. Some may even be different from others in the same layer. Not all screws always have to be removed; some only grip small internal parts, and unscrewing them may drop a nut or a washer deep into the works. Especially in devices with motors or speakers, both of which use strong magnets, a lost metal part can get pulled in and cause real trouble later.

As you take out screws from a layer, count them and put them in the empty cup that's on top of the one containing the outside screws. Don't forget to include a record of the number of screws, along with a diagram if different lengths are used. Now put another cup into that one. When you start on the next layer, put its screws into the open cup, and so on. If the device is especially complicated or has many layers, take a photo of each layer, with the stack of cups visible in the picture. That way, you'll know which set of screws goes with which layer—just count the number of cups in the photo. You might be surprised at how easy it is to lose track of that after the unit has sat in pieces on your bench for days or weeks. When the disassembly operation is complete, you'll have a stack of cups, with the screws from the last layer in the top cup. Place an empty cup in the top one to keep the screws securely inside, and put the stack somewhere safe and out of your way.

Opening a Shut Case

Let's take a look at some case-opening procedures for common products, starting with the easiest and working up to the really challenging adventures.

Receivers and Amplifiers

Most shelf-style audio gear opens up with no hassle. You'll find four screws, two on each side, two or three smaller ones at the top edge of the back panel, and perhaps one to three on top, just rear of the front panel. Unscrew them all, put them in a cup, and the top should slide off. Often, you'll have to spread the sides slightly while lifting the back edge because the front edge is under the top of the front panel.

Disc Players and DVRs

Most disc players and DVRs open the same way, but there are some variations. Some have screws on the bottom edges instead of the sides, and you may find some on the back's upper edge. Plus, the front edge may have a lip fitting into a groove on the front panel.

TVs and LCD Monitors

Today's flat-panel TVs and monitors usually unscrew from the back. The panels are recessed from the bezel around them, so you should be able to lay the set face down gently on your bench, after sweeping the table's surface and checking for anything that could stick up and put pressure on the screen. You may find lots of screws of various lengths. Be very careful to note which go where, because under those screws is the back of the display panel! You *really* don't want to mix up the lengths and drive a screw into that when you put it back together.

Turntables

Turntables are an old technology, but they enjoy a following among audiophiles, so there are plenty of them still around. In fact, vinyl has been making a comeback, and new turntables are being manufactured. Turntables are uniquely shaped and somewhat delicate, making them awkward to service.

A turntable's platter may be driven by a rubber wheel, a belt or a direct-drive motor turning at the same rate as the record. Most better turntables are belt- or direct-driven. To change the belt, first lock down the arm so it won't flop around and damage the stylus. Then, lift the rubber mat on the platter, and you'll see the motor spindle somewhere on the left side. Putting the new belt on requires lifting the platter straight up and out. Most come off without removing anything, but some have a retaining clip around the spindle.

For any other repairs, you'll need to get to the underside of the turntable, which involves laying it on its face. To do so, put it on a pillow arranged such that the weight of the unit won't be on the arm. Never put pressure on the arm assembly; the arm probably won't survive.

Before you flip the unit over, it's wise to take off the stylus and put it aside because it's the most fragile, easily damaged element. Whack anything into it, and it'll get trashed. The quickest and safest way to get the stylus out of harm's way is to pull the entire cartridge. Many later turntables use a *P-mount* cartridge that unplugs easily, with no individual wires and connectors to deal with. Some with the old-style screw-in mount have removable head shells. If you see a sleeve where the head meets the arm, it's probably a removable shell. Unscrew the sleeve a few turns, and the shell should pop right off.

Once the stylus and/or cartridge has been removed and placed out of harm's way, secure the arm with its retaining clip. Take a look at the back of the arm. If you see a little antiskate weight hanging down on a wire, make a note of its setting, and then remove it so it won't get damaged when you flip the turntable over. If the primary counterweight slides on and off, slide it off after noting its setting as well.

Remove the mat. Unless the platter is held on with a retaining clip, remove the platter, too, so it doesn't fall off.

Many turntables are mounted on springs, so you need to hold the corners of the chassis as you turn it over or the machine can fall out of its base. Hold those corners, turn the unit over slowly and place the turntable face down on the pillow, making sure none of the weight is on the arm. If there's a bottom plate, remove the screws securing it, and it should come off.

Video Projectors

If your projector has a lamp (in other words, it's not lit with LEDs), be sure the lamp is completely cooled down, and take it out *first!* The bulb represents most of the cost of the projector. Plus, it's fragile and contains mercury. Put the lamp assembly aside, far enough from the work that you won't drop a tool on it or knock it off the bench.

Most video projectors have screws on the bottom. After their removal, the top half will lift off. Some projectors are entirely snapped together, with no screws. Even with screws, there may be hidden plastic snaps.

On some units, the lens has plastic rings for focus and zoom that must be pulled off before the case can be separated. Pry the rings off with your fingers, avoiding the use of tools. If you must use a screwdriver, do so especially carefully. I've seen a few units that required prying plastic tabs away from the lens. Do it gently so you don't break the tabs.

There may be ribbon cables between halves to connect the control buttons and indicator lights. Separate the halves slowly to avoid tearing them.

As you remove the case from a lamp-based DLP projector, keep your fingers away from the front of the unit because the color wheel is just inside, and it's fragile. Putting any pressure on it is likely to result in its destruction. Those darned color wheels can cost more than $100, so don't break one! LCD units and LED-lit DLPs have no color wheels to worry about.

Portable DVD Players with LCD Screens

These usually have screws of varying lengths in the back. After you remove those, the back should come off, but make sure to have the unit lying on its face because the laser sled assembly can fall out and tear its ribbon cables if you hold it in any other orientation. The assembly sits on rubber bumpers, and the back holds it in place. It's supposed to be loose, for vibration damping and skip resistance (Figure 9-3).

If you need to get to the LCD monitor and its associated circuitry, its screws are probably under the rubber bumpers on each side at the top. Check the back of the LCD before peeling off the bumpers, though, in case the screws are back there. On many players, the plastic bezel will come off the front, with the LCD anchored to the back, but some are the other way around. Often, the speakers are on the bezel, connected by wires, so remove the bezel carefully to avoid tearing them. If the bezel won't come off after you pull the screws, it either has internal snaps or is glued at the seam. Many of them are glued to prevent rattling from speaker vibrations. Use the snap-popping procedure described at the start of this chapter. If you find no snaps, try peeling up the bezel gently, one edge at a time. The feel of separating glue will be unmistakable. Just remember not to let the bezel pop off hard or you'll probably rip out those speaker wires (Figure 9-4).

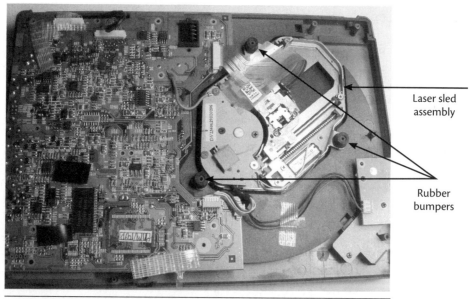

Laser sled assembly

Rubber bumpers

FIGURE 9-3   **Inside a portable DVD player.**

Inverter

FIGURE 9-4   LCD with bezel off.

Smartphones and Tablets

Every model of phone or tablet has a specific order of disassembly that must be followed for successful repair. There is no way to cover the multitude of product models here. I *strongly* urge you to search online for a video tutorial for whatever device you want to fix, and to follow it carefully as you proceed.

Depending on the brand, your phone or tablet may have screws in the back and a separable case. Expect hidden snaps all around the edges. After removing the battery (if accessible), SIM card and any micro-SD card you may have installed, unscrew the screws and carefully work your fingernail or a thin piece of plastic around the edges, starting in a corner, to pop the snaps. The case should come apart. The PC board and the screen will still be on the front.

If you're taking a phone or tablet apart to change the battery, it's likely that the replacement cell came with a *spudger*. This is a plastic tool with a fine edge you use to pry apart case halves and pull up screens (Figure 9-5). Spudgers break easily, but they're much safer for the product than any other prying method, both physically and because they're made of plastic, so they can't cause short circuits. You can buy them separately, and they're plentiful online and pretty cheap. If you'll be doing a lot of work on phones and tablets, I recommend you get a few.

If the back is sealed, access to the unit's insides is through the front, which means taking off the screen first. The darned things are glued on, and you have to use heat to

FIGURE 9-5 Spudger tool.

soften the adhesive enough to separate the screen from the case. Most hobbyists use a heat gun or a hair dryer for this, but it's very easy to ruin the screen that way. Too much heat and the screen will get burned. Too little and you can bend it slightly while pulling it up, destroying it. LCDs won't survive being bent even slightly! You have to heat up the edges of the screen to just the right temperature to soften the glue and then lift it off oh so gingerly by working around the edges with a spudger.

There's a better, safer way. As mentioned in Chapter 2, LCD separators are very affordable these days. They feature a carefully controlled heating surface that lets you soften the glue without applying too much pinpoint heat that might wreck the screen. If you plan to do much phone or tablet repair, it's well worth getting one of these.

There are some great online tutorials to help you through the screen removal process.

Camcorders

Camcorders that record to memory cards are like digital cameras, so see that section (next) for advice on taking them apart. Tape-based camcorders, analog or digital, are a whole 'nother story. They're mostly gone now, but some people are keeping them to dub precious home movies to digital formats, so I'm including them here. Tape-based camcorders can rival laptop computers for complexity. In some ways, they're worse because they are so oddly shaped that boards and mechanisms are crammed into nooks and crannies, making them hard to extricate (Figure 9-6). Plus, the mechanism is delicate and easily damaged during disassembly. Drag out the cups and the digital camera.

If you only need to clean the heads, open the cassette door and then remove its two screws at the top. They may be under rubber covers. Take off the door, and you should be all set.

For more extensive repairs, the machine will have to come apart. The typical camcorder body is in two pieces. Before you try to separate them, it's best to remove the cassette door

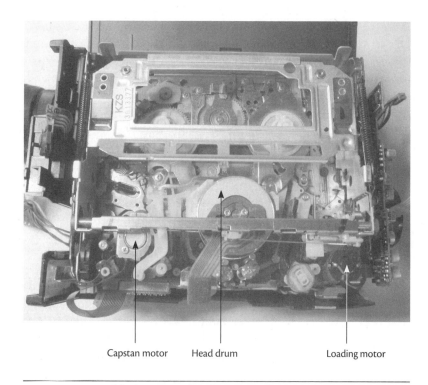

Capstan motor   Head drum   Loading motor

FIGURE 9-6   Mechanism side of a camcorder.

because it probably prevents getting that side of the body off. If the machine works enough to get the door open, pop it open, take out its screws and remove the door. If the door is stuck closed, remove the screws anyway and see if you can slide off the door. Most likely, it won't budge, but you might be able to remove it once the case is loose.

Look for arrows on the case indicating which screws need to come out. You may find screws all over the case, and most of them will have to go. On some cameras, various covers on the front and top have to come off because there are screws under them securing the shell to the chassis. Once the case is loose, gently pull the two halves apart, being careful not to get your fingers inside, where they could damage the mechanism. You'll see ribbon cables all over the place, and you'll have to disconnect a few once the machine is open.

Digital Cameras

Along with camcorders, digital cameras are some of the hardest items to service. Most of today's cameras are larger, semipro models because people just use their phones for the kind of basic picture taking that used to be done by small cameras, wiping out the market for those. Still, even in bigger cameras, the works are crammed in there tightly. Plus, cameras have lots of buttons, and some have sliding switches with plastic cradles that fall off into oblivion as the case comes apart. We'll talk more about switch cradles in Chapter 13.

The case halves on many digital cameras are three-dimensional puzzles. To get them apart, you may have to bend them around the edges slightly. On some, there's a plastic sidepiece surrounding the two halves, with tabs from each half fitting into it.

Generally, the back comes off, with all of the circuitry and the LCD remaining on the front. Be careful not to press on the LCD once the protective plastic window lifts off with the back of the case.

While LEDs act as camera flashes in phones, digital cameras may still use flash tubes because of their better color characteristics and potentially higher brightness. Such cameras store the energy for the flash tube in a large electrolytic capacitor. That baby can hold its charge of several hundred volts or more for weeks after its last use. Usually, the cap is stuffed under the main circuit board, next to the optical assembly (Figure 9-7). Its leads, however, may join the board just about anywhere. As you get the case apart, keep in mind that the connection to the flash cap could be right under your fingers. I've gotten zapped by digital cameras more often than by anything else. In addition to the danger to you, discharging the cap through your finger or a tool can leak high voltage into the camera's sensitive circuits, causing instant, silent damage.

FIGURE 9-7 The evil flash capacitor. Note the voltage rating!

Laptop Computers

Laptops are among the most complex consumer devices and the toughest to take apart. Talk about layers! You'll want to use the cups and camera for these. Before you begin, do an internet search for disassembly instructions. Laptops are pretty trouble-prone, and sites abound with help. Very often, the disassembly sequence must be followed exactly or the machine can be damaged. Plus, where you begin depends on what area you need to reach. Changing the hard drive might require a different procedure and degree of disassembly than would resoldering the power jack or replacing the keyboard.

If the battery is removable, take it off before doing anything else. To remove the keyboard, look for snaps at the top. If you don't see any, check for screws on the bottom of the machine. They'll nearly always be placed such that they screw into the back of the keyboard near its top. I've seen a few near the middle, but none at the bottom, which usually has slots fitting into grooves on the top half of the case. Once the keyboard is loose, pull it up gently, keeping in mind that a ribbon cable connects it.

If you're trying to replace the screen or repair a backlight problem, you should be able to open the screen assembly and not need to get into the rest of the laptop at all. Most LCD housings are screwed together. Look for screws along the edges of the housing. If you don't find any, check for cosmetic covers or bumpers on the front, near the bottom of the LCD. I've seen a few cases where screws were under the rubber bumpers at the upper corners, but not many. Often, those bumpers are pretty permanently attached and will tear if you try to pull them out. Then you find there's nothing under them anyway. Before going that route, exhaust all other possibilities. Hidden snaps are common here, too. Just avoid pressing on the screen; it's easy to do while pushing the seam along the edges, feeling for snaps. Once LCDs are pressed on hard enough to deform them and affect the picture, they don't come back to normal.

Probably the most common failure in laptops is a loose, intermittent power jack. Repair is simple: Just resolder the jack to the motherboard. Alas, getting to it isn't always so easy. For this one, you'll need to take the case apart. Look for screws all over the back, and keep track of their lengths when you take them out. Use your cups! Watch for hidden snaps along the sides, and don't bend the back too hard or you can break the internal frame or the motherboard. Go easy.

Some models mount the board to an internal frame, while others have it screwed to the plastic case. Slim, lightweight notebooks typically don't have frames. Instead, the major components are simply screwed to the back.

If you're lucky, the jack's solder connections will be visible, and you can resolder them without further disassembling the machine. If not, you may have to remove the top half of the case. Be very careful of ribbon connectors going to the track pad and other items on the top half.

I recommend not trying to take apart a valuable laptop if you've never done it before because the chance of wrecking it is substantial. Outdated machines are available for very little, or even for free, from neighborhood websites and local computer recyclers. Get one and practice on it.

AC Adapters

Despite the title of this chapter, with most AC adapters, you will have to do some breaking! There are adapters with screws holding them together, but most of them are ultrasonically welded. Some use both! The only way you'll get in is by cracking them open, and it takes some work. The process definitely will mar the casing, but it's necessary if you want to get inside these things.

First, look for screws on the back. They could be under rubber feet or labels. Always feel labels for indentations that could indicate screws underneath. Look for the seam between the two halves. Take a flat-blade screwdriver that's fairly thin along the front edge (in other words, it's somewhat sharp), and place it in the seam. Tap gently with a hammer until the screwdriver just goes through the plastic. Try not to let it go in beyond that, or it may damage components or the circuit board. Pull out the screwdriver, and you'll be left with a slot in the plastic.

Tap in another slot a short distance along the seam from the first one (Figure 9-8). Then put the screwdriver into the first slot and twist it. If you're lucky, the seam will break and the case will pop open. If not, make some more holes, and keep trying until it pops apart. If twisting in one slot doesn't work, try another one. I've run into some seams that popped open easily and others that wouldn't yield until I'd made a mess almost all the way around them. Most give way after a few tries.

FIGURE 9-8   Do this carefully!

Chapter 10

What the Heck Is That? Recognizing Major Features

When you open a modern electronic product, the apparent complexity may seem overwhelming. Today's pocket-sized gadgets can sport a surprising complement of goodies inside. Tablets, smartphones, GPS units and smart watches, for instance, are miniature computers, with RAM (random access memory), ROM (read-only memory) and a microprocessor. Most of 'em offer Bluetooth and/or WiFi as well.

Older products, including some with double-sided circuit boards, typically had through-hole components, and only on one side. Not anymore! Thanks to the complexity and size of today's gadgetry, surface-mount parts are all over the place on both sides. Components that stick up, like transformers and can-style electrolytic capacitors, are often relegated to one side so the board can fit flush against the case. Everything else is fair game. Transistors, chips, resistors and small inductors and capacitors may be anywhere. As you wend your way through a circuit's path, you can expect to flip the board over numerous times.

To find your way around in a box crammed full of parts, cables and boards, you need to become acquainted with what the major sections of the product look like, how components tend to be laid out, and how to follow connections from recognizable features back to those less obvious. Though the features vary a great deal depending on the product's function, pretty much every device has a power supply section, an input section (or several), some kind of signal processing and one or more output sections. Let's look at some common circuit sections and how to locate them.

Power to the Circuit: Power Supplies

Everything has a power supply of some kind. It could be a pair of AA cells, a simple linear supply or a complex switching supply with multiple voltage outputs. Power supply problems account for many repairs, so recognizing the power supply section is vital.

Batteries, obviously, are hard to miss. Battery-powered devices may have other power supply components as well, though, such as a switching converter to step up a single AA cell's 1.5 volts to a level high enough to run the product, or a battery management system (BMS) for safe charging of a lithium pack. Even when the battery voltage would be adequate on its own, a device may include voltage conversion and regulation to ensure that a weak battery doesn't affect the reliability of writing to memory cards, which can be seriously scrambled by insufficient voltage during write operations. Digital cameras and laptops have such systems so they can operate properly until the battery is nearly dead and then shut the device down gracefully.

To find voltage converters and regulation systems in battery-powered gear, look for small inductors and transformers. There are many varieties of them, but the telltale sign is metal. These things don't look anything like transistors and chips. They may be round or square, but most are made from ferrite material, which is a darkish metal with a matte surface. Toroid (doughnut-shaped) cores are common. You may be able to see the wire wrapped around the core, but don't count on it (Figure 10-1).

Along with these, keep an eye out for electrolytic capacitors. Although 'lytics can be sprinkled throughout any circuit, the larger ones tend to be congregated in or near power supply sections. Diodes and voltage regulators will be found there too. Most very small products don't use enough current to require heatsinks on regulators and switching transistors, but larger devices usually do.

It might seem natural that you could follow the wires from the battery compartment straight to the supply section, but it doesn't always work out that way. Most of today's gear uses transistor switching driven by a microprocessor to turn itself on and off, rather than a real power switch actually interrupting the current between the batteries and the circuit. There may be some distance between where the batteries connect to the board and the

FIGURE 10-1 Miniature inductors and transformers.

location of the voltage conversion and regulation circuitry. Those little transformers are your best landmarks.

In AC–powered products, finding the power supply is a lot easier. Follow the AC cord, and it'll get you there! Hard switches are still used in some AC devices, but those with remote controls, like DVRs, TVs and most projectors, use the same microprocessor-controlled soft switching found in battery-operated gadgets, so you can't assume that the on/off switch is in line with the AC cord. Look for a transformer. Switching supplies are pretty much standard now, but some products, especially high-end audio amps and receivers, still use linear supplies for their essentially noiseless operation. The transformer in a linear supply (Figure 10-2) is a lot larger than the one in a switcher. In a switcher, look for a transformer like that shown in Figure 10-3.

The power supply sections of pocket-sized devices are almost certain to be on the main circuit board, while those in AC-powered devices are more likely located on a separate board, with a cable feeding the output to the rest of the circuitry (Figure 10-4). When they are on the main board, expect them to be separated from the rest of the circuitry in a clear manner. On boards with AC line–powered switching supplies, look for slits or holes in the board separating the AC and DC sides of the supply. Those are put there so a voltage surge from a nearby lightning strike can't jump easily from the AC line–connected parts to the output side, which at some moment or another could be connected to you.

FIGURE 10-2   Linear power supply transformer.

FIGURE 10-3 Switching power supply transformer.

FIGURE 10-4 Separate power supply board.

Once you've located the power supply, its major features are easy to spot. In a switcher, the AC line will go through a line filter that looks like a small transformer and then through a fuse and on to the rectifiers, which may be separate diodes, a bridge rectifier or a double diode, depending on the design. Usually, it's a bridge rectifier. After that comes the chopper transistor, typically mounted on a heatsink. Near it will be a large electrolytic capacitor with a voltage rating in the hundreds. Then comes the transformer, followed by the low-voltage rectification and regulation circuitry. At the very end, right by the output wires, you'll see the electrolytic filter caps rated at around 16 to 25 volts.

In a linear supply, which has much less complexity, the AC line will go through the fuse and to the transformer, which is larger than a switcher's version. You won't see a heatsinked transistor or a big electrolytic cap on the AC line side of the transformer. On the low-voltage side, you'll find the rectifiers, regulators and filters.

A special type of power supply is the voltage inverter, a step-up supply driven from the main power supply's DC output. The inverter takes that low voltage and produces the high voltages required by the fluorescent backlights used in older LCD screens that were not lit by LEDs, or the vacuum-fluorescent displays used in home theater audio receivers. Inverters look like miniature switching supplies, which is basically what they are. The parts are much smaller, though, and you'll see two transformers and two output cables in bigger LCDs with two fluorescent lamps behind the screen. Most designs put the transformers at opposite ends of the board (Figure 10-5).

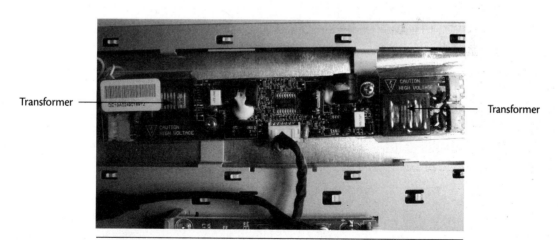

FIGURE 10-5 LCD backlight inverter. Note the plastic insulation over the transformers.

Follow the Copper-Lined Road: Input

The input sections collect signals and feed them to the signal-processing areas. The type of input circuitry present depends on the nature of the incoming signals. In radio and TV gear, input comes from an antenna or cable in the form of radio-frequency (RF) signals with strengths ranging from millionths to thousandths of a volt. The function of the input section is to amplify those very weak signals and then separate out the desired one, also called *tuning* it. Until recently, the most common approach to tuning involved inductors (coils) resonating at the desired frequency. Analog-controlled tuners used a mechanically variable capacitor in parallel with the coil to change the frequency. Digitally tuned, non–software-defined radio (non-SDR) setups also may use coils, but the tuning is controlled by the digital circuitry and accomplished in a phase-locked loop (PLL) system with a *varactor*, which is a voltage-variable capacitor, controlling an oscillator, or via direct digital synthesis (DDS) of the required oscillator frequencies. This has changed as SDR has overtaken analog techniques, but some sort of amplification still has to occur before the signals can be applied to the digital tuner's digital signal processor (DSP) chips.

Today's TVs use HDMI (high-definition multimedia interface) for digital input from external devices. Some still accept DVI (digital visual interface), which uses a different connector but carries the same signal format, except without audio data. A few sets still accept VGA (video graphics array), which is analog but can carry high-resolution signals and was commonplace on computers before DVI and HDMI took over. Many modern digital TVs still have inputs for *baseband* video, which is the analog video signal from a non-computer source without an RF carrier. Various flavors of baseband video include *composite*, in which the entire signal is carried on one wire; *S-video*, in which the chroma (color) information is carried on separate wires from the luminance (brightness) signal; and *component*, which provides separate connections for red, green and blue. The input circuitry for each style of video is somewhat different.

The easiest way to find video input circuitry is to follow the lines coming from the input jacks. These lines will go to some sort of switching circuitry first so the set can choose the desired signal source. After that, analog signals will be sent to the appropriate stages for amplification and preparation for further processing. Digital inputs will go directly to specialized chips that can decode them.

Input to a device can also be from a *transducer*, such as a laser optical head, phono cartridge, phototransistor, tape head or microphone. The signals from transducers may be very weak, as with magnetic phono cartridges, video heads and hard drive heads, or somewhat stronger, as with laser optical heads and phototransistors. Most of the time, the input circuitry will involve low-level amplification to boost the signal for the signal-processing sections. Because of their sensitivity to weak signals, input stages for transducers are often hidden under metal shields to mitigate interference from other parts of the circuit or outside sources. Sometimes, antenna or cable input stages are also shielded for the same reason. The shields are a good giveaway that what's under them handles tiny signals.

Shake, Bake, Slice and Dice: Signal Processing

Most products process a signal of some sort, be it analog or digital. Some kind of information is taken in or retrieved and massaged into whatever it is you want to hear, see, record, play back, send or receive. Much of the circuitry in any device is dedicated to signal processing. This is the little stuff, with lots of resistors, capacitors, transistors and chips used in analog processing. Digital circuitry is mostly chips, with a few bypass capacitors and other small components that set operating parameters (Figure 10-6).

The chips employed in analog circuitry tend to be small-scale, with around a dozen leads, not a hundred. Analog sections also use more transistors than do digital types, and you may see variable capacitors, trimpots and adjustable signal transformers, especially in radio and TV receivers (Figure 10-7).

FIGURE 10-6   **Digital signal-processing section. No adjustments anywhere!**

Trimpots

FIGURE 10-7 Analog signal-processing section.

While most modern products are primarily digital, plenty of them combine analog and digital functions, with a digital control system operating the analog sections. Frequency-synthesized superhet radios use analog stages to pick up, amplify and detect radio signals, but their tuning is entirely digitally derived. CD players process everything in the digital domain and then convert the results to analog for output. Many DVD players offer analog output along with HDMI. The sound sections of TVs typically provide both analog and digital outputs as well.

Digital control sections are recognizable by the large microprocessor chip with lots of leads. Look for a crystal or resonator very close to the micro. If the product has a display, it'll probably be near the heart of the control section, as will a keypad or a series of control buttons (Figure 10-8). Even in all-digital devices, signal-processing, output and power supply sections look quite different from the control circuits.

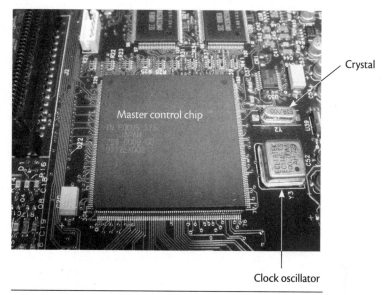

FIGURE 10-8   Digital control section. Look at all the leads on that microprocessor!

Out You Go: Output Stages

Output stages prepare the processed signal for display, a speaker, headphones, a transmitting antenna, a print head, a motor and so on. In many cases, a large part of this preparation is current amplification to give the signal the oomph to drive a speaker, move a motor or push a transmitted RF signal for miles. The primary identifying characteristic of output stages is that they are larger than most of what's around them. In some cases, like headphone amplifiers and small speaker drivers used in pocket radios and cell phones, the current required is low enough that the output stage may be quite small. Usually, however, more current is needed, so the parts are larger and better equipped to dissipate the higher heat.

The style of output stage depends on what is being driven. Analog audio amplifiers driving large speakers with many watts will have sizable heatsinks for the output transistors or modules. If modules are used, they will be much larger than power transistors, with more leads. If discrete transistors are employed, they'll range in size from around a postage stamp to perhaps 1½ inches long (Figure 10-9).

Output components used to drive motors may have heatsinks, but they might not if the motor is small and doesn't carry much load. Some motor drivers are just transistors on the circuit board. They're likely to be a little larger than the others, though.

Today, even some small products are using all-digital audio amplifier chips so efficient that they can be tiny and require no heatsinks. Portable Bluetooth speakers loud enough to drive you out of the room can be driven by small chips that run cool. Eventually, all but

FIGURE 10-9 These output transistors in a stereo receiver are larger than average, but they have only three leads, and their 2SC and 2SA part numbers prove they're transistors, not modules.

audiophile-grade audio amps may go this way, but for now, you'll still see plenty of analog amplifiers out there in larger products like home theater audio receivers.

The output circuitry for dot-matrix LCDs of the sort used for frequency and operating status displays is integrated into the display module itself, along the edges. This stuff isn't serviceable; if a row or column goes bad, in most cases the LCD must be replaced. Simple numeric LCDs don't carry their own driver circuits. Instead, they're driven directly by the micro or by an external driver chip.

A Moving Tale: Mechanisms

We're heading toward a time when electronic products no longer include mechanical elements. Tape recording is already a dead technology. At this point, just about everything records to memory chips, with hard drives and optical discs being the last mechanical media. Not long from now, even those will fade into obscurity.

While it draws ever nearer, that time has not quite come yet, and some of today's products still have mechanisms. Those in hard drives are sealed to prevent even the tiniest dust mote from crashing the heads into the platter, but the mechanics in optical disc players are readily accessible, and they cause enough trouble that you should get familiar with them. Digital light processor (DLP) video projectors that have hot lamps use color wheel assemblies rotating at high speed. And, if you're servicing older technology like VCRs, tape-based camcorders, audio tape recorders and turntables, you're going to get well acquainted with mechanisms and their peculiarities. Figures 10-10 through 10-13 show a few mechanical sections you might encounter. For more details on specific mechanisms, see Chapter 15.

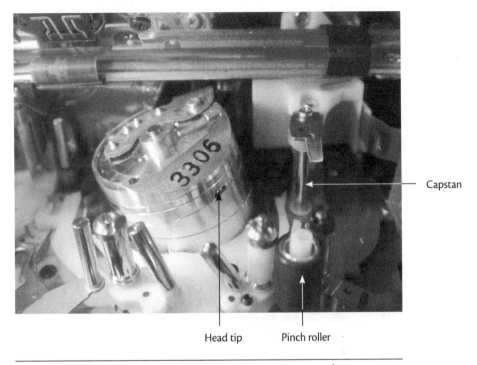

Capstan

Head tip   Pinch roller

FIGURE 10-10   Video head drum in a mini-DV camcorder.

FIGURE 10-11 Camcorder capstan motor.

Back of spindle motor   Back of laser head

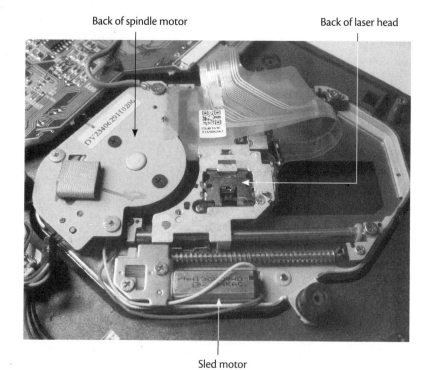

Sled motor

FIGURE 10-12 Laser optical head sled assembly.

FIGURE 10-13   Color wheel assembly in a DLP video projector.

Danger Points

As discussed in Chapter 3, there are dangerous spots in many electronic products. While it seems obvious that AC-powered circuitry would be the most hazardous, don't discount battery-operated gadgets as being harmless. Some products step up the battery voltage to levels that can give you a nasty jolt. As mentioned in Chapter 9, digital cameras, especially, generate hundreds of volts for their flash tubes, and the energy gets stored in a big capacitor capable of biting you weeks after being charged.

Watch out for any exposed points connected to the AC line. Heatsinks are usually grounded, especially in audio amplifier output stages, but those in switching power supplies may not be. The heatsink on a switcher's chopper transistor can be at hundreds of volts, and its ground reference is the AC power line's neutral or ground, making it quite hazardous. *Never* touch it if the AC cord is plugged in. Even when the supply is unplugged, the heatsink may carry a full charge from the electrolytic capacitor. Unplug the product, and measure from the heatsink to the negative side of the big capacitor to see if any voltage is present. Remember, the AC side of a switcher is not connected to circuit ground (the ground at the output side, where the rest of the product, including its chassis, connects), so measuring from the heatsink to circuit ground will show 0 volts regardless of what's actually there!

The cases of metal power transistors in output sections can carry significant voltage as well, even if the heatsinks on which they're mounted are grounded, because there is usually a thin, hard-to-see insulator between the case and the heatsink. Avoid touching those transistors without first measuring from the case to circuit ground. Even 50 volts can do you harm if it's applied across your hands, especially if they are wet or sweaty. If you've ever touched your tongue to the terminals of a 9-volt battery, you know how little it takes.

The output connections of fluorescent backlight inverters can be at 1,000 volts or more. Keep away from them when the device is powered on. VCRs, DVD players, home theater receivers and some other products feature small fluorescent displays. Lighting those up requires a few hundred volts, so beware of their connections. Think safety at all times!

Chapter **11**

A-Hunting We Will Go: Signal Tracing and Diagnosis

Now that you have the unit open and ready for diagnosis, it's time to apply the ideas we've been examining and put that oscilloscope to good use. Locating the trouble is the heart of the matter and much of the battle. The general approach is to reduce your variables to eliminate as much circuitry as possible, concentrate on what seems a likely problem area, take some measurements, apply a little logic and gradually narrow your focus until you reach the bad component.

Where to begin? That depends on what symptoms are being displayed. In order from least functionality to most, here are some good ways to pick a starting point.

Dead

As we discussed a while back, dead means nothing at all happens when you try to turn the unit on. If this is the case, head straight for the power supply. Check the fuse first. If it's blown, assume something shorted and blew it. The short could be nearby, in the rectifiers, the chopper transistor or its support components, or it could be somewhere in the circuitry being powered by the supply, far from where you're looking.

If the supply feeds the circuitry through a cable, disconnect it. Replace the fuse, and try applying power. Does the fuse blow again? If so, the problem is in the supply. If not, the fault could still be there, but more likely a short in the circuit being powered is drawing too much current and popping the fuse.

Many power supplies have small, low-current sub-supplies for standby operation so they can keep enough circuitry alive to respond to remote control commands or soft switches. The main supply turns on only when commanded to do so by the product's microprocessor. With the cable disconnected between the supply and the rest of the circuitry, the micro can't command the supply's main section to start up. If the short is in a part of the supply not running while in standby, the supply can appear to be okay and not blow the fuse, confusing the matter of where the short lies.

If the fuse is not blown but the unit still does nothing, take a good look at the electrolytic capacitors in the supply and also in other areas of the unit. See any with even slight bulges on top or leakage around them? If so, forget about continuing your exploration until you've replaced them. By the time a cap bulges, it's pretty far gone, with perhaps 10 percent of its original capacitance left. Its equivalent series resistance (ESR) will be way up as well. Very likely, replacement of the bulging parts will restore operation of the unit.

Hard brown stuff around a capacitor that looks like glue is, in fact, glue! Manufacturers put it there to keep larger caps from breaking loose during shipment. There's no need to replace these parts unless you're sure what you're seeing is leakage.

Even if you see no bulges or obvious leaking, get out your ESR meter any time a power supply isn't working. Make sure power is removed and the caps are discharged. Be extra careful when discharging that high-voltage cap on the primary side of a switching supply! It's charged up directly from the AC line, through the rectifiers. Even if the chopper or the rest of the supply is dead, the cap can still have hundreds of volts on it. No matter how you discharge it, expect a spark. But don't do it with a direct short, as the current dump will be tremendous. Use a resistor of around 100 ohms and keep it connected for 30 seconds or so. Then, check the charge with your DMM to see if any is left, repeating the process if necessary.

One nice thing about ESR meters is that they can be used for in-circuit tests, so it's quick and easy to check every electrolytic in the supply, at least for a preliminary test. As we discussed in Chapter 6, keep an eye out for paralleled capacitors that can make a bad one look good. Also, don't miss the caps right at the supply's DC output lines; quite often those are the bad ones, and small electrolytics you might easily overlook can be the culprits. As a matter of fact, I've found high ESR in far more of the smaller caps than the big ones.

If all the electrolytics are good, check the supply's output voltages with your DMM. Find circuit ground on the output side (*never* on the AC input side of a switcher!), and hook the black lead to it. In a unipolar design, the negative output lead is almost always ground. If the supply is bipolar, there will be positive, negative and ground. Even if the supply provides several output voltages and both polarities, one ground serves them all, although multiple wires may be connected to it.

In a product like a disc player or an audio receiver, the metal chassis should suffice as long as you can find a spot that's not painted over. In a pinch, you can usually use the outer rings of RCA jacks in audio/video gear. Choose an input jack, not an output jack, so you don't risk shorting an output if your alligator clip makes contact with the jack's inner conductor. All the jacks will have the same ground anyway.

Many supplies have markings for the voltages on the board, right next to where the output cable plugs in. If so, see if the voltages are there and are fairly close to their rated values. Don't worry if a line marked 5 volts reads 5.1. If it reads 4 or 6, then something's out of whack. When the voltage is too high, the problem will be in the supply's regulation.

When it's too low, it could still be a regulator issue or a failing capacitor, but a short elsewhere in the circuitry might be pulling it down. If the voltages are okay, the supply is probably fine. If they read zero, it might still be fine and just isn't being turned on, as described, but it's quite possible it isn't working. If the supply is turned on and off by the unit's microprocessor, there has to be some voltage from the standby supply to run the micro or it couldn't send a signal to start the main supply. So, one of those lines should have the standby voltage on it.

A product running off an external AC adapter might not blow its fuse even when seriously shorted. Most modern AC adapters are switching supplies. A well-designed one will go into self-protect mode, sensing the excessive current draw and shutting down. Usually, it'll restart every second or so, pumping some current into the device and then stopping again because the load is outside the normal range, never staying on long enough to melt the wire inside the fuse. Even the primary-side fuse in the adapter may survive, for the same reason. I once fixed a laptop computer's AC adapter that had a shorted output cable but never blew its fuse. The adapter's self-protect mode saved the fuse and the rest of the supply as well. The good fuse confused my diagnosis attempts until I considered the self-protect mode, checked the cable and found the short. After cutting off the bad section of cable and resoldering the DC output plug to the remaining good length, I plugged in the supply, and it worked fine. Internal power supplies in AC-operated devices also may survive shorts without blowing their fuses, but they usually aren't as well protected as external adapters, and the fuse blows. After all, that's why it's there!

If you have a working supply but no operation, head for the product's microprocessor and check for an oscillating clock crystal or resonator. If you find no voltage at all, there could be a little subregulator (a voltage regulator fed from a regulated source) on the board to power the micro, and it might be bad. If you see voltage there (typically 5 volts, but possibly less and very occasionally more) but no oscillation, the crystal may be dead. Without a clock to drive it, the micro will sit there like a rock. If you do see oscillation, check that its peak-to-peak (p-p) value is fairly close to the total power supply voltage running the micro. If it's a 5-volt micro and the oscillation is 1 volt p-p, the micro won't get clocked. This isn't likely when the crystal is connected directly to the micro, but it can occur when the clock oscillator is separate. If you have power and a running microprocessor, you should see some life someplace.

Not very long ago, lots of products included small backup batteries on their boards (Figure 11-1). They're less common today, but some laptops still use them. These batteries keep the time-and-date clock running and preserve user preferences. Loss of backup battery power causes resetting of data to the default states but usually doesn't prevent the product from working. In some cases, though, a bad battery can indeed stop the unit from turning on. I've seen laptop computers that wouldn't start up unless the bad backup battery was disconnected.

The batteries may be *primary* (nonrechargeable) lithium coin cells or *secondary* (rechargeable) types. Often, they're soldered to the board. Primary types can be replaced with a standard lithium coin cell and a holder as long as the arrangement will fit. You can

FIGURE 11-1 Soldered rechargeable backup battery in a digital camera.

even use bigger or smaller cells, since they're all 3 volts anyway; smaller cells just won't last as long. Secondary cells need to be replaced with the same type as the original, and those are not easy to find. Most likely, you'll have to try to order one from the manufacturer. Don't replace a rechargeable cell with a primary type, because the applied charging voltage will cause the nonrechargeable cell to burst.

If you're suspicious of a bad backup battery, measure its voltage. Lithium coin cells are 3 volts nominally, and they spend most of their service life pretty close to that voltage. A new one measures about 3.2 volts, and a weak one is around 2.9 volts. Anything lower than that can result in data loss or unreliable operation. Should the voltage be low, disconnect the battery and see if the product comes to life. Sometimes, the data loss from removing the battery is exactly what you need; it clears the memory of corrupted data, and the device resets itself and starts up with factory defaults.

There are some products, especially laptop computers, that use NiMH (nickel–metal hydride) rechargeable coin cells instead of lithiums. The giveaway is that the cells are in a series stack. These cells have a nominal voltage of 1.2 volts each. Most stacks have three cells, for a final voltage of 3.6 volts.

A rechargeable lithium cell will have a nominal voltage of 3.6 or 3.7 volts. At full charge, expect 4.2 volts, and a minimum of 3 volts when discharged. Anything much under that indicates a bad cell. As mentioned earlier, be sure to replace it with another rechargeable cell.

Comatose or Crazy

This situation is trickier. When the unit turns on but is completely bonkers, with random segments on the LCD and improper or no response to control buttons, this usually indicates one of three things: Power supply voltages are way off (probably too low), there's lots of noise on the power supply lines, or the digital control system is seriously whacked out. Check the supply voltages first. If they look close to what they should be, scope the noise. Using your scope's AC coupling, look at the supply's DC output lines. There shouldn't be more than 100 millivolts (mV) or so of junk on them. If you see much more than that, you can expect to find bad electrolytic caps in the supply.

The most likely culprits are the caps right on the output lines. A good electrolytic will smooth out that noise, so its presence tells you the cap is not doing its job. Even if the part doesn't bulge or exhibit obvious leakage, check it with your ESR meter. If you're at all suspicious of the reading, try replacing the cap or temporarily putting another cap of the same value across it. Be sure to get the polarity right when you do this! And, of course, shut everything down first and verify that the existing cap is discharged. Don't worry about the two caps adding up to more than the correct value; extra capacitance on a power supply line will only provide better filtering. If the noise drops dramatically, change the cap, regardless of whether proper operation was restored when you jumped it with the good one. It might still be bad, but there could be others that will have to be changed before the unit will work again. Electrolytic capacitors die from age and hours of use, so multiple bad ones are not only common but expected after you find one. If the noise drops only a small amount, then the original cap is probably okay, and you're just seeing the added capacitance smoothing things over a little bit. The trouble is elsewhere.

When you're sure the power supply is working properly, go to the microprocessor. Check it for clocking, just as with a dead unit. If the clock looks normal, it's possible that the reset circuit, which applies a pulse to the micro's reset pin when power is first applied, isn't working, so the micro isn't starting up from the beginning of its program. Many reset circuits are nothing more than an electrolytic cap between the positive rail and the reset pin, with a resistor going from there to ground. When the cap is discharged and the unit is turned on, the change in the cap's charge state momentarily lets a spike through to the reset pin until the cap charges up and blocks the rail's voltage. If that cap has dried out or leaked, the reset pin won't get tripped, and the micro will start up at some random place in its firmware, resulting in digital dementia. If you can find the reset pin, disconnect power, scope the line going to the pin and then reconnect power. You should see a pulse. If you don't, try turning on the unit. Still no pulse? The reset system isn't working. Try replacing that capacitor. Or, if there's a transistor, diode or other circuit generating the reset pulse, work backward from its output to its input, scoping your way until you either find a pulse or locate its missing source.

Alive and Awake but Not Quite Kicking

This is where your sleuthing skills really get a workout. The unit powers up and responds properly, but some function doesn't work. Perhaps a backlight is out, or a disc player has trouble reading discs, or a TV plays only in black and white. Maybe a projector turns on but the lamp won't strike or a color is missing. Figuring out these kinds of failures can take much more work than it does to troubleshoot the dead and semiconscious types.

After checking for the usual power supply issues, take an especially careful look on the board for bulging or leaky electrolytics. Change any that don't look normal. Make good use of your ESR meter. It also helps to scope the caps while the unit runs. As a rule, any electrolytic with one end tied to ground should have very little besides DC on the other end. Especially if you see high-frequency elements to the noise—it's fast or has spiky edges—jump that cap with another one, and look at the noise content again.

If these simple checks don't turn up anything, it's time to go snooping. Is the problem at the input side, the signal-processing midsection or the output stages? With audio gear, listen for a slight hiss from the speaker. If it's absent, the problem is likely to be in the output stages, because they'll produce a little noise if they're working, even with no signal input. At least this is true with traditional analog amplifiers. You can't count on it with the digital ones. The speaker itself could be bad too; check with headphones or scope the output lines going to the speaker. With older video equipment, there might be some noise on the screen, indicating that the stages driving it are working. With digital video, the screen will probably just be blank or have random blocks on it.

Items that move, like laser heads, print heads and swing-out LCD viewfinders on camcorders, have plenty of problems with broken conductors in their ribbon cables. If a moving part misbehaves, look at its ribbon with a magnifier. Even if you can't see a break, disconnect the ribbon and check the integrity of each line on it using your DMM. The very thin, flexible ribbons with black printed conductors rarely break, but the slightly thicker green ones with copper conductors are quite prone to fractured lines. Even if the conductor side looks black, check the fingers on the ends, where they make contact with the connector's pins. If they're copper colored or look like they're coated with solder, check the ribbon carefully.

In older gear, dirty trimpots (small variable resistors used to set voltages and signal levels) are prime troublemakers. They depend on the mechanical connection between a small spot on the rotating contact and the resistive substrate. In time, oxidation works its ugly magic, and the connection opens, resulting in no signal flow. Depending on the trimpot's function, that can mean no audio, a servo unlocking or a host of other ills. The fix is easy: Spray the trimpot with contact cleaner. To get it to work requires moving the trimpot's wiper back and forth, which means loss of where it was set. Marking the trimpot with a line can help you set it back where it was, but in critical circuits like voltage regulators, this might not be close enough. It's far better to reset the wiper by observing the circuit function with your scope or DMM and adjusting the pot to match the original performance. After all, once you clean the trimpot, the optimal wiper position may not be exactly where it started anyway.

A lot of newer products use flash RAM for data backup instead of a battery. Flash is basically the same stuff that's in a thumb drive, but it's a small chip soldered to the board someplace. In most applications, flash RAM lasts for the life of the device. It does wear out from repeated writes (but not reads), though, and in some products this causes it to fail while the rest of the device has plenty of life left. The tipoff is that the unit functions but doesn't retain user data after being turned off. In a TV, that includes channels, brightness, volume settings and so on. A DVR might lose all your program schedules. If the device starts up with factory defaults or old settings every time you turn it on, regardless of what you've saved, bad flash RAM is the likely cause.

Sometimes Yes, Sometimes No

Want to give a tech nightmares? Sneak up behind the poor sap and whisper the word "intermittent." Watch for neck shivers and twitching muscles. Nothing is more difficult to find than a problem that comes and goes. Naturally, it goes when you look for it and comes back after you're done.

Thermal intermittents represent the easier-to-cure members of the genre. If a product works when cold but quits after warm-up, or the other way around, at least you can control those conditions while you hunt for the trouble. Your most powerful weapon here is a can of component cooler spray. After the operating status changes state, grab that can and get ready to spray. You don't want to spray the entire unit part by part, so concentrate on the kinds of components most likely to cause thermal problems: those that get warm. Voltage regulators, power transistors, graphics chips and CPUs all generate lots of heat and should show sudden change when you blast them with the spray, if they're the troublemakers. Spray small parts for about a second. Big ones may require as long as 5 seconds. Avoid hitting your skin, as the spray can cause frostbite. And, of course, keep your eyes away! I keep my face at least 12 inches from any part I'm spraying.

Now and then, thermal intermittents occur in small-signal parts too. If a transistor is leaky, it may get warmer than would a properly functioning part, driving itself into thermal weirdness. I remember one radio transceiver (transmitter/receiver) that would peg its signal strength meter to the far right when turned on, and no signals could be heard. As it warmed up, the meter would slowly drop to normal, and signals would gradually rise until they came in loud and clear. Knowing that the meters in radios are driven by the automatic gain control (AGC) circuit, I went hunting in that area. Spray can in hand, I finally found it: a small, garden-variety NPN transistor, leaky as could be. Oddly, it was shorting when cold and started working properly as the current through the short warmed it. Twenty-five cents and three solder joints later, the receiver was back to normal.

Electrolytics can be thermal as well. Of course, a cap shouldn't get hot in the first place. Some get a little warm from their normal internal resistance, especially in switching supplies, but a really hot cap means there's a DC current path through it, so it's acting like an actual resistor. In other words, it's leaking.

If you blast any part and the unit starts or stops working, that's a pretty good sign you've found the bad component. As mentioned in Chapter 6, when parts are crammed together, you may think you're hitting the right one, but a little bit of the spray is splattering on another component that's actually the culprit. You might have to spray a few times from different angles to be sure which part you're really affecting, letting the suspect and the components around it warm up again between sprays.

Solder joints can be thermal too. Sometimes, when you spray a component and it starts (or stops) working, the real problem is at its joints. Before changing the part, always check the soldering, and touch it up if you're not sure. Test again before replacing the component. Spraying the joints themselves to find a bad one doesn't work, though. Assessing solder joints is best done visually.

Mechanically influenced intermittents are the hardest problems of all to find. When a machine exhibits symptoms by being tapped on, turned or tilted, there goes your night. And the next, and the next, probably.

Vibration- or position-sensitive intermittents are caused by bad connections. They could be cold solder joints, circuit board cracks, dirty connectors, bad layer interconnects or, rarely, fractures inside components. Tap around, see what trips the symptoms, fix it, done. Seems simple enough, right? How hard could this be? Plenty. These kinds of intermittents tend to be very sensitive, causing malfunction no matter where you tap or flex. The basic search technique is to press and tap ever more gently as you home in on the problem area, hoping to localize the effect until you get down to one spot. Alas, even when you barely touch the board, the part that's flexing or vibrating may be far from your point of contact.

Circuit board cracks are rare these days, except when a product has been dropped. Most cracked boards stop working completely, but now and then a cracked trace will have its edges touching just enough to cause a vibration-sensitive intermittent. Far more common are bad layer interconnects, or *vias*, especially the conductive-glue variety. Even copper-plated vias can cause intermittents, but not very often. Conductive glue may look fine but not be making a solid connection with the upper or lower foil traces.

If you suspect a bad via or a cracked trace, jumping with wire, even temporarily, will settle the question. If an interconnect isn't solid with an inner layer of the board, it can be tough to figure out where the jumper should go unless you have a schematic. Because the connection isn't totally lost, though, you can use your DMM to trace to other components. Keep an eye on the actual resistance to avoid reading through other parts and thinking they're connected when they're not. Expect to see some resistance. After all, that's the problem, right? If you see what looks like a connection, tap on the board and see if the reading changes. Remember that a DMM doesn't respond very fast. The needle of an analog VOM will bounce, which is more useful in this case.

Many products use the chassis or case as circuit ground, with grounding pads on the board making contact when it's screwed down. As the device ages, loosened screws and oxidation degrade these critical connections, leading to intermittent behavior. In a unit more than 5 years old, check those pads even if the screws are tight. If the pads are dirty or oxidized, clean them up and see if that cures the symptoms. In a newer item, all should be well unless the screws are loose or the unit has lived in an especially corrosive environment like a boat or a beach house.

Probably the most frustrating intermittent of them all is when the unit works just fine until you close up the case. Then it either won't work at all or it becomes motion-sensitive. You open the case back up again, and the little monster works perfectly. Arggh!

To get to the root of one of these seemingly intractable dilemmas, consider what's happening when the case is closed. Look at the inside of the case and visualize where it will press on the board, on wiring and on ribbon cables. Some cases hold down the corners of the circuit board, flexing it when the screws are tightened. Experiment while the case is open, trying to re-create the problematic conditions. Most of these can be solved, but I've run into a couple I couldn't straighten out.

If the board isn't too sensitive to probe without altering the symptoms, use normal signal tracing techniques to locate the intermittent. If everything you touch disturbs the intermittent, it's very difficult to make sense of what you see on the scope.

To and Fro

Some techs like to work backward most of the time, starting at the output stages and hunting back toward the input area, looking for where the signals are disappearing. Others prefer to start at the input and see where things get lost. What's the best method?

Either way may be appropriate. The output-to-input approach is especially useful when there is an output signal but it's not normal. Very often, such problems arise in the output stages and their drivers, so why start way back at the input and scope through stage after stage to get there? If you see a normal signal feeding the output stage, you've pretty much nailed it without a lot of hunting.

Digital devices with time displays, such as disc players, offer a powerful clue to help you decide your direction of attack: Is the time counter moving? If so, the device is receiving data, be it off a disc, a memory card or internal memory, and at least the heart of the digital section is working. So, start at the output and work your way back. If the counter is not progressing, head for the input and digital control areas and find out why not.

With items like radio-frequency (RF) receivers, you may have normal audio hiss or random video blocks but no reception. Because the path through a receiver is fairly complex, with oscillators, tuned amplifiers and demodulation stages in superhets and DSP systems in software-defined radios, it makes more sense to start at the input and work forward. At some point, you'll discover a missing oscillator or a dead intermediate-frequency (IF) or demodulator stage, with corresponding loss of signal.

If you are going to work backward, be certain you have a valid input before you start looking for it way down the line! Just because you plug an audio source into a stereo receiver, for example, doesn't mean it's getting to the amplifier board. There could be an issue with the input switching, or your connecting cable might be bad. Check the input signal *at the board* to be sure. That old tape-based camcorder won't play in color? Are you sure the tape you're trying to play *has* color? If the recording got made on the same machine, it could be that the fault is in record, not play. Use a known good tape, or verify the existing one by playing it on another machine.

In many cases, a hybrid approach is the most effective. Start at the output and work back a few stages. If you can't find the signal, go to the input and work forward. In complex systems with multiple inputs, such as servos, check the inputs to be sure they're all there, since one missing signal will turn the whole thing into a mess.

All the World's a Stage

Always remember the all-important organizational concept of the circuit stage. You're not going to scope every darned component in the device. Instead, you'll focus on a particular area in the unit and look at it stage by stage.

Test points are very handy. With a schematic, you can look up TP204 and find out what it's supposed to show. Even without a diagram, you can often guess the signal being tapped from the waveform when you scope it. Sometimes you really get lucky, and the test points are labeled for function, in addition to their call numbers. You might see "Reset" or "Trk gain" (tracking gain). Checking those points and interpreting their signals can save a heck of a lot of work. If a test point at the end of a chain of stages shows the expected behavior, there's no need to scope each stage—they all have to be working for it to be there.

Test points for digital signals like "Reset" may show a line above the word. This means "*Not* reset," which is tech-ese for "the signal goes low to initiate the reset, not high." When there's no line, the signal should go high, but don't count on it. Some manufacturers don't bother adding the line. If you see one, though, the signal definitely goes low.

Only when you find a nonfunctional stage is it worth trying to discern what part in the stage is preventing it from working. In the vast majority of cases, that part will be either an electrolytic capacitor or an active element—a transistor or a chip. With a few exceptions, like crystals and high-voltage transformers, other components that may have gone bad are probably victims of having had too much current pulled through them, and are not the perps themselves.

Diodes, rectifiers and zeners represent a special case. Though they're not active in the sense of having gain, they are semiconductors susceptible to the same kinds of failures found in transistors. Most techs think of them as active elements and check them before looking at more reliable components like resistors, coils and ceramic capacitors.

Zeners, which dissipate excess power as heat, are particularly prone to being open. Replacing a marked zener is no big deal because you can look up the value by its part number. Unmarked zeners present a much bigger problem if you don't have the schematic. What was the zener voltage (the voltage at which the diode conducts backward) supposed to be? You'll never know for sure, but you can make an educated guess.

First, the zener voltage will be less than what you're measuring at the blown zener, because the whole point of a zener is to reduce the voltage to the diode's breakdown rating; the part does nothing when the voltage is below that value. Theoretically, the zener voltage could be as little as 1 volt less than the applied voltage, but expect it to be at least a few volts less. Look for electrolytic caps in whatever circuit the zener regulates. The zener voltage will be less than the caps' voltage ratings. Again, it should be at least a couple of volts less, since few designers are foolish enough to run 'lytics all the way up at their ratings.

Though there's a wide range of zener values, many circuits operate on 5, 6, 9 or 12 volts, and it's reasonable to expect most of the zeners you find to be at or near one of those values. Microprocessor circuits commonly use 5 volts. In audio power amplifiers, zeners are used to establish bias on the transistors (some current to keep them turned on), and calculating the correct value isn't simple. Luckily, you should have another channel in which to measure the voltage across its good zener.

Check, Please

When you find a stage that isn't functioning, don't be too quick to indict it. First, be sure it's receiving the power and signals it needs to do its job. You really can't blame the poor transistor if it's not getting voltage, if its bias is way off or some other stage isn't turning it on or providing proper input.

Unfortunately, cause and effect aren't always so clear. When a signal or a voltage appears to be missing, it could be that the stage is receiving it but a bad part is shorting it to ground. Or, a coupling component could be open, preventing the signal from getting to the active element. How to tell?

If there's a resistor between the source of the signal and the stage you're examining, check on the other side of it. The current limiting of the resistor isolates the far side from anything happening at the suspicious stage. The bigger the resistance value, the more isolation you can expect. A few ohms won't give you much isolation—signals will be about the same on both sides—but a kilohm (1,000 ohms) or more sure will. You should be able to see something of the original signal on the other side, even if it's reduced in amplitude. If not, then the stage on that side isn't sending it, and you need to move your hunt to that part of the circuit.

A coupling capacitor can provide isolation for AC signals, but how much depends not only on the size of the capacitor but also on the frequencies involved. The higher the frequency, the smaller a capacitor it takes to pass it, so the less isolation you get for a given capacitance value.

The quickest, easiest way to probe through a series of stages is by looking at their outputs. If the output is correct, the input also has to be okay, right? The reverse is not true, though; the input may be fine while the output is not. Also, input signal levels to some stages are so small that even the high impedance of a scope probe can reduce or otherwise alter them, leading you to think the signal isn't correct when it really is. If a stage's output looks wrong, then you can work back toward the input of the stage to find out why.

There is a peculiar, uncommon circuit configuration that can confuse the heck out of you. While most circuits feed the output of one stage to the input of the next, things that synchronize each other, such as stages of servos and phase-locked loops (PLLs), sometimes accomplish synchronization by feeding the synchronizing signal to the *output* of the stage requiring the sync. This forces it to operate in lockstep, but it also mixes the sync signal with the signal being synced! When you scope the output of the stage providing the sync signal, it looks like it's there, but you're really seeing the next stage's output being fed back through the coupling element, typically a capacitor. You think the earlier stage is working even when it's not, and you hunt elsewhere.

I ran into this very problem while troubleshooting an old video camera that was driven by external sync signals. The incoming sync was fed through an amplifier chip to a square wave oscillator. The oscillator wasn't syncing to the incoming signal, causing all kinds of interference in the resulting video signal. When I scoped the output of the sync amplifier chip, though, the darned sync signal appeared to be present. I was baffled as to why the oscillator wouldn't sync to it. Then I noticed that it was fed through a capacitor to the oscillator's output. I disconnected one end of the coupling cap and checked again. Aha! There was no signal on the output of the amplifier chip after all. That was the trouble. I replaced the chip, and the camera synced up properly. This kind of oddball circuit configuration is rare in modern gear, but keep it in mind when you see an output signal from a suspected faulty stage but its presence makes no sense, based on the malfunction.

When tracing along the signal path in either direction, also keep in mind that bad connections are a prime cause of signal loss. It might seem obvious that a connection is either there or it's not, but that just isn't the case. Especially where weak signals are concerned, oxidation and cold solder joints can act in a nonlinear fashion, passing no signal until the voltage rises above a certain level. Above that level, the electron motion pushes its way through the oxidation barrier and conduction occurs. So, you check a suspicious connection with your ohmmeter, and it looks good, but it won't pass a millivolt-level signal. I've seen this happen many times with perfectly normal-looking solder joints where they connected to tape heads, microphones, phono cartridges and audio input switches on stereo receivers, all of which generate or handle seriously small signals. It also happens where the heads connect to the board in hard drives (see Chapter 15). Measuring resistance is certainly useful, but your scope is your best bet to see if a signal is getting through. The rub is that most scopes can't see a signal of a few millivolts very clearly, if at all. Sometimes you have to scope a little farther down the signal chain, after more amplification has occurred, and infer whether the signal might be getting lost in the earlier stages.

Once you're sure the correct conditions have been met and input signals to the stage are present while the output is missing or wrong, it is reasonable to conclude that you have a bad component, and it's time to start checking them. Unless you see a leaking cap, head for the active element first. Even if you find a burned resistor, you can bet the active element pulled too much current through it. Resistors don't char themselves during normal operation.

Sometimes parts can be tested while they're in the circuit, but usually the effects of other components will confuse the measurement, and you'll need to pull the suspect part before you check it. Surface-mount components will have to be desoldered and removed. It's a little easier with through-hole parts, and those are often the trouble because they're the larger items that handle significant current and get hot.

Two-legged parts need only one lead disconnected. If one lead goes to ground, remove the other one and leave the ground side connected. It's easier because ground lands are typically the biggest and hardest to desolder. Also, you can leave one test lead connected to circuit ground and will have to connect only one lead to the component—except for ESR tests. For those, you'll need to connect the meter directly across the capacitor being tested, even if the part's ground lead remains connected.

Wick the solder out of the hole and bend the part up on its other lead. For a three-leaded component like a transistor, pull two leads, and be sure one of them is the base or gate lead so that other parts can't influence the sensitive terminal with added capacitance or noise pickup via the rest of the circuitry.

With some parts, especially electrolytics, it might not be possible to bend the component on one lead if the leads are too short. If you can wick the hole thoroughly enough that the stub of the lead moves freely within it, testing is possible without pulling the component. With your test probe, push the stub away from the walls of the hole, watching the test results as you do so. The readings should make it apparent when contact with the rest of the circuit has been broken.

You'll be amazed at how many times you're absolutely *sure* you've found the problem, you pull the part, and it tests out fine. It can be frustrating, but that's just the nature of the repair game. Eventually, you'll nail it. It feels really great when you pull the fifth part you're certain has to be the culprit, and it actually is!

When All Else Fails: Desperate Measures

No matter how good a sleuth you are, sometimes nothing works. You pull part after suspicious part, and they're all good. You've been at it for hours, you're out of ideas, and desperation sets in. Welcome to the technicians' club! It happens to all of us once in a while. Here are some desperation techniques to try. They may seem crazy, but they're better than giving up and tossing the unit on the ol' junque pile. Now and then, they actually save the day.

Shotgunning

This is as old as electronics itself. When you have an intermittent connection you just can't find, solder them all! Back in the days when circuits had a few dozen parts, shotgunning was easy and quick. Today, with hundreds of joints on every board, the prospect of redoing so many can be daunting, and it's not feasible with laptops and other extremely dense, complex products.

Start with an area you think is causing the trouble, and hit every joint in it. If it doesn't work, keep on going. Don't be surprised if you wind up redoing every joint in the entire product, and it still won't work. Frankly, shotgunning is rarely successful; the real problem always seems to be something obscure that gets missed. Now and then, though, luck prevails and the symptoms disappear. Don't get too excited—you might have only wiggled the actual bad connection, and the problem will return . . . typically right after you tighten the final case screw. Once in a great while, I've seen shotgunning result in a real repair.

Current Blasting

This one has a little more basis in sanity. It's useful only when you have a dead short across the power supply rails somewhere on the board, but you can't find it. Especially on today's digital boards, there are lots of little bypass capacitors from Vcc (the positive supply rail) to ground. Now and then, one of them shorts. You see the short no matter where on the rail you probe with your ohmmeter, so it's impossible to deduce which of the 50 little caps might have become a 0-ohm nightmare, and pulling them all to test them presents too much risk to the board. Plus, it'd take hours, and you can't be sure the short isn't in some other component anyway.

Because ESR meters can read fractions of an ohm, you might be able to find the shorted part by following the traces and looking for the lowest resistance. This can be next to impossible with dense boards, though, because you can't get your probes where you need them, and even locating ground points on different parts of the board might prove frustrating.

There's a faster, easier way. You probably have a high-current power supply, either on your bench or perhaps in a discarded desktop computer. To perform current blasting, you need a supply of the same voltage as the product's supply. Many of these direct-short situations involve 5-volt digital boards, so a computer supply is a good choice. You need a *lot* of current—perhaps 20 amps or so. Your little 2-amp variable bench supply won't do it, but a PC supply has the required oomph.

If at all possible, disconnect the product's own power supply so you won't be feeding voltage into its output. That's usually okay, but some voltage regulators can be damaged when their output voltage exceeds their input voltage (which will be zero in this case), so it's best to avoid having them connected during this maneuver. If there's a fuse between the supply's output and the rest of the board, removing it should do the trick. Otherwise, pull the connector or unsolder the positive wire.

With the hefty supply turned off, connect its +5V and ground wires to the board's supply rail and ground traces. Naturally, + goes to +. Make sure you're past any fuses, because this procedure will blow them. This is one time you don't want protection. Turn on the supply and wait. After perhaps 30 seconds, the shorted part will start smoking and burning because pretty much all the supply current is going through it. As soon as you see the smoke, kill the supply; you've found the bad component. If the part is something nonessential, like a bypass cap, the product may start working as the current cooks the cap.

I've seen the voltage rise high enough to start up a device even before the short clears. It's amazing what enough current will do.

This procedure will work with other voltages, too, of course, as long as you have a supply that can source a lot of current.

Some caveats: It's possible the board's traces could melt before the shorted part gets hot enough to smoke. Traces that feed power supply current to the various sections of the circuitry tend to be wide, reducing the likelihood of such a mishap, but they're not made to handle 20 amps! I haven't seen one melt, but it could. On a dense or multilayer board, a melted trace might prove disastrous. Also, the big supply must not have self-protection, or it'll refuse to dump lots of current into a short. I've had good luck with desktop PC supplies; they are very sturdy and don't mind the overcurrent, at least not for the period of time required. They also don't seem to have self-protection circuitry.

If nothing gets damaged, current blasting pretty much always works when the shorted part is a capacitor, and it's usually effective with transistors and diodes. Sometimes, the short is in some microscopic element inside a chip, and the high current instantly blows the short open. Nothing smokes, you don't know where the short was, the device still doesn't work, and you're no better off than when you started. Still, current blasting is a useful technique, and it beats just giving up.

I fixed a really nice little hard-drive MP3 player this way. It had a dead short across the power supply input jack, and the board was too small and dense for me to consider trying to pull parts and test them. It was a 5-volt unit, so I hooked up a computer supply and hit the switch. In 10 seconds, a surface-mount electrolytic cap right next to the power jack lit up like a tiny light bulb. I changed it, and the unit came back to life. I saved a rather expensive digital piano with current blasting too. The shorted component was a tiny 0.1-µF bypass cap near the microprocessor. I'd never have found it any other way.

LAP Method

This is the craziest last-ditch method of them all, but it has worked for me on rare occasions. LAP stands for "least accessible place." Where's the hardest place to reach in the entire product? If every other option has been exhausted, head there and suffer through whatever it takes to examine that difficult area. After a few LAP successes, I used to wonder how this could possibly be real. Was some cosmic force hiding things from me? Was there a ghost in the machine with a bad sense of humor?

The more reasonable explanation is that it was the one place I hadn't been yet! It seems like no matter how hard we try to check everything, there's always some forgotten nook so inaccessible that we don't even notice it or we subconsciously avoid it. And, if it's the last possible place, the trouble just might be there.

Chapter 12

Presto Change-O: Circuit Boards and Replacing Components

Once you've found a component you want to test, or one that's obviously blown, you need to remove it from the board. Back when all components were mounted on leads pushed through holes in single- or double-sided circuit boards, removal was easy. A little solder wick or a pump of the solder sucker, and the holes would clear. After that, all you had to do was pull.

Sometimes the process is still like this, but now there's much more variety of component styles requiring different removal techniques, and multilayer boards have complicated the situation. Component removal ranges from trivial to maddening, and it's easy to destroy the circuit board when a recalcitrant part simply refuses to budge.

Unless both sides of the board are accessible, you'll have to remove it from the unit before you can desolder anything with leads poking through the board. Either way, first make sure all power is disconnected. I always look at the AC plug before beginning to unscrew a board or desolder components, just so I know the plug is definitely lying loose. Even if I remember having pulled it, I take another look.

Removing Through-Hole Parts

Many larger components still use the old wire-through-the-hole mounting technique. To remove power transistors and other through-hole parts, the solder must be sucked out of the hole, or the lead has to be pulled out while the solder is molten. Clearing the hole is preferable because pulling a hot lead usually leaves the hole clogged with solder anyway. For small joints, use solder wick, as described in Chapter 6. Place the end of the wick on the joint you want to desolder, and then press the iron's tip against the wick. Hold it there for about 20 seconds, and the solder should flow up into the wick (Figure 12-1).

FIGURE 12-1 Using solder wick.

This doesn't always work, though. Sometimes the solder won't flow well enough to clear out the hole. The usual reason is insufficient heat, but transferring the heat to the joint is an issue, too, as is thermal absorption by large copper lands. If you can't get a *small* land's hole to clear, try adding some fresh solder, and then wick it out again. Boards manufactured with lead-free solder don't desolder well. Adding leaded solder to a lead-free joint lowers the existing solder's melting temperature, making removal easier. For desoldering a multi-legged beast like a microprocessor or memory chip, try ChipQuik (see Chapter 2).

Solder wick absorbs some heat, so it takes a hotter iron to desolder a lead than it does to solder it. Plus, to remove a lead requires wicking out all of the solder in the hole. With thick or multilayer boards, some of it may be a millimeter or more away from the heat source, making the solder hard to melt.

Desoldering is complicated by the increased thickness of multilayer boards and their extra heat-sinking effect from internal foils contacting the copper coating inside the holes. Applying enough heat to wick the solder out can destroy the board. To remove a stubborn lead from a multilayer board, it's best to heat one side while pulling the lead out on the other, and then clear the hole after the lead is gone. Even then, you may struggle with it and be tempted to reach for the big soldering gun. Those produce too much heat for small boards, and can deform them and break internal connections in multilayer boards, wrecking the device (Figure 12-2).

Large lands used for power supply and ground buses create a heck of a heat-sinking effect. It can be quite frustrating trying to get them hot enough to melt and clear the solder. The big gun might be called for here, but it's still possible to trash the board because there may be other lines running over the big land inside the layers. Heating up the big land to an extreme level can break those other lines or short them to their adjacent layers.

If a part won't come out no matter how hard you try, it's a lot safer to clip the leads and solder in the new part without clearing the holes. Clip the new part's leads close, and solder them to the residual solder in the holes. You should be able to heat a hole enough to make a good joint, even if the solder at the far end of the hole never melts (Figure 12-3). If there's room, you can leave a little of the old part's leads and solder to those.

FIGURE 12-2   How not to do it! Excessive heat destroyed this multilayer board.

FIGURE 12-3   Attaching new parts to a board with stubborn ground lands.

Sometimes, you can't get to the leads to clip them. On most electrolytic caps, the leads are under the parts, unreachable with any tool. The easiest way out is just to chop off the component near its base with a pair of wire cutters. Then you can clip or desolder the leads easily.

Bigger joints with lots of solder can overwhelm solder wick, saturating it before much of the total solder is removed. To clean out an entire large joint might require a foot of wick, which isn't cheap. These are jobs for solder suckers. After applying the sucker a few times, you should be left with only a coating of solder on the joint. A sucker will not remove that, so finish up with wick.

As mentioned in Chapter 6, avoid using a spring-loaded solder sucker on static-sensitive components like CMOS chips and MOSFETs. The rapid release of the plunger can generate static charges capable of damaging those parts.

Removing Surface-Mount Components

Wicking surface-mount parts is easier because all of the solder is touching the wick, and many of the lands are very small and readily heated. Most surface-mount pads will desolder without incident. If the solder on a small land won't flow into the wick, try the same trick I described earlier: Add some fresh solder to the joint before trying to desolder it.

Large lands on power supply and ground buses may still be hard to heat, but a normal iron will take care of most of them. Using a big gun on a surface-mount component is asking to destroy the part and quite possibly the board. Tiny surface-mount resistors and capacitors have sputtered-on solder pads. Too much heat can delaminate them, making reconnection to the parts' bodies impossible to achieve. If you delaminate a pad, you'll have to replace the part.

Larger surface-mount device (SMD) components are more easily desoldered with soldering tweezers. Put the two heated tweezer tips on the pads, and the part should pop right off in a few seconds. It helps to tin the tips with a little tin-lead solder first.

If you have a hot-air rework station with a narrow nozzle attachment, you can desolder SMD parts of any size with it pretty readily. A pair of normal, unheated tweezers helps here. Grab the center of the part and apply the hot air, alternating between pads. When they're both obviously melted, the part should come off. Be careful not to get enough hot air on the surrounding parts to loosen them up as well! Also watch out for pulling too hard on the component before both ends are thoroughly melted because you could tear the copper traces off the board, which is a *nasty* nightmare at the size scale of traces on modern boards.

Most SMD components are glued in place before being machine-soldered at the factory. Very often, desoldering the ends of a part will break the glue and free the part, but not always. If you see a red shellac-like blob around the edges of the component, it's glued on and may not budge after desoldering. To dislodge it, wick both ends, and then heat one (or both, if you have the tools to do so) while pushing on the component's body with a small screwdriver or pulling it up with tweezers. Again, be careful not to tear the copper from the board, should one end of the component still be too well attached. And beware, the tiny part may pop off suddenly and blast away into oblivion. Somewhere in the universe there must be a room full of sad, homeless SMD parts that flew off circuit boards, never to be found. Plenty of 'em came from my workbench.

Replacements and Substitutes

Any time you need a new part, you just breeze on down to your local electronics supply store, buy the exact replacement and pop it in. Um, right, sure you do. Ah, if only real life could be like that! Few towns even have parts stores anymore. And, while lots of standardized components are available via mail order, many newer consumer electronics products are stuffed with all kinds of obscure and specialized components nobody but the manufacturer can provide. Luckily, in most cases, you have a few options.

Buying New and Used

Proprietary components have to be procured from the manufacturer (unlikely these days, but worth a try), the component maker who supplied them (possible) or a parts unit. For popular products, check eBay for broken units of the same model being sold for parts.

Standard components are widely available through online mail order, but many parts houses have minimums, so you might have to spend a lot more than the part is worth. The shipping is usually more than the cost of the parts as well. Oh well, you can always stock your components supply with other goodies you might use later. Or, you can save up your parts needs until you have a big enough order. That'll delay your repair work a long time, though.

In the United States, these are the two largest, most popular parts houses:

- DigiKey (digikey.com)
- Mouser Electronics (mouser.com)

Do an online search, and you'll turn up dozens more sources for both prime and surplus components. Be careful about buying semiconductors (transistors and chips) from unknown online sources! The market is flooded with counterfeit Chinese parts that don't live up to their specs. These components may have the logos of major manufacturers on them, but it means nothing. The only way to be sure you're getting a genuine device that will work as intended is to buy it from an established, domestic parts house.

Parts Drawers

If you've stockpiled components, see if what you have is a close enough match. When using parts that have been sitting around for a long time, take some fine sandpaper or an X-Acto blade to the leads to scrape off oxidation that will have built up. Otherwise, soldering to those leads won't be successful. This applies to SMD pads as well. Just don't overdo it on those or you could file them off, ruining the component.

Scrap Boards

There's a reason I've encouraged you to save boards from dead machines. Those from the same manufacturer as the unit you're repairing might use the same component, even if they're a different model. Manufacturers save costs by reusing sections of their designs and techniques in lots of models. If you can't find an exact replacement, you still might locate something close enough to work. Check all your parts machines, even those made by other companies. You're more likely to find a compatible part from the same type of product because the function is similar. So, if you need a part for a camcorder, check boards from those; you probably won't find what you need in a DVD player. If you locate something you can use but the leads are too short, solder on a little wire to extend them.

Choosing Components

Now that you've tracked down what needs to be replaced, it's time to put in the new component. Especially if you're using your stash or pulling parts from old boards, you might not have an exact replacement. Substituting a part with something close but not an exact match requires consideration of how the part is being used, what parameters are critical and what you can get away with in a particular application. The general idea is that a part with better specs can sub for a lesser one, but not the other way around. Even then, there are exceptions. Different component types have varying requirements. Let's look at the most common parts and how to pick replacements and substitutes.

Fuses

A replacement fuse must have the same current rating as the original. Using one with a higher rating risks allowing too much current into whatever short blew the original fuse. This could cause further circuit damage, power supply damage or even a fire. Putting in a fuse with too low a rating will cause it to blow even when the circuitry is working properly, confounding your troubleshooting efforts.

Always replace a slow-blow fuse with the same type. Otherwise, the fuse is likely to blow when the device starts up and pulls momentarily high current. If the fuse is a high-speed type, you can be sure the manufacturer spent the extra money for one because the circuitry is especially sensitive. Replacing a high-speed fuse with a standard-speed version will work, but you may lose your product later on when a very short overcurrent condition damages whatever the high-speed fuse was meant to protect.

Fuses have voltage ratings. While many glass fuses are rated at 250 volts AC (VAC), some are rated for 32 VAC. This will be shown on the end cap, and you should replace it with the same type.

Any fuse used on the AC power line in the United States will be rated for 250 VAC. The lower-voltage fuses are usually used for DC power outputs from power supplies. If all you can find is the current rating, replace the fuse with a standard-speed fuse of the same rating; base your estimate of the voltage on the application (AC power line or low-voltage DC) and you should be fine.

MOVs

Your best bet is to look up the part number and get an exact replacement. If the bad MOV (metal-oxide varistor) is burned to the point that you can't read the number, get a replacement rated for standard AC line voltage and of approximately the same size, and it should do the trick. Generally, the bigger the part, the more joules it can handle, so putting in a larger MOV is okay as long as the operating and varistor voltage ratings are correct.

Safety Capacitors

To replace a safety capacitor, get one with the same capacitance, class and numbers. Never replace a Y-class part with an X-class substitute! And don't even think of subbing in a standard capacitor. Using incorrect parts could cause you to get shocked later on, especially when the original was a Y-class part connected between the power line and ground.

SCRs, TRIACs and Thyristors

The important characteristics of these components are the gate turn-on voltage, the breakdown voltage and the amount of current the part can handle. While you can substitute types by matching these requirements, the easiest path is just to get an exact substitute. SCRs and TRIACs are readily available from mail-order parts houses.

BMS for Lithium Batteries

For a BMS (battery management system) that's part of a battery, just replace the entire battery, being certain that the replacement also has a BMS! Some supposedly exact replacements for BMS-protected batteries omit the thing, which can be dangerous. When the BMS is part of the product, you'll have to troubleshoot it at the component level, like you would with any other circuit. The transistor sending the charge current to the battery is a common failure point. It'll be a MOSFET (see "Transistors" later in this chapter) and probably will look like an eight-pin IC chip.

For more on lithium batteries and other types, check out my other book, *How to Get the Most from Your Home Entertainment Electronics*. It has detailed info on the care and feeding of all kinds of batteries commonly used in consumer electronic devices.

Capacitors

While it's possible to replace one type of capacitor with another, the easiest way to be sure of not adding to your problems is just to get the same type with the same ratings. Mostly, this means matching the capacitance value and voltage rating.

The vast majority of the capacitors you'll replace are electrolytics, of which tantalum capacitors are a subset. The old candy drop–style tantalum caps fail pretty often, too, but they aren't used much anymore, so you may never run across one. SMD tantalum capacitors are still found in modern products, but they're more reliable than the candy drops. The major factors in any electrolytic are its size, capacitance, voltage rating and temperature rating.

The most important consideration after size—it does, after all, have to fit on the board in the allotted space—is voltage rating. Electrolytic capacitors simply won't withstand voltages higher than their ratings, at least not for long; they fail catastrophically by shorting. Don't try putting in an underrated capacitor, not even for a quick test. In general, though, you can improve reliability with a part rated for higher voltage, especially when the original capacitor has swelled. There are several reasons for swelled caps, but running them close to their rated voltage accelerates wear, reducing lifetime. Designers know this and usually build in lots of headroom by specifying a capacitor whose voltage rating is well above what will be applied to it. Don't count on it, though. This ability to withstand higher voltages comes with two costs: the literal cost of the part and its size. Keeping things small and cheap is a prime goal in manufacturing in this age of miniature, disposable electronics.

If there's room for one, put in a cap rated for around 10 volts more than it'll be subjected to, and the part will last a lot longer. As a side benefit, electrolytics with higher voltage ratings also tend to have lower ESR (equivalent series resistance), which is always a good thing.

Some circuits require caps with especially low ESR. They're pretty much the rule in switching power supplies and computer motherboards. Replacing them with standard caps is not wise. The device may work, but it could behave unpredictably due to the less effective filtering of the inferior capacitors. Even if it works fine, it'll probably start showing flakiness as the caps age a little and their ESR rises. All electrolytic capacitors exhibit rising ESR as they age, but if you're starting out with an ESR that's too high in the first place, it'll get out of the acceptable range in a hurry.

The capacitance rating is not as critical as you might suppose. Most 'lytics have rather wide tolerances, in the range of −20 to +80 percent. If the capacitor is being used to couple signals from one stage to another, the capacitance value is more important than it is when the part is a bypass or filtering cap, where one side goes to ground. In audio amplifier stages that use caps of a few microfarads from a transistor's emitter to ground, however, it pays to keep the value close to the original because a higher value might increase low-frequency response, upsetting the audio quality.

For filtering use, if your available replacement's value is no more than 50 percent higher than the original, go ahead and use it. A little extra filtering never hurt anything, and +50 percent is likely within the stated tolerance of the original part anyway. To be sure that the new part isn't at its maximum tolerance value of, say, 80 percent over the stated value, measure it with a capacitance meter. Despite their stated wide tolerances, most brand-new electrolytics I've measured have been within +20 and −10 percent or so of their printed values.

Combining capacitors to get near the needed value is fine in most applications. I don't recommend it, though, for the big storage cap at the input of a switching power supply (near the chopper) or in other high-voltage circuits. Putting caps in parallel adds their values, and putting them in series drops the final value according to the following formula:

$$\frac{1}{\dfrac{1}{C_1} + \dfrac{1}{C_2} + \dfrac{1}{C_3} \cdots}$$

Capacitors combine exactly opposite to how resistors do. For more info, see the section "Resistors" a bit later in this chapter.

When putting polarized capacitors in series, be sure they connect + to − so you wind up with one + and one − at the ends of the string. When you parallel them, connect all the + terminals to each other and all the − terminals to each other. In either case, be sure each capacitor's voltage rating is equal to the entire applied voltage. When in series, the individual caps won't really be subjected to the full voltage during normal operation, but a voltage spike can occur when power is first applied and they haven't all charged up yet, so it's a smart safety move to be certain every one of them can handle it.

Especially in power supply applications, the cap's temperature rating matters. Electrolytics that get charged and discharged very fast, as they do in a switcher, can become a bit warm from the power dissipation of their internal resistance. Standard 'lytics are rated to operate at 85°C, with higher-temperature caps rated as high as 150°C. Manufacturers hate paying for things they don't need, so respect the temperature ratings of the original parts if you want the repair to last; the higher-temp, more expensive component is there for a reason. It's fine, though, to replace an 85°C part with one rated for higher temperatures. For quick testing purposes while troubleshooting, you can disregard the ratings because the trial capacitor won't be in use long enough to fail from overheating.

Tantalum capacitors should always be replaced with the same type. They have lower impedance at high frequencies than standard electrolytics and are used only where that matters. Replacing a tantalum with a garden-variety electrolytic will result in performance degradation or circuit malfunction. The capacitance tolerance of tantalums is much tighter than that of standard electrolytics, so use a part with the same value. An increased voltage rating is fine, however.

Yellow SMD capacitors are tantalum. Black caps usually aren't, but there are exceptions. If the part offers a lot of capacitance in a very small size, it might be tantalum. You can replace a non-tantalum part with a tantalum. It'll just cost a little more.

Replacing other types of capacitors, such as ceramic and plastic film, involves consideration not only of capacitance and voltage rating but also thermal drift, if the cap is part of a critical timing circuit. Luckily, most aren't, but be on the lookout for "NPO" printed on the part or for black paint at the top of a small ceramic type. This stands for "negative-positive zero" and means it doesn't drift in either direction (more or less capacitance) as the ambient temperature varies. To avoid thermal variability, be sure to replace an NPO cap with another NPO. These parts are usually ceramic, though, and a bad one is rare. Sometimes, you'll see some cracking of the coating around the leads, but that doesn't indicate a bad capacitor. It cracks easily during mounting on the board and during desoldering, and the cracks typically don't affect anything.

High-end audio equipment may contain some rather expensive, esoteric coupling capacitors in the signal chain, the path through which the audio signals travel from one stage to the next. They're used because they are perceived to be especially transparent to the audio signals, not causing any sort of coloration or degradation of the sound. To preserve those pristine qualities, be sure to replace those parts with the same types.

If you're restoring antique electronics from the 1940s through the 1960s, you'll run into paper and oil capacitors. No such thing is made anymore, but you can replace them with modern varieties. Work on antiques is outside the scope of this book, but there are online guides to such restoration work that will help you pick out the correct replacements.

Crystals and Resonators

Crystals are specified for frequency and tolerance, or how far off that frequency is allowed to be. Most likely, though, all you'll see printed or stamped on the crystal is its frequency. You can infer the tolerance from the number of digits to the right of the decimal point. The more digits, the more precisely the frequency is specified.

For instance, if the original crystal was 15.00000 MHz and you replace it with one that's 15.000 MHz, it might not be precise enough to work correctly. Manufacturers don't like to spend extra for high-precision crystals. If the original was very precise, there was a reason, and you should replace it with one at least as precise. If your replacement has *more* digits after the decimal, that's fine; it'll just be more likely to be closer to the intended frequency.

Generally, digital clock oscillators aren't required to be ultra-precise, but in applications like TV tuners and radio equipment, getting very close to the desired frequency can be critical. When looking for a replacement, you might see it specified in PPM, or parts per million. Finding the tolerance is easy: Just divide the frequency by a million and then multiply that by the PPM rating. So, a 14-MHz crystal rated at ±5 PPM could be off in either direction from its marked frequency by 14 × 5 Hz, which is 70 Hz.

Crystal Clock Oscillators

Get one of the same size and style, with the same number of pins, and it'll almost certainly have the same power supply voltage rating. Most of these things run on 5 or 3.3 volts, and some will run on either. Be sure to get one with the same frequency tolerance or number of digits to the right of the decimal point, and it should work fine.

If there's room for it, you can use a bigger oscillator than the original, if that's all you happen to have. I did this when replacing the jumpy one described in Chapter 7. Use short lengths of wire to get the pins to the appropriate points on the board, keep the output and ground leads as short as possible, and you should be good to go.

Diodes, Rectifiers and LEDs

Small-signal diodes, those tiny glass types, don't handle much current. Replacing one with anything other than the exact part number requires understanding what the diode is supposed to do, but in many cases, subbing in a suitable part is easy. If it's just directing DC voltage to various circuit stages to turn them on and off, pretty much any old silicon diode will do. Common types are 1N914 and 1N4148, and they have voltage ratings well above what you'll encounter in most products, so they make great generic replacements. If you have room, you can replace a surface-mount diode with a part that has leads, and it'll work fine.

When the diode is used in some high-speed application like detecting radio and TV signals, or it's a special type like a *varactor*, which tunes radio stages depending on the voltage applied to the diode, you must replace it with the same type. There are also very specialized diodes made from gallium arsenide (GaAs), Schottky diodes, zener diodes, and even the old technology of germanium diodes. All of these specialized parts have to be replaced with the same kinds.

Rectifiers also come in various types. For a basic AC power rectifier or a bridge rectifier package, as you might find at the AC power input of a switching power supply, the issues to consider are how much current it can handle and the peak inverse voltage (PIV), which is how much voltage the diode can withstand in its nonconducting direction. Exceed the PIV, and the part will arc over inside and be destroyed. Most rectifiers connected to the AC power line have around a 450-volt PIV rating.

The current rating indicates how much current can pass in the conducting direction without overheating the part and burning it out. It should be matched to the original part or exceeded. If there's no number on the old one or it's too burned to read, estimate the current from the size of the rectifier and from what the label on the back of the product says it draws. If it's specified in amps, use a rectifier that can handle a bit more. If it's specified in watts, divide the watts by the voltage (120 in the United States) to get the amps. Most consumer electronics products are fine with a 4- or 5-amp rectifier. A power supply in a desktop computer or a multichannel home theater receiver might need something a little bigger.

Fast-recovery (high-speed) rectifiers are used in switching power supplies. The standard low-speed rectifiers or bridge rectifier connected to the incoming AC power line feed power to the chopper circuit, as shown in Figure 8-5, which breaks the incoming power into fast pulses. After those pulses pass magnetically through the transformer, they are converted into DC current with high-speed rectifiers. These parts can accept pulses of current many times faster than what's on the AC power line and must be replaced with the same type. If you can't find an exact replacement, look up the part's spec sheet and be careful to match or exceed its breakdown voltage, current and speed ratings. Take the speed rating seriously—a standard, low-speed rectifier simply won't work.

LED indicators can be matched with just about any small LED of similar brightness. The voltage required to light an LED can vary from type to type, but most of the small LEDs used as indicators are pretty similar. Current through these things is low, too, typically around 15 milliamps or less, so it's unlikely anything will burn out due to a difference in the parts. The worst that could happen might be that your replacement is dimmer or brighter than the original. Today's LEDs are much brighter for a given amount of current passing through them than LEDs of just a few years ago. If your replacement is too bright, you can put a resistor of around 100 to 450 ohms in series with one leg of the LED to cut down the current and brightness. Either lead will do. You'll have to experiment to find the right value to match the brightness of the other indicators.

Bright white LEDs used in TVs are another story! In most sets, the LEDs are wired in series strings, so current has to pass through one to get to the next, just like in old-fashioned Christmas lights. Backlit LED TVs typically have multiple strings, each driven off a separate connection to the power supply. In that supply is a regulator that keeps the current through the LED string constant, rather than the voltage. It's done that way because the amount of current passing through an LED for a given applied voltage varies with temperature, and it's the current that determines the brightness. Also, the current increases as the LED gets hotter, which increases it even more until the LED is destroyed, in a process called *thermal runaway*. Constant current supply stops this from happening.

If you replace such an LED with something that can't handle the current, it'll burn out in a hurry. If it restricts the current, the other LEDs will be dimmed. If it's too dim or bright, you'll see that on the screen. So, you're best off matching the original LED as closely as possible. LEDs aren't marked with part numbers, so if you can't get the service data on the TV, finding a suitable replacement is basically hit or miss. Don't try to replace a TV's backlight LED with a small indicator type. Look for a similar TV for scrap and pull an LED out of it. People give away dead LCD TVs all the time on neighborhood websites. Even if it's not the same model TV you're trying to fix, a set from the same maker is likely to have compatible LEDs. I recommend you harvest them all before recycling the rest of the set. LED failure is one of the most common faults in LCD TVs, so you just might need some more one of these days.

Some video projectors use super-bright LEDs instead of a hot lamp. Except in very cheap, toy-level projectors, which might have just a single bright white LED, there will be three: one red, one blue and one green.

A lot of current passes through these LEDs to generate the required brightness, and they get hot, so they're mounted to heatsinks. Manufacturers promise 20,000 or more hours from these things, but I've seen plenty of them die at around 4,000 hours. Because of the high current and brightness, you will have to find an exact replacement. It's unlikely anything else will fit or work properly. Be sure to mount the new part properly on the heatsink, using thermal grease to ensure good heat transfer.

Inductors and Transformers

The primary characteristics of an inductor are the inductance value and the current rating, which depends mostly on the size of the wire used to wind the coil. This also affects the DC resistance of the inductor, which can be a factor in some circuits, especially those that pass significant current through the coil. It can matter in a power supply, but probably not so much in a small-signal circuit like a radio or TV's receiver stages.

Without a schematic, guessing the value of an unmarked or burned-beyond-recognition inductor isn't easy. It requires a fairly deep understanding of the circuit and some informed guesswork, and is beyond the scope of this book. Luckily, you'll probably never have to do it.

Replacing transformers is even trickier. With those used for signal transfer or driving a speaker in an older audio amplifier, the impedance (opposition to AC current) of the input and output coils is the main thing to look for, rather than inductance. Like resistance, impedance is specified in ohms. A speaker-driving transformer might be rated at 2,000 ohms on one side and 8 ohms on the other. Current capacity matters as well, but a similar-sized transformer should be in the ballpark.

When the transformer is in a switching power supply, you'll need an exact replacement. These transformers typically have many windings and are custom-made for the specific product. There's nothing generic that will do.

Integrated Circuits

ICs are specific to the functions they perform. There are thousands of types, but few can be interchanged. Common ones are easily available at parts houses. Get an exact replacement.

Some products incorporate chips not available in the United States, and you're out of luck. I ran into that with the LCD driver in an electronic air cleaner. There was just no way to get a replacement part. Also, ROM (read-only memory) chips and some microcontrollers (simple microprocessors used to control many products) have firmware burned into them. You can get another IC, but without the code, it's not gonna do anything. Only the manufacturer or a parts machine will be able to provide a replacement.

Replacing a flash RAM chip means losing whatever data was on the old one. Most of the time, that's okay. You might lose all the channels and picture settings programmed into a TV, but you can always rescan the channels and readjust the settings. In some products, though, factory adjustments not accessible to us mere mortals are stored on those chips. If so, the unit will probably still work after you replace the part, but not properly.

Op Amps

The main characteristics of small-signal op amps are speed, noise and power supply type. Most op amps require both positive and negative power supplies, while some operate from a single polarity. You can't substitute one for the other. Replacement power op amps that drive heavy loads need to have similar output capabilities to avoid overheating and self-destruction.

The whole concept of an op amp is that it is generic, with most of its characteristics set by external components. Consequently, lots of small op amps can be replaced with similar types as long as the pinout (arrangement of leads or solder pads) is the same. Especially in audio applications, many types are interchangeable. Their differing amounts of inherent noise, though, can matter a great deal depending on the application. If you replace an op amp with a noisier type, you'll probably regret it. The noise specs can be found on the op amps' data sheets.

Op amps used in higher-frequency applications like radio receivers have to be at least as fast as the original to pass the signals they're intended to process. Be careful, though, not to use parts a whole lot faster, or the replacement may become unstable and go into oscillation, where it generates unwanted signals on its own.

Resistors

Match the resistance value (ohms) and the power-handling capability, specified in watts. Unless a resistor is a special kind, such as a metal-film or wire-wound type, a standard carbon-film resistor should work properly. Don't replace a metal-film resistor with a standard carbon type, but replacing a carbon part with metal film is fine; it's just a better component.

Metal-film resistors are found in circuits that handle very tiny signals, such as microphone and phono preamplifiers, because the resistors generate less noise, so replacement with a carbon resistor will degrade the performance. Wire-wound resistors are used in high-current circuits in things like power supplies. I've seen some in audio amplifier output stages as well. Never use a wire-wound part unless the original was the same type. Because they are basically coils, they have some inductance that can make a mess out of circuits not designed to use them.

Common resistors have a 5 percent tolerance. That is, the true resistance is within ±5 percent of the marked value. In critical circuits, 1-percent resistors may be used, and it's important to replace those with the same type. However, in a pinch, you can measure your 5-percent parts and find one that's within the 1 percent tolerance. As long as the circuit it's in doesn't heat it up with lots of current, it should stay close enough to work properly.

If you can't find the exact value you need, consider the original part's tolerance (see Chapter 7), and try to combine a few other resistors to get to a value well within the original part's specs. For instance, if you need a 3.3-kΩ resistor, you could put a 2.2-kΩ and a 1-kΩ resistor in series. Resistor values in series add together, so that'd get you to 3.2 kΩ. If the original resistor had a 5 percent tolerance, as most do, it could vary by ±165 ohms and still be within spec. So, 3.2 kΩ would be fine as long as the combined resistors' own tolerances didn't push their total value outside the tolerance range of the original part. Check the real value of the combination with your DMM to be sure.

Resistors in parallel combine opposite to how capacitors do. The resistance value goes *down* according to the following formula:

$$\frac{1}{\frac{1}{R_1} + \frac{1}{R_2} + \frac{1}{R_3} \,\cdots}$$

Two resistors of the same value will produce half the resistance. The larger the resistance of the second resistor, compared to the first, the less effect it has on it. Play around with a few resistors by combining them in parallel and measuring them, and you'll get the hang of it.

Potentiometers

The most important characteristic is the resistance value across the ends of the pot. How much current the pot can handle usually isn't an issue because only small currents go through these parts in most circuits. If in doubt, figure that a similar-sized replacement should handle about the same current as the original.

If you're stuck for an exact replacement, you can replace a linear or log-taper pot with the opposite type. The penalty will be that the pot's effect on the signal won't be the same at a given spot on the knob. Especially in audio applications, this can be annoying because the increase in volume, bass or treble will bunch up at one end of the rotation.

Switches

Match or exceed the current capability and voltage rating of the switch, especially when the switch is connected to the AC power line. In small, low-current and low-voltage devices, though, a vastly overrated switch may have been used for convenience. They really don't make switches that can handle only 5 volts at 200 milliamps, so the original may have been rated much higher, and you don't need to worry about it. Use whatever you have. You'll need as many poles and throws as the original, of course, but a switch with some extra poles can be used as long as it fits in the available space. Just ignore the extras.

Relays

Relays are failure-prone, especially with age, and finding an exact replacement can be tough. When subbing relays, the most important specs are the coil voltage and the current-handling capacity of the contacts. Naturally, you need the relay to have the same number of poles and throws as well.

The current required for the coil to pull in the relay matters, but there's some leeway. Sometimes it won't be specified on the relay, but you might be able to find it from the part number by looking it up on the internet. You may see a resistance specified on the body of the relay. This is the coil's resistance, and you can calculate the current using good ol' Ohm's law. If the voltage is 12 volts, for instance, and the resistance is 500 ohms, the current that will pass through the coil is 12/500, or 0.024 amps. In other words, 24 milliamps. As long as that's somewhere near what the original part used, it should work fine. It's also okay to use a relay that takes any amount less coil current than the original. If the new one takes a lot more, it could overheat the transistor driving it.

Relay coils pretty much always have diodes connected across them in the reverse direction to prevent the reverse current spikes they generate from damaging the driving transistor. Sometimes the diode is inside the relay, and sometimes it's on the board. If it's inside, you can replace the relay with a diodeless version as long as you add a diode outside. Just about any silicon diode like a 1N914 will do. If the diode is already outside and all you have is a relay with a diode, you can use it. You don't even have to remove the extra diode on the board. Just be certain that the coil's connections are oriented so that the cathode of the diode connects to +. Relays with diodes usually have + and − marked on their coil connections.

Transistors

Transistors are the most complicated parts to substitute. Major semiconductor manufacturers used to give away large transistor substitution books filled with hundreds of pages of transistor types and their brands' appropriate cross-referenced substitute part numbers. Because many transistors have similar characteristics, a few hundred parts can sub for thousands of transistor types.

These days, you can look up this stuff online, but you may run into numbers for which you can't find a cross, or there might be a valid sub but you can't get one. Alas, some parts are made of *unobtainium*. Even when a substitute component is available, you may prefer to speed up the repair process by using a part you already have. Small transistors like 2N2222, 2N3904 and 2N3906 can replace a wide variety of bipolar parts in non-critical circuits, and are worth stocking in your parts drawers.

To choose your own substitute requires some understanding of the transistor's application and how a change in characteristics might affect circuit performance. Some functions, like simple switching of voltage to direct it to various circuit stages or turn an indicator on and off, will work with just about any transistor of the same basic construction (bipolar or field-effect transistor) and polarity. Others, such as high-frequency signal processing or current amplification in complementary audio amplifier output stages, often require stringent adherence to the original part's specs.

All this assumes that you know the old part's number. Usually, you will, but at times you might have to fly blind. If the original transistor literally blew apart, which happens occasionally when a heck of a lot of current has been pulled through one, there may not be a number to read! I've seen surface-mount output transistors in LCD backlight inverters blow so hard that there was little left between the solder pads. Even when the number is visible, it could be a proprietary house number with no cross-reference to a sub. And some transistors, especially tiny SMDs, sport no numbers in the first place.

If you're lucky, the board will be marked with "ECB" or "GDS," showing what terminal goes to which pad. ECB indicates "emitter, collector and base" and thus a bipolar transistor. GDS means "gate, drain and source," the terminals of a field-effect transistor (FET). These markings also give you strong clues to the part's polarity. If C goes to the positive side of things, it's an NPN. If E does, it's PNP. With a FET, if D is positive, it's an N-channel part. If S is positive, it's a P-channel part. Usually, anyway. Some FETs are symmetrical and can be used either way.

Without board markings or a part number, the transistor is a total mystery. Use your scope and understanding of basic transistor operation to deduce the part's polarity and layout of connections. Start by looking for the power supply voltage feeding it. If it's positive and fed through a resistor or a transformer, you've probably found the collector of an NPN transistor or the drain of an N-channel FET. Find the stage's input by looking for whatever signal operates the transistor. If it's a continuous signal, you should see it. If it's something that happens only when you press a switch or some other operation signals that area of the circuit, create those conditions and find the signal. When you find it, you've found the base or the gate. Whatever's left will be the emitter or source. The exception is the grounded-base amplifier, which uses the emitter or collector for input. Without a diagram, that oddball configuration can confuse the heck out of you. Luckily, you won't see one very often.

Most bipolar transistors are NPN. If the connection to the positive supply line is direct, without a resistor, or there is a resistor but it's of very low value, the transistor could be PNP, and that connection would be its emitter. Find what looks like the base by scoping for signals. See if there's a resistor from the base to the transistor terminal connected to the supply voltage. PNPs are used to turn on and pass current from the supply to some other circuit when the input signal goes low, toward ground. The resistor going up toward the supply keeps the base high and the transistor turned off until the input signal pulls it low. You'll find PNP circuits of this sort in power switching sections of battery-operated products.

Assume the part is an NPN bipolar transistor or an N-channel MOSFET, and you'll be right most of the time. In small-signal stages, depletion-mode FETS are common, but the ones used for switching power on and off are likely to be enhancement-mode types. Replacing a depletion-mode FET with an enhancement-mode type won't work, but it isn't likely to hurt anything. The other way around, however, could keep the new part turned on all the time, which will allow current to pass when it shouldn't, possibly causing damage if much current is involved. Finally, if your replacement turns out to be the wrong polarity, such as NPN for PNP or N-channel for P-channel, the circuit won't work, but trying it shouldn't do any damage.

All bets are off if the transistor is part of a complementary push-pull amplifier. This configuration uses NPNs and PNPs in more complicated, hard-to-deduce ways. And, if the original part was a FET, the issues of enhancement and depletion modes and JFET versus MOSFET make the whole thing very tough to fathom. By far, most push-pull amps use bipolar output transistors, but some MOSFET designs are in use. Luckily, it's pretty unlikely for a blown output transistor to be so destroyed that you can't read the part number.

Don't try to sub chopper transistors in switching power supplies without knowing the correct part number and finding a legitimate sub from a cross-reference. Most choppers are power MOSFETs with specs that must be closely matched for reliable operation. Even if a sort-of-close sub works, it probably won't run for long before failing. (See Chapter 8 for what happened when I ignored my own advice!) Sometimes, even a legit sub will die in a hurry, and the only part that will work is an exact replacement of the original part number. The same was true of horizontal output transistors in CRT (picture-tube) TVs, another application involving fast pulses at fairly high voltages and currents.

In some cases, the original and replacement transistors are electrically compatible, but their arrangement of leads, called *pin basing*, is different. Most small-signal American bipolar transistors are EBC, left to right, while Japanese parts are usually ECB. You can replace one layout with the other as long as you switch the two leads, being careful not to let them touch as they rise from the board toward the transistor. Small FETs are usually SGD. Power bipolar transistors are usually BCE, with C connected to the metal tab (if there is one), but check to make sure. Power FETs typically use GDS, with D connected to the tab.

Once you've figured out what should go where, whether from the original part or from scoping and deducing, you can proceed with trying out a new part. The primary characteristics to be concerned with are gain, high-frequency cutoff point and, with larger parts, power dissipation capability. Secondary characteristics, but still very important, are the maximum voltages permitted from base to emitter and from collector to emitter.

Very often you'll find a transistor that matches pretty well but has a little more or less gain. Depending on the application, the part might work. If the circuit is linear, producing output proportional to the input, the transistor isn't normally saturated (fully turned on), so a slight gain difference may not cause a problem. In switching circuits like backlight inverters, though, inadequate gain can result in lots of heat from the transistor's not-fully-turned-on resistance, burning out the part in a hurry. Too much gain in a linear circuit like a small-signal amplifier may cause distortion, increased output or spurious signals. Not enough usually just results in a bit less output.

The high-frequency cutoff point specifies at what frequency the transistor's gain will have decreased to one. In other words, it won't be amplifying at or above that frequency. In low-frequency applications such as audio, any transistor will be more than fast enough. At radio frequencies, the situation can be quite different, requiring a transistor whose cutoff frequency is approximately equal to the original part's spec. Too little might result in low or no output, while too much could result in unwanted harmonics or spurious signals riding on the desired one. When in doubt, go for too much, as long as the difference isn't excessive; at least the thing will try to work.

Power dissipation is very important. The new part should be able to dissipate at least as much power as the old one. A better dissipation spec is fine. Rarely is this an issue with small-signal transistors, but it's a biggie with the larger ones that handle significant current in power supplies and audio amplifiers.

Maximum permissible inter-electrode voltages must be respected. Exceed them, and the transistor might emit some of that magic smoke. Most transistors' collector-to-emitter specs are well beyond what a small-signal circuit produces. The circuit's base-to-emitter voltage, however, could exceed the capabilities of some replacement parts, so keep an eye on that. Large parts used in output stages can have pretty high voltages applied from collector to emitter, so don't take that spec for granted either.

If all this seems overwhelming, stick to replacement part numbers from a cross-reference book or online source, and you'll be fine. Even with expertise, matching up transistors can be very much a roll of the dice.

Voltage Regulators

The industry-standard three-terminal regulators, such as the 78XX and 79XX parts, are used often enough that it pays to keep some around. LM317, a variable regulator whose output voltage is set by external resistors, is also still in common use. It's fine to replace a lower-power 78XX or 79XX part with a higher-power version if you can fit it in there, as long as the output voltage is the same. Just beware that the *pinout*, or layout of the leads, may not match up, requiring you to cross one over the other when you install the part. Beyond these standard kinds of part numbers, it's best to get an exact replacement.

Zener Diodes

The purpose of a zener diode is to break down nondestructively in the reverse direction and conduct when the part's reverse voltage spec, or *zener voltage*, is reached. The important specs are the zener voltage and the power dissipation. Unlike normal diodes, zeners' dissipation limits are specified in watts, not amps or milliamps. Always replace a zener with one of the same zener voltage and at least as much dissipation capability. A higher dissipation spec is fine and can increase product reliability when the design is marginal.

My electric blanket died a year after I bought it. Its identical replacement did the same thing. When the third blanket died, I got tired of buying new ones and opened up the control box. It turned out that one of the zener diodes in it had gone open. There were two ½-watt zeners, one 5.1 volts and one 9.1 volts, both of which were standard values I had in my parts drawers. I replaced them with 1-watt versions, and the blanket has worked for 5 years. More dissipation capability never hurts with these things.

You can put zeners in series to add their voltages, but don't parallel them to increase dissipation capability; even zeners with the same zener voltage won't start conducting at exactly the same voltage, so one will always take more current than the other, resulting in its premature failure. When combining them in series, be sure that the wattage of each zener is at least as high as the original part's rating, and watch the polarity. Each zener should feed the next one cathode to anode, so you wind up with one anode and one cathode at the ends of the string.

Installing the New Parts

Once you've procured or substituted components, it's time to put them in! Proper installation is crucial for successful long-term repair. Let's look at some issues specific to various kinds of parts.

Through-Hole

Replacing a through-hole component is pretty easy, requiring nothing more than pushing the leads through the holes, bending the ends a little so the part doesn't fall out, soldering the leads and then clipping off the excess. If the part is attached to a heatsink, it's a little more complicated, but not much. For a free-floating heatsink bolted or clamped to the top of the component, install the heatsink before soldering the part to the board. When the part mounts on a heatsink attached to the chassis, put the leads through the board's holes without soldering them, and then screw or clip the component to the heatsink. If the original part had a mica or plastic insulator between it and the heatsink, don't forget to install it on the new one!

If the old part used *heatsink grease*, also called heatsink compound, you need to do the same with the new one. The grease used is a special silicone compound formulated for maximum heat transfer. You can get it from online parts houses, and computer supply shops that carry CPU upgrades and bare motherboards also carry it. Most heatsinks, including those with mica or thin plastic insulators, do require the grease. Those with rubber separators usually don't, though, because the special rubber is adequately heat-conductive and also conforms well to its mating surfaces. Figure 12-4 shows typical insulator setups requiring thermal grease.

A thin smear of the special grease on one of the mating surfaces helps heat transfer across the less-than-perfect contact area, filling in tiny gaps and increasing effective surface area. Too much grease can separate and insulate the surfaces, *reducing* heat flow, so don't overdo it. Smear on the grease with a swab, and be careful not to put bending pressure on the insulator or it may break. Mica insulators are especially brittle, and even a single crack

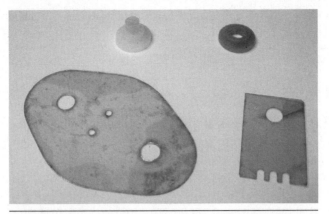

FIGURE 12-4   Transistor mounting hardware with screw sleeves.

can lead to a short later on. To avoid bending it, place the insulator on your workbench before applying the grease.

The insulator's job is to isolate electrical contact between the component and the heatsink while facilitating heat transfer. If there's no insulator, either the part has no metal contact area on its case, as with an all-plastic transistor, or it's okay for the contact area to be connected to the heatsink. Voltage regulators sometimes have their ground connections on the metal tab, so contact with a grounded heatsink is a good thing. Many power transistors, though, have their collectors or drains at the tab, and really big ones with all-metal cases use the case for that connection. The cases are usually connected to voltage sources, and contact with ground would be a short. Insulators are used to avoid the connection.

When there is an insulator and the component has a metal tab, the mounting screw will pass through a plastic washer with a sleeve. Be certain to use it, and watch its orientation. The sleeve should fit into the hole on the transistor's tab, preventing the screw from touching the inside of the hole.

Tighten the mounting screw more than you would a screw holding a board down or a case together. You want good heat transfer, and that takes some pressure. Don't overdo it to the point of breaking the insulator or stripping the screw, of course. After you have the part mounted, check with an ohmmeter to be certain there's no unintended connection between the metal case or tab and the heatsink. Solder the leads to the board only after the mounting procedure is complete. And, of course, ATE!

Occasionally, you will find a thermistor (a heat-sensitive resistor) glued to the case of a power transistor, especially in the output stage of a push-pull audio amplifier. Thermistors are used to adjust the bias of bipolar power transistors as the parts heat up, because their gain and optimal bias current drift with temperature. If you can get the thermistor off without destroying it, glue it with epoxy to the new part. If you can't remove it, you'll need a new thermistor. Look up its part number, and order one just like it.

You may run into a thermistor stuck on a transistor with a glob of heatsink compound instead of glue. This makes it easy! You can just peel off the thermistor and affix it to the replacement transistor the same way.

SMD

Putting in a new SMD part is a bit tougher than installing a through-hole component, thanks to the size scale. How do you hold it in place long enough for soldering? Gluing is not recommended. Sure, the manufacturers do it, but they have special glue made for the purpose, and we don't. More than likely, some other line runs underneath the component, and a later attempt to remove the glue will tear the copper off the board. Also, the electrical properties of the glue you might use are a wildcard; you have no idea how its presence might affect circuit performance. It could exhibit capacitance or even conduct current. Some glues do!

To get an SMD component in place using a traditional soldering iron, first use wick to clean the board's solder pads so that there are no raised bumps of solder on them. You want the SMD part to lie flat. Melt a little solder onto your iron's tip. Now, place the part on the board and line it up carefully with a tiny screwdriver. Center the part between the pads so it can't create a short across two lands. Hold down the body of the SMD component with the screwdriver while touching the iron's tip to one end of the part. The solder on the tip should flow onto the board, making a joint at that end. Don't worry about getting a good joint; all you want to do is prevent the part from moving.

Once the component is held in place by the solder on one end, solder the other end properly. Then go back and redo the messy end. Don't keep heat on the part any longer than necessary; if you delaminate the component's contact pads, you've ruined it and will have to get another one. When you've finished soldering, take a good, close look with your magnifier or microscope to be sure you haven't created any solder bridges to adjacent pads or parts.

The newer, easier way to solder SMD components is with solder paste and a hot-air soldering station, as discussed in Chapter 2. All you have to do is put a little of this on the board's pads, line up your part, hold it down in the middle with a small screwdriver and hit both ends of the part with the hot air.

Solder paste won't stick to anything but the pads, so shorts between them are unlikely. Still, it pays to inspect the results carefully, just in case. ATE, after all.

Sometimes, you can get away with replacing an SMD part with a slightly smaller one, as long as it's not a resistor that has to handle enough current to burn out the smaller one. Usually, this is not the case, because most of these parts handle very little power. If the size difference isn't too great, the pads might reach the board's contact points. Just make sure there's no spot between them that might get shorted by the smaller part's pads.

More likely, you'll want to replace a smaller part with a larger one because the original size is too tiny to handle. I had to do this not long ago when an incredibly minuscule resistor broke off a board and I couldn't hold it in place well enough to solder the darned thing back on. After a few attempts, it flew off into another dimension, never to be found.

When putting in a larger part, there probably won't be room on the board, or the size mismatch will create a short to a nearby pad. Assuming there's a little space vertically, solder a strand of wire to one of the pads on the board. If you're lucky enough that there's a nearby pad connected to one of them, use that instead of one that will be under the part after it's attached. Leave the other end of the wire hanging.

Now, hold the part with needlenose pliers or a hemostat, with one side of the component angled up so one of its pads touches the board's pad opposite to the one with the wire. If you're using a hemostat, don't lock it, because you won't be able to unlock it without tearing the part off the board.

Solder the pad and take away the pliers. Then, touch the wire to the floating pad on the component and solder that one. Do it quickly, or the other end will unsolder. One second is about the maximum you'll get away with. It might desolder anyway, and this is why it's better to use an adjacent pad if you can. If you had to use one under the part, you'll need to desolder it to reattach the wire, which is a pain. Plus, an adjacent pad will be a little farther away, giving you slightly more time during soldering before the solder melts and it detaches (Figure 12-5).

FIGURE 12-5   The angle trick. One under, one adjacent.

Saving Damaged Boards

When you desolder a through-hole component, one unfortunate result of failing to get the hole hot enough is that its copper lining comes out with the component lead. If you see what looks like a sleeve around the lead, you've torn out the copper. On a double-sided board, it's not a catastrophe. When you replace the part, be sure to solder both the top and bottom contact points, and all will be well. You might have to scrape some of the green *solder mask* coating off the top area to get contact between the lead and the foil. That's best done with the tip of an X-Acto knife.

Pulling the sleeve out of a multilayer board can destroy it because you have no way to reconnect with interior foil layers that were in contact with the sleeve. If you're lucky, a particular hole might not have had inner contacts, and soldering to the top and bottom may save the day, so it's worth a try. Don't be surprised, though, if the device no longer works.

If you can figure out where they go, broken connections can be jumped with wire. On double-sided boards, it's not too hard to trace the lines visually, though you may have to flip the board over a few times as you follow the path. When you find where a broken trace went, verify continuity with your DMM, from the end back to the break, just to be certain you're in the right place. Don't forget to scrape off the solder mask where you want to contact the broken line.

Wire jumping can help save boards with bad conductive glue interconnects, too. On a double-sided board, you can scrape out the glue, or use a very fine bit to drill it out if it's too solid to scrape, and run a strand of bare wire through the hole, soldering it to either side. Forget about trying this on a multilayer board, however; you'll probably trash it while trying to clean the hole. On those, it's best to run an insulated wire around the board from one side to the other. That adds extra length to the conductive path, which could cause problems in some critical circuits, especially those operating at high frequencies. At audio frequencies, it should be fine. If some interior layers are no longer making contact with the glue, this won't work. Most conductive-glue boards I've seen have been double-sided, making them suitable for wire jumping.

If the board is cracked from, say, having taken a fall, scrape the ends of the copper lines at the crack. It's possible to simply solder over them, bridging the crack, but that technique tends to be less reliable than placing very fine wire over the break and soldering on either side. To get wire fine enough, look through your stash of parts machines for some small-gauge stranded wire. Skin it, untwist it and remove a single strand. You can also use "wire-wrap" wire, which is thin, solid-conductor (in other words, not stranded), insulated wire used for prototyping circuits with a wire-wrapping gun. If that's not fine enough, cannibalize a small speaker or a pair of headphones. The voice coils suspended around the magnets are made of ultra-fine, enamel-coated "magnet wire" that you can unwind. To remove its insulation for soldering, bathe the wire ends in solder on your iron's tip until the enamel comes off and the wire accepts solder.

Sometimes, there are multiple broken lines too close to each other for soldering without creating shorts between them. To save boards like this, scrape the solder mask off close to the crack on every other line. Then, scrape the in-between lines farther away from the crack. Use bare wire strands to fix the close set, and insulated wire, like the wire-wrap or enamel-coated stuff, to jump the farther set.

Wire-wrap and magnet wire are especially good for this kind of work because their insulation doesn't melt very easily, so it won't crawl up the wire when you solder close to it, exposing bare wire that could short to the repaired lines nearby. Plus, these kinds of wire are thin enough to fit in pretty small spaces.

It's possible to repair broken ribbon cables in stationary applications (the ribbon doesn't move or flex during the product's operation). Ribbons with copper conductors can be soldered across or jumped with tiny wires, just as with a circuit board. The thinner, printed kind can't be soldered, but you can bridge across bad lines with conductive paint. With copper ribbons, you'll have to scrape off the masking where you need to solder. Printed ribbons also may have masking, but some don't. You can test for this by checking the resistance along one of the lines. If it's open, there's masking you'll need to remove before applying the paint.

On multilayer boards, cracks and torn sleeves are extremely difficult to bypass. If you have a schematic, you may be able to find the path and jump with wire. Without one, it's pretty much impossible when the tracks are inside the board.

Bridging broken conductors is a tedious, time-consuming technique, but it works. Accomplishing it without causing shorts takes practice and isn't always possible with very small, dense boards and ribbons.

LSI and Other Dirty Words

Back in Chapter 7, I promised to describe a trick for resoldering big ICs with very close lead spacing. Those large-scale integrated (LSI) chips with 100 leads are in just about everything these days. It's not likely you could find a replacement chip, so why would you want to resolder one?

With so many leads, an intermittent connection to an LSI chip is not uncommon. Surface-mount boards are factory-assembled with *reflow soldering*, in which solder is applied to the pads and then reflowed onto the component leads with hot air, infrared lamps or in a special oven. Reflow soldering relies on low-temperature solder that can break after the numerous heating and cooling cycles encountered in a product's normal use. Now and then, one connection out of an LSI chip's long row of them will go flaky. The leads are so close together that there's no way to apply solder to one without causing a short to the adjacent leads.

Here's the trick: Go ahead and short them! With all power removed, including from any backup batteries mounted on the board, solder away, and let as many leads get shorted together as you want. Once you have good solder on the problem lead, lay solder wick across the leads where the excess solder is shorting them. Heat it up and wick off the excess, but don't wait until the leads are bone dry. Pull the wick off a little sooner. If you get the timing right, you'll be left with a perfectly soldered row with no shorts. The wick soaks up the solder in between leads faster than what's underneath them, where you want it to stay.

If you wait too long and wind up removing so much solder that the connections to the board aren't solid anymore, resolder the area and do the wick trick again. After you try this procedure a few times, you'll get the hang of how long to wait before pulling the wick. I've had tremendous success with this approach. The one caveat is that it's hard to wick out solder if it gets under the edges of the chip. To avoid that problem, solder as far from the body of the IC as possible. This helps prevent overheating the chip, too.

When you're all done, use your magnifier to verify that the contact points with the board are soldered, and that no bridges exist between leads. Honest, it really does work! I've even replaced a few LSI chips this way using chips from parts units. Getting those babies lined up accurately on all four sides . . . well, that's another story. I solder two leads diagonally across the chip at the corners to hold the part in place, and then do the rest.

You can use solder paste and hot air to rework these bad connections. But, since you can't apply it under the chip's leads because the part is already in place, the paste doesn't always flow well where you need it. I've found the ol' short-'em-and-wick trick to be more effective in these situations.

Chapter 13

That's a Wrap: Reverse-Order Reassembly

You fixed it. Congratulations! Now it's time to put everything back together. Just screw the boards down, plug in the connectors, close the lid and you're done, right? Well, sometimes, but not performing the reassembly methodically can lead to all kinds of trouble, from failure of your repair to new damage, and even to danger for the product's user. You're sewing up your patient after the operation, and it's important to put in the stitches carefully and avoid leaving a scar.

Common Errors

It might seem absurd to think that one could reassemble a machine and have parts left over, but it happens all the time. You snap that final case part into place, breathe a sigh of relief, glance at the back of your workbench, and there it is: some widget you know belongs inside the unit, but you forgot all about it. Maybe it's a screw, a spacer or a heatsink insulator. Hmm, is it really worth all the trouble to backtrack, just for one little, seemingly nonessential item? Sigh . . . Time to pull the whole mess apart again. It's very easy to forget to replace a bracket, a washer, a shield, a cover or even a cable. The unit might function without one of those pieces, but it's not going to work completely right.

Putting the wrong screws in the wrong places may have no consequences, but it also could seriously damage the device. If a screw that's too long presses against a circuit board, it might short whatever it touches to the chassis, hence to ground. You flip on the power, and voilà, you have a new repair job on your hands. See the definition of *magic smoke* in the Glossary.

Overtightening screws can strip their heads, making it very hard to get them out again. It can also break plastic assemblies and cause cracks in the case. Undertightening screws may lead to their falling out later, possibly jamming mechanisms or shorting out circuitry if they're internal screws.

Our memories can really fool us sometimes. You're certain that part went over here, but now it doesn't quite fit. So, you press a little harder because you *know* that's where it goes. Snap! Oh, right, it went over there, not over here. If something doesn't want to fit together, it probably belongs somewhere else. This kind of error happens more with mechanical parts than with electronic components.

Manufacturers try to make internal connectors different enough from each other that only the correct cables will fit. In products with lots of connectors, though, there may be ambiguity. CRT projection TVs, for instance, had so many connectors that there was just no way to make them all distinct. LCD and DLP rear-projection sets were similar, and even some modern flat-screen TVs have an awful lot of similar-looking connectors. Sometimes, a color code is all you have to guide your reassembly. Sometimes, not even that. It's not hard to get one wrong. Guess what happens when you turn on the set.

Even when you put ribbon cables in the right connectors, it's easy to damage them. They're delicate and easily torn or folded hard enough to snap their printed copper conductors. The connectors are fairly breakable, too, especially the tiny ones in pocket-sized products. Those plastic latches that hold the cables in place snap off without much pressure. Worse, the entire connector can break its solder joints and fall right off the board, just from the stress of being pushed on while the cable is inserted and clamped down. I've seen this happen on digital cameras and pocket video gear. It is *not* easy to correct!

Getting Started

To begin reassembly, reverse the order in which you took the machine apart. When the unit has multiple boards, you'll need to get the inner ones reinstalled first. If a board that'll wind up under another board has connectors, put the cables in before covering up that board.

Ground lands are contact areas on circuit boards that press against brackets sticking up from a metal chassis, where a screw will hold down the board. You won't find these lands in plastic products, but they're quite common when there's a metal chassis.

In anything more than a few years old, take a look at them before you put the boards back in place. Those lands connect parts of the circuitry to ground via the chassis, and a poor connection due to oxidation or corrosion of either the land or the bracket can seriously affect the product's performance. Clean them up with some contact spray or, in extreme cases, fine sandpaper. After you place the board, tighten the screws firmly so contact will be reliable, but don't overtighten to the point that you might crack the board. Heating and cooling in larger items, and physical stress in portable devices from being bounced around, can cause cracks later on if the screws are inordinately tight.

A little sealant on a screw head is better than pushing the limits of tightness. Manufacturers and pro shops use a type of paint called *glyptal* to keep screws from loosening. Swabbed around the edges of the screw head, it is highly effective. You can use nail polish. Don't glob it on; just a little smear will do fine. Be careful not to cover components or their leads, and let it dry before closing the case so the outgassing won't remain inside. I use red polish so I can see where I've been, should I open the unit again later on.

Placement of wires and cables is called *lead dress*, and it can be surprisingly important. When you took the unit apart, you may have noticed that some wires were tacked down with hot-melt glue or silicone sealer. If the manufacturer went to the trouble to do this, there was a reason. Maybe the wire needed to be kept away from a hot heatsink that could melt its insulation. Perhaps a cable carries a weak, delicate signal that would receive interference if it got too close to some other element of the machine. This was sometimes the case with the cables going to video head drums in VCRs. Or maybe the reverse is true: The cable would cause interference to other sensitive circuits. Wires carrying high voltages, like those used to run projector lamps and older fluorescent LCD backlights, may need to be kept away from all other circuitry to avoid not only interference but also the possibility of arcing. The closer a high-voltage wire is to ground, the more those devious electrons want to punch through the insulation and get there. Give 'em time, and they just might, especially if the wire actually touches the metal chassis.

If the manufacturer tacked wiring down, put it back the way you found it. Hot-melt glue is somewhat flammable, and it melts with heat, of course, so it isn't used much in larger products. Now and then, you may find it in smaller items that don't carry much voltage or produce significant heat. To tack wires back down into it, you can melt the existing blob with your plastic-melting iron, avoiding any other wires, or drip a little more glue on top from your glue gun.

More often, you'll find silicone sealer used to secure wires. The type used is called *RTV*, and it's best to replace it with the same kind because it offers the correct insulating strength. RTV is available at most hardware stores, and electronics supply houses carry it too.

Now and then, and especially in small-signal radio-frequency (RF) stages operating at very high frequencies, you'll see blobs of wax covering transformers and capacitors. The wax holds the parts to the board, dampening vibrations that can cause howling or other noises in signals or frequency instability in oscillators. In some circuits, the capacitance of the wax may affect circuit operation, so it's best to remelt and reuse the original wax. It's not the same stuff that's in the candles on your dinner table.

Even if the manufacturer took no extra care with wiring, you should pay attention to the issue to achieve maximum product reliability and safety. Could a power supply lead touch a heatsink? Is the cable from a tape head going right by the power supply? And, perhaps most important, is any wire or cable placed such that it'll get crimped by a circuit board or part of the case when you close up the unit? The sharp ends of a board's component leads can go right through a wire's insulation, with disastrous results. Crimping caused by the case can break the wire or cut through the insulation and short it to the chassis. With a wire carrying unisolated AC power, a crimp could even present a shock hazard. If the case halves don't mate properly, a wire is probably in the way. Don't just squish them together and go on.

These scenarios may sound farfetched, but they're really pretty common. I always make it a point to watch the wiring as I close up the case, imagining where things will be and how they'll press on each other before I actually snap the halves together or tighten the final screws.

Reconnecting Ribbons

Insert ribbon cables into their sockets carefully. It isn't hard to put them in wrong, which can lead to anything from no operation to circuit damage. Most ribbons have bare conductor fingers on only one side. If you get one of those in upside down, the product won't work, but it's unlikely to cause damage unless the socket has U-shaped contacts that touch both sides of the ribbon. There are some like that.

Double-sided ribbons offer more opportunity for calamity. How do you know when they're in the right way? Many are keyed with a notch at one end so they can't be inserted upside down. Some are not, though. I sure hope you heeded my advice in Chapter 9 to mark the darned things! If not, see if the cable has a bend or curvature suggesting its original position. In most cases, the correct orientation should be apparent.

As discussed in Chapter 9, some ribbon connectors have no latches, and the ribbon just slides in. This type requires some force for proper insertion. With such a connection style, the ribbon cable will have a stiff reinforcement layer at the end. Even when you find one, take a good look at the connector to be sure it has no latch that slides in or flips up, because some latch-type connectors accept reinforced ribbons too. If you see no latch, grasp the cable's reinforced tab, carefully line up the ribbon with the connector and press it in firmly. The tab should protrude from the back of the connector equally on both sides. If it's crooked, pull and reinsert the ribbon.

Connectors with latches are easier to manipulate. First, be sure the latch is open. Slide it out or flip it up. The flip-up kind will stay up while you insert the ribbon, but the slide style has an annoying habit of going partway in before you want it to, blocking full reinsertion. One end may slide in, resulting in a crooked latch. If this happens, pull that end back out, remove the ribbon and try again. You might have to hold the slider's ends with one hand while you insert the ribbon with the other.

Latch-style connectors require almost no insertion force. Gently slide in the ribbon until it stops. Don't press firmly here. Look down at the top of the connector and verify that the ribbon isn't crooked. Then, close the latch while holding the ribbon in place with your other hand. This is easy with flip-up latches and a little harder with sliders. Occasionally, I've had to close sliders one end at a time with a thumbnail or a screwdriver while holding the ribbon in my other hand. To avoid a crooked result, it's better to close the ends at the same time, but now and then you gotta do what you gotta do. Close each side a little at a time, alternating between sides, rather than pushing one all the way in while the other is fully out.

When the latch is closed, look again at the exit point to be sure the ribbon is straight. Leaving a ribbon crooked can cause crossed connections and damage. You may see the edges of the bare conductors sticking out. That's fine as long as they're even and you're certain the ribbon is in all the way. Many of them are designed this way.

Special ribbons used for hard drives, laser heads, projector imagers and other very dense applications can have two sets of fingers, one behind the other. These will always have latch-style connectors. Be absolutely positive that the ribbon is fully inserted so there's no chance the wrong set of fingers could make contact with the mating pins in the connector.

Oops! Broken Latch

If you were unlucky enough to break the latch on a sliding-style connector when you removed the ribbon, don't despair. The object is to get pressure on the conductive fingers so they make good connection with the socket. Find some thin, soft plastic from, say, the bottom of one of those little pudding cups in which you keep screws. Cut the plastic into a rectangle that just fits into the socket and sticks out a few millimeters. Trim carefully so the edges line up well with the edges of the socket, without a gap. Now put the ribbon in and wedge the plastic piece in to replace the broken latch. Be sure to insert it on the side of the ribbon that does *not* have the conductive fingers, or the plastic will block the connection. I got this wrong once and went around in circles for hours trying to figure out why that confounded shortwave radio wouldn't turn on anymore!

If you get the thickness right, it'll take a little pressure to slide in the plastic, but not a lot. If it slides in very easily, it may not put enough pressure on the ribbon to make good connections. The few times I've had to do this, I used forceps to push in the plastic piece. Needlenose pliers will work as well.

Flip-up latches are much harder to repair. It might be possible to modify the socket by melting a piece of plastic over it and converting it to a sliding arrangement, and then using the plastic insert approach, but it'd be a difficult modification to pull off, considering the size scale of some of these connectors. Trivial as it may seem, a broken flip-up latch often means the end of the product unless you can scrounge a latch from a parts unit.

Switch in the Cradle

Many small products feature slide switches, or at least they did up until a few years ago. It's likely you'll run into them. The switches are tiny things mounted on the circuit board, typically at the edges, and each one fits into a plastic cradle when you put the board back into the case. The other side of that cradle faces the outside and is what your finger pushes on when you operate the switch (Figure 13-1).

When inserting a device's innards back into the case, it's imperative that you line up the tiny switch with the notch in the cradle! Otherwise, you are highly likely to break the

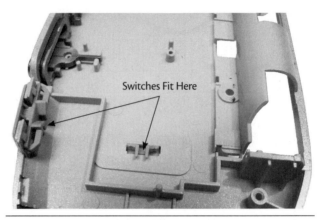

Switches Fit Here

FIGURE 13-1   Switch cradles.

switch when you screw down the board or snap the case closed. One trick is to see which way gravity will pull the cradle while you're doing the reassembly, and move the switch to the same position. This way, you don't have to hold the cradle in place while getting the board in position. If there are multiple switches, get them all set up so they'll align with their cradles. After putting the board in place, hold it steady and operate all the switches to be sure they're properly seated in the cradles before you screw it down. If any aren't, lift the board slightly and move the cradle until you feel it grab the switch. Then, retest them all to be sure none of the others slipped out of place. ATE!

Layers and Cups and Screws, Oh My!

Here's where those pudding cups come into play. If you used them as I suggested back in Chapter 9, your innermost layer's screws will be in the cup second to the top, just under the empty protective cup. Put that layer's boards, shields and assemblies into place and fasten them with those screws. If the screws are of different sizes or styles, you should have taken digital photos or made a drawing and placed it in the cup with the screws. Be certain to use all the loose parts in the cup because you won't be able to reach that layer once you reinstall the layers covering it.

Sometimes the screws came from difficult spots where they will be harder to put back in than they were to take out. Magnetizing the tip of your screwdriver with a speaker or headphone magnet can help keep a screw in place as you guide it to a difficult location. But even this might not work, especially if the screw will have to pass another magnet nearby in a motor, laser head or speaker. One of my favorite methods for keeping screws attached to the screwdriver is to position the screw on the tool and then wrap it at the head with a turn or two of adhesive tape. That'll hold it for sure. Squeeze the tape firmly around the screwdriver's shaft but not so firmly around the screw head. Insert the screw a couple of

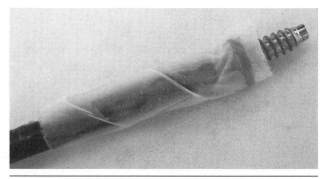

FIGURE 13-2   **How to get a screw back into a difficult spot.**

turns but don't tighten it. Then, pull out the screwdriver slowly. The tape will pull away with the tool, and the screw will be right where you want it, ready for tightening (Figure 13-2).

For now, insert each screw loosely. Don't tighten them until you have them all in, because you might have to shift the board or shield slightly to make the holes line up. If a screw falls into the machine, you may even have to back up a few steps and remove a board to get at it.

There's a trick to making sure you get all the screws back in and tightened. As you insert them, count aloud, advancing the count once for each screw. If you made note of how many there were and put it in the cup, you'll know you got them all and none was lost along the way. I've had the count come up short, only to find that the missing screw got knocked out of the cup while I was digging out another one and was lying halfway across my workbench. A missing screw may seem trivial, but if it connects a ground land to the chassis and it's not there, product performance can be significantly affected.

The counting technique has another important benefit: Once you have all the screws in loosely, you're going to go back and tighten them, and you want to make sure you don't miss any. If you've counted to eight, you'll know you've tightened them all when you reach eight again. This might seem silly or overly cautious, but it really helps, and I still do it after all these years. More than a few times, I've finished at seven and thought, "Wait, wasn't there another one?" Sure enough, there was one hiding between a couple of big capacitors or behind a heatsink, and I'd have left it loose had I not counted, simply because I didn't notice it. Loose lips may sink ships, but loose screws can destroy electronics when they fall out and cause shorts.

When you're ready for the next layer, pull the empty cup and put it aside, exposing the next set of screws. Continue on with the next layer, and so on. When you're all done, you should have a nice set of empty cups ready for the next project. If you have screws left over in the final cup, remember to check under labels and rubber feet for hidden holes. Those screws go somewhere!

Oh, Snap!

Those nasty hidden snaps are much easier to close than they were to open! Line them up carefully and apply pressure until they pop into place. The edges of the case should fit smoothly. If there's a bulge, either the snap isn't all the way in or a wire is caught underneath.

Sometimes, a case has to be snapped together at one end before the other, even if it didn't come apart that way. Look at the style of snap, and how it fits together should be apparent. If you get it wrong, you might break a snap, but it's not a big deal. Heck, they can break even when everything is done right. Often, you can live without one or two. If a snap's loss makes the case wobbly, it might be worth some careful repair with your plastic-melting iron. Be sure to pop the case apart first; trying to melt plastic near the outside will almost certainly result in very visible damage. Even from a half-inch away, the iron puts out enough heat to soften and deform many plastics. A repaired snap is weak, so you may get only one chance to close up the case properly. Still, it's better than nothing. If you hear something floating around inside the unit after you finish putting it together, the snap has broken off. Open the case and remove the plastic piece.

Screwing It Up Without Screwing It Up

As I mentioned, screws should be reinstalled carefully to avoid damaging them or the plastic into which they are screwed. Phillips screw heads are especially easy to strip, and trying to remove one is mighty frustrating once you do. How tight is right? Hold the screwdriver with your fingers, not in your palm. Turn the screws just until they stop, and snug them in ever so slightly. That's it. Don't twist until you can't twist any further.

When You Had to Break In

When you had to break into a device like an AC adapter, closing it up again is simple. Just put the two halves together and use your plastic-melting iron to rejoin the seam. Be sure to have adequate ventilation, and avoid breathing in plastic fumes! Or, you can glue the device back together if the mating surfaces aren't too damaged.

Done!

If everything fits together well, you should be ready to fire up the unit and consider your repair complete. Don't forget to ATE! Be sure to bench test receivers, projectors and other heat-generating products if the work you did could possibly make them run too hot. You'll want to bench test a projector whose fan or ballast circuit you repaired or replaced, or a receiver that needed new output transistors. A digital camera or other pocket product won't require that extra step, of course, but it could have other problems like buttons that don't work because they aren't lined up properly with the board or switches out of their cradles. Test every button and every function. ATE! Now, go show off your work and bask in the glory of a job well done. You've earned it!

Chapter 14

Pesky Parts and Persnickety Problems

While electrolytic capacitors are the bane of modern electronics, there are other pesky parts and scenarios that cause frequent trouble. Let's look at a few you might encounter.

Automotive-Related Problems

Equipment for use in cars has to operate in a hostile environment, both physically and electrically. The extremes of heat and cold in a vehicle are tremendously greater than what is found in a home. In warm conditions, a closed car's interior can approach 140°F. In winter, it'll be just as cold inside the car as outside. If you live in New England or North Dakota, well, you know what that means. Plus, there are wide swings in humidity, and dirt gets into everything. Let's not forget vibration, too! Car gear gets bounced around a whole lot.

How do these things affect automotive gear? For electronic components, heat is far worse than cold. High heat cooks electrolytic capacitors, greatly shortening their life spans. Automotive gear is made with parts rated for high temperatures, and that helps, but caps still go bad quite often, even in systems built into the car. I've seen melted caps causing failure in speedometers and tachometers, and the entire vehicle can stop or run poorly when the caps in the car's ECU (electronic control unit, a.k.a. computer) go bad. It's even worse under the hood. My own car came to an abrupt stop when the radio noise-suppression condenser (the automotive term for a capacitor) inside the distributor shorted out from age and heat, blowing the ignition fuses. The darned thing was there just to keep static out of the radio, and its failure shut down the whole car!

CD players are pretty obsolete now, replaced by satellite radio services and Bluetooth connections from phones, but plenty of the players are still out there in older cars, even if they're not used much. The players' optical heads can be damaged by elevated temperatures, especially the laser diode, if operated when very hot. Rubber belts used in the loading mechanisms and sled motor drives can get stretched prematurely from heat. The resulting slippage causes skipping, difficulty in finding tracks and inability to load or eject CDs.

High humidity and exposure to salt air lead to corrosion of connectors, and flaky connections ensue. I fixed a factory-installed car stereo that howled intermittently out of one speaker and made random noises the rest of the time. The problem was nothing more than corrosion in the connectors from the main board to the power amplifier board on the back of the unit, thanks to the years the car had spent in Florida's humid, saline environment.

Vibration usually won't cause problems with circuitry. Bouncing around can exacerbate the symptoms of corroded connectors and other failing connections, though. That's a good clue; if the problem comes and goes as the car hits bumps in the road, go hunting for a bad connection.

When you work on automotive electronics, keep all of these factors in mind. As we discussed in Chapter 4, knowing the history of a device helps you diagnose its ills. Did the fault crop up in winter? After a heat wave? In a car that stays parked near the ocean? Is the car older or newer? How long has the equipment been installed?

You're most likely to see problems with internal connectors in the stereo and the wiring leading to the speakers. Check connectors for oxidation and corrosion. Any wires that get flexed, like the ones in the door hinges, are highly suspect as well. You can bend a wire only so many times before it breaks inside its insulation and makes intermittent contact across the break.

LCDs in car stereos tend to go black from leaky seals around their glass panes, due to the wide temperature range to which they get subjected. We'll talk more about darkened LCDs later in this chapter.

The electrical system in a car is another source of stress. Automotive gear has to contend with significant voltage fluctuations and spikes from the alternator. The situation is even worse in an older car when one of the alternator's rectifiers is burned out. Alternators generate alternating current (hence the name), usually in three phases, or parts of the rotation where the output voltage peaks. Each phase has its own coil winding and rectifier. When a rectifier goes bad and one phase is lost, there's enough energy left from the other two that the system still keeps the battery charged. The car runs fine, but there are lots of spikes on the output after the AC gets turned into DC by the remaining rectifiers. One easily recognizable symptom of a bad alternator rectifier is a whine in the stereo's amplifiers, no matter what the audio source. If you hear a whine that varies with engine speed, turn the volume all the way down. If the whine is still there, the noise is coming through the DC power wiring, and a failing alternator is likely. Another cause is a weak car battery, because the battery is the filter capacitor, so to speak, for the entire electrical system. As the battery nears the end of its life, its internal resistance goes up, just like in a capacitor with high ESR, and filtering of alternator whine declines.

Even with a good alternator, spikes can be huge at the moment the engine is started or shut down. Most modern auto gear uses soft on/off switching, so power is applied to the device's capacitors and protective diode even when the unit is turned off. Those spikes can punch holes in the dielectric layers of electrolytic caps, shorting them out. To protect against this, the reverse-polarity protection diode in a lot of auto electronics is a zener connected directly across the 12-volt line that cuts the spikes way down by shorting them out when they exceed the zener's breakdown voltage. (See Chapters 3 and 7 for more about zener diodes.) With age, and especially if the car's electrical system is overly noisy, the zener can short, blowing whatever fuse feeds power to the device.

An automotive product with a direct short across the DC power input lines is fairly common. The clue is the blown fuse someplace in the car or the device. Typically, the short is in a 'lytic right across the power wires or in the protective diode. If a fuse is blown, always check those first. If they're good, take a look at the audio output transistors or modules because they can be connected directly across the incoming power, and spikes can take them out too.

Some factory-installed stereos use a special theft-prevention code. Once power has been removed, there's no way to operate the device again without entering the code. Often, it's printed in the car's owner's manual. You'll need that code to work on the stereo and, of course, to reinstall it in the car. It's wise not to pull the stereo for service unless you have the code or are sure the unit doesn't require one.

Condenser and MEMS Microphones

Virtually every small gadget that has a microphone uses a condenser mic or its ultra-tiny modern variety, the MEMS (microelectromechanical systems) mic. These are the tiny mics you'll find in your phone, tablet, laptop, Bluetooth headset, digital camera and walkie talkies.

Condenser and MEMS mics employ a permanently charged diaphragm called an *electret* to pick up sound vibrations. Inside the mic is a tiny *preamplifier* to boost the signal to something usable. MEMS types can output an analog signal, like a standard condenser mic, and some convert the audio to a pulse train easily processed by digital circuitry. Others output actual digital bytes for direct input to a microprocessor, in the same way that a USB podcast mic does. Because of the preamp or other processing circuits, condenser and MEMS mics need a power supply. The supply voltage for a condenser mic ranges from 3 to 6 volts or so, with very little current required, while a MEMS type is likely to be around 1.5 to 3.6 volts.

Condenser mics are sealed in an aluminum cylinder, typically about the size of your pinkie nail or smaller. MEMS types are usually square or rectangular, and much tinier. Some are less than 1 millimeter across! On a condenser mic, the sound pickup hole is in the front, often covered with a circle of black felt. On the back are the solder terminals. The cylinder is crimped around the edges of the back plate, where the terminals are, and the ground connection to the metal casing is provided through that crimp via a conductive ring on the plate (Figure 14-1).

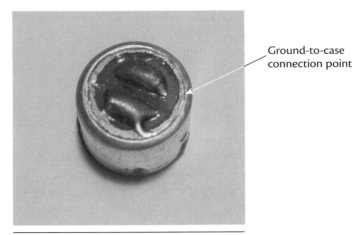

Ground-to-case
connection point

FIGURE 14-1 **Condenser microphone back plate.**

Both condenser and MEMS microphones are pretty reliable, but condenser types are prone to a peculiar failure. After a few years, and especially if the mic has been in a harsh environment like, say, your back pocket, the contact through that crimped seam gets oxidized, and resistance develops between the ground connection and the metal can. The loss of shielding lets the mic pick up AC hum and other electrical noise from outside sources.

If your device uses a cylindrical condenser mic and has distorted audio pickup, hum or other audio anomalies, suspect the mic before digging into the circuitry! To verify the problem, press your X-Acto knife blade or a small screwdriver against that seam at the back, making sure it contacts both the ground connection from the terminal and the outer casing. If the problem disappears, you've found the trouble.

It's tempting to simply solder across the seam to restore the ground, but it won't work because aluminum can't be soldered without special techniques that are not practical with something as small and delicate as a microphone. The ring on the plate will accept solder, so it looks like you've accomplished it, but there's no real solder bond to the aluminum can, and the repair will fail pretty quickly.

There is a way to fix these mics by using conductive paint. Scrape the seam clean and then paint across it, and the ground connection will be restored. The caveat is that a tube of conductive paint costs considerably more than a new mic. But, if you plan to do ongoing repair work, it's worth buying some and keeping it around. Sometimes the cost of the mic isn't the issue; it's the availability. And conductive paint comes in handy occasionally for other hard-to-restore connections that don't handle much current. A small tube will keep you supplied for years.

The audio and electrical characteristics of little condenser mics are all pretty similar, so you can use any condenser mic that will fit. Parts supply houses have them, and you can find them in junked phones, pocket recorders and such.

The mics' internal preamps are simple field-effect transistor (FET) inverting amplifiers, similar to the bipolar-transistor inverting amplifier described in Chapter 8, but with a resistor going up to the power supply instead of a transformer. The audio signal is picked off from where the resistor meets the FET's drain. A capacitor is used to block the DC and couple the audio to the rest of the device (Figure 14-2).

There are two styles: two-wire and three-wire (Figure 14-3). They both work the same way, but there's a slight difference in how they're connected. In a two-wire mic, the resistor R2 and capacitor C1 are on the product's circuit board. This reduces the wiring required and is the most common scheme in use today. In a three-wire arrangement, resistor R2 is internal to the mic, and only the capacitor is on the board.

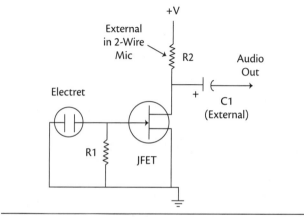

FIGURE 14-2   Condenser mic schematic.

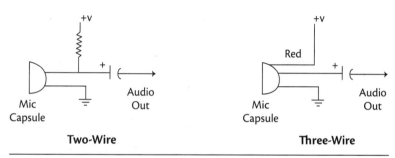

FIGURE 14-3   Two- and three-wire condenser mic diagrams.

You can replace the three-wire mic in an older device with a modern two-wire type by adding a 2.2-kΩ to 4.7-kΩ resistor between the audio output solder pad (the one that *isn't* the case-to-ground connection point shown in Figure 14-1) and the third wire coming from the board, which should be red. Now, connect the audio output wire to that same pad, where the resistor meets the mic.

The higher the resistor's value, the less current flows, and the easier it is for the JFET inside the mic to pull it up and down in step with the sound striking the electret diaphragm. Thus, the mic will be more sensitive to softer sounds. Excess sensitivity can cause overloading and distortion with mics that go right up against your mouth, so start with a 2.2-kΩ resistor and see how that works.

To replace a two-wire mic with a three-wire type, connect the ground and audio output wires, and ignore the third terminal on the mic.

It's unlikely you'll have to do any of this, but it comes in handy now and then to know how, just in case you're stuck for a mic and have only the wrong type. I've saved a few products this way over the years without having to order new mics and wait for them to come in the mail.

Push My Buttons

These days, pushbuttons have taken over for nearly all functions, at least on devices without touch screens. As discussed in Chapter 7, there are two styles. Hard-contact switches cause two pieces of metal to touch. These switches click when pressed and should have resistance close to 0 ohms when engaged. Soft switches, which have no click, use conductive rubber and can exhibit some resistance without being bad.

There also are two ways pushbuttons can be connected. In a matrix layout, the switches are arranged in an *X-Y* grid (Figure 14-4). The device's microprocessor scans the lines quickly, looking for an intersection. When it finds one, it executes the correct command, based on which lines were crossed. The resistance of the buttons doesn't matter a whole lot as long as it's low enough for the micro to sense the connection. This is the way devices with lots of buttons, like remote controls, are organized, because it requires far fewer connections than there are switches. Just be aware that the physical layout doesn't always match a grid, depending on the shape of the device. As long as the switches cross common lines, it's still an *X-Y* grid system.

When there are few buttons, a simpler method requiring only one connection to the microprocessor can be used (Figure 14-5). Each button is connected along a ladder of resistors. The arrangement is a voltage divider, with the ratio of R1 to the other resistors varying, depending on which switch is closed. The micro reads the voltage at the junction between R1 and the lower string and knows which button has been pressed. The resistor values are chosen so that if you press more than one button at a time, the voltage falls between what you get from any one button, and the micro can be programmed to ignore it.

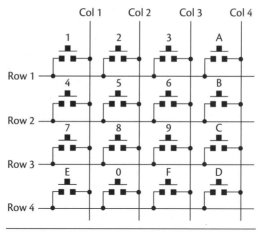

FIGURE 14-4   *X-Y* grid button diagram.

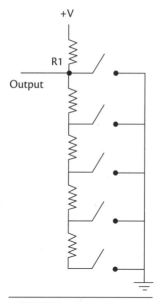

FIGURE 14-5
Resistive ladder button
diagram.

   This method has a pitfall that leads to more frequent repairs. Because their resistance is unpredictable, conductive rubber buttons aren't likely to be found in a resistive ladder. So, the metal contact types are used instead. After some time, the contacts get oxidized and their resistance goes up. Generally, designers leave a pretty wide gap between expected voltages so small resistance differences won't foul up the scheme, but sometimes the oxidation is bad enough that the micro thinks a different button has been pushed, resulting in unexpected operation. Or, the button's resistance gets high enough that the micro doesn't see the switch closure at all. If your device has a resistive-ladder switch arrangement and unexpected things happen when you press a button, suspect an oxidized switch.

With matrix arrangements, the usual failure mode is that one button or an entire row or column doesn't work. If it's just one button that quits, the problem has to be the button itself, because the same lines are used for others, and they're still working. If an entire row or column goes out, either the connection to the micro or the micro itself (or an interface chip between the micro and the buttons) is bad. Almost always, it's a connection problem. Especially in remote controls, flexing of the board from years of button pressing causes flaky vias (interlayer connects) that disconnect several buttons at once. They may not be on the same row physically, but they are electrically.

Conductive-rubber buttons wear out over time. The conductive coating erodes away, and the micro sees no connection. That's the most common failure in remote controls. In Chapter 15, we'll explore how to fix them.

LCDs: Liquid-Crystal Disasters

LCDs have been around for decades. They've gotten increasingly reliable, but they still go bad often enough to make the list of pesky parts.

The concept behind all LCDs is simple: A liquid that has crystalline properties is sandwiched between two layers of glass. When voltage is applied, its crystals line up and twist the polarization of light passing through the panes by 90 degrees. A fixed optical polarizer in the front pane blocks the shifted light because its polarization no longer matches. This makes the segment or pixel go dark, in the same way that light is blocked by two polarized sunglass lenses when you rotate one in front of the other.

There are two basic styles of LCDs. Simple ones used for frequency displays and time counters are segmented, with numbers made up of seven individual segments, and other indicators, such as "Stereo," "AM" and "FM," being fed by a single line that turns on the entire word at once. Dot-matrix LCDs are the ones used for video screens on all of our modern devices like phones, tablets, laptops and flat-screen TVs. Letters and numbers are formed by turning on and off individual dots. We'll discuss dot-matrix displays in Chapter 15.

Segmented LCDs, which are still common in non-smart wristwatches, digital clocks, radios, car stereos, walkie talkies and DMMs, tend to develop blackened areas when they get old or have been exposed to temperature extremes. A dark blob oozes across the display, obscuring the segments and rendering the LCD unreadable (Figure 14-6). The cause is flowing of the liquid-crystal material from its assigned spots, leaving some areas bare. Why? The glue holding the front and rear glass panes together starts to fail, letting air in and separating the two glass panes slightly so there's not enough liquid crystal to fill all the volume. I've experimented with restoring these by carefully squeezing the LCD from both sides to push the material back where it belonged. It was successful sometimes, but

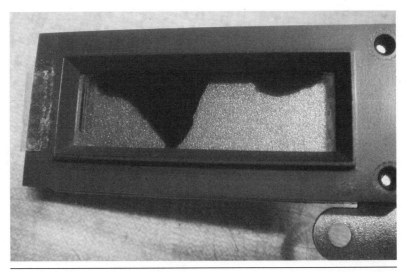

FIGURE 14-6   Blackened LCD.

eventually the liquid flowed back to where it started, and the display returned to being unreadable. The fixes did last a few months, though, which is better than nothing, I suppose.

A warning: The liquid-crystal material is toxic and shouldn't be handled. You'd probably only come in contact with it if you broke the display, but that could happen while you try to repair it. Also, it's possible some of the material could ooze out of the display's edges. I've never seen it happen, but I recommend that you wear protective gloves while handling a bare LCD with oozing problems, just in case. If you do touch the stuff, wash your hands pronto. You don't want to ingest liquid crystal or get it into a cut.

The other big issue with segmented LCDs is loss of one or more segments. Usually, this isn't a problem inside the glass panel, but is caused by a bad connection to the board. Most of the numeric LCDs you'll encounter are in a frame that presses them against one or two conductive rubber *elastomeric strips*, the other sides of which contact the board (Figure 14-7). The strips aren't conductive along their lengths, though. They have a bunch of vertical channels, each of which is insulated from the others. The channels are much smaller than the contacts they're pressed against, so they don't have to be aligned with them. At least a couple of channels will touch each contact on the glass edge of the LCD and the circuit board, and they're spaced so that shorts between adjacent contacts can't occur. Over time, oxidation causes loss of a few of those contacts, and some of the display's segments stop working.

To remove the LCD, you have to get it out of its frame, which typically involves untwisting four metal tabs that go through the circuit board to hold it in place. Sometimes the tabs are soldered to the board's ground traces, and you'll have to desolder them first. Now and then, you may find a frame that's screwed on instead. Once you pull off the frame, you'll notice that the strip seems glued to the LCD. Peel away the LCD slowly, being careful to keep the strip glued to the LCD, not the PC board. Once those strips peel off the

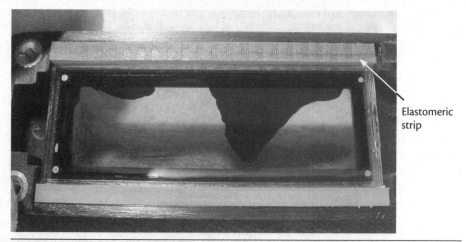

FIGURE 14-7 LCD with elastomeric strip.

LCD, it's hard to get all of the display's segments working again after reassembly. Typically, the bad connection is at the board side, not where the strip contacts the glass.

Clean the strip's contact edge, as well as the PC board's mating pads, using a clean, dry cotton swab. Don't use alcohol on the strip. You can use it on the board, though, if it looks like it needs it.

When you rejoin the display to the board and fire it up, you may still find missing segments on the display, indicating that the strip is making poor contact. Try poking on its edges with something non-conductive, if the edges are accessible. I like to cut off the end of a cotton swab and form the stick into a point. Sometimes, moving the strip a little bit rubs the contacts and makes them connect. If it doesn't work, or you can't reach them, a little extra pressure on the frame might get things connected. In one device, I had to put a foam shim behind the board to get it to push just a tad harder on the display. Other times, I've had to take the display off and clean the strip a few times before it made good contact. Most of the time, you can get it to work eventually. If you have no luck restoring contact, peel off the elastomeric strips and squeeze them to make their vertical span slightly larger. That might help them reach the contacts and exert a little more pressure when you put them back in place. This involves removing the strips from the LCD, so it's a last-ditch effort to be tried only when all else has failed.

Lithium Batteries

Thanks to its light weight and high energy density, the lithium battery has supplanted other types of rechargeable cells in just about all consumer electronics. Without lithiums, we wouldn't have tablets, smartphones, smart watches and small laptops that run for an entire workday on a charge.

Lithium batteries come in varieties of chemistry and construction, but their basic characteristics are similar. The native voltage of a rechargeable lithium cell is 3.6 to 3.7 volts, depending on the chemistry. Discharging them below 3 volts can damage them, and charging them past 4.2 volts can make them swell, burst and catch fire. And what a fire it is! Go to YouTube and watch some videos of exploding lithium batteries. You don't want this to happen in your house!

Recharging lithium batteries requires a special charging circuit called a *BMS* (battery management system), as discussed in Chapter 7, that monitors the voltage and switches to a low-level charge rate as the voltage nears 4.2 volts. Then, the charger shuts off. Charging a lithium with any other kind of charger is just asking for a fire.

Even when used properly, lithium cells can outgas as they get old, swelling to proportions several times their original size. Around the cell is a soft plastic casing designed to accommodate the swelling without bursting. Swelling has been a problem with laptop batteries for years. The battery chemistry has improved, and swelling occurs less often, but now and then those batteries still inflate like balloons. If the battery is sealed in a flat product like a smartphone or a tablet, the device can be destroyed by the pressure (Figure 14-8).

As with all batteries, the energy capacity of a lithium cell declines with age. Eventually, the operating time of the device gets so short that you want to replace the battery.

Batteries for popular items are available online, and most of them are not expensive. The hard part is getting the unit open to replace the cell. See Chapter 15 for a discussion of how to open tablets and smartphones. There's lots online for specific models, too.

When replacing cells, be sure the new one is an exact replacement. Some lithiums have an onboard BMS to prevent overcharging, and some don't. Unfortunately, unscrupulous sellers are happy to save a few pennies by omitting the protection circuit while claiming the battery is an exact replacement. Because most products have their own BMS boards, few people experience catastrophe due to the missing protection. A fault in the controller, or a badly designed product that relies on the battery's BMS to stop the charging cycle, can

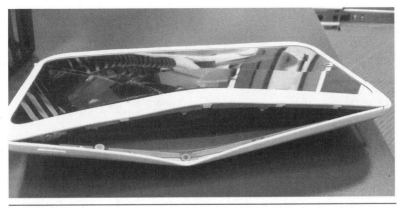

FIGURE 14-8   E-reader destroyed by a swollen battery. Look how big it got!

lead to a real disaster when there is no such circuit in the new battery. If your product's original battery has the protection circuit, make sure the replacement does too! With the flat lithium-polymer cells used in most products, the circuit is on a strip at the top, where the leads connect. See Figure 7-6 for a picture of one of those. Cylindrical lithiums of the sort used in high-power flashlights that include a BMS put it on a little round board on the end of the cell. Many of these cells have no BMS, though, because they're meant to be recharged in external lithium chargers made for them, and the space taken up by the board can be used for more battery chemistry, increasing the battery's milliamp-hour rating. With these, the BMS is in the charger.

Chapter 15

Aces Up Your Sleeve: Tips and Tricks for Specific Circuits and Products

Although the principles we've covered apply to pretty much all electronic devices, various product categories are different enough that they benefit from specific troubleshooting techniques. Let's look at some of the most common gadgets, their typical problems, and how to approach their repair.

Linear Power Supplies

Linear power supplies are no longer used in most consumer electronics, but there are still some out there, especially in high-end audio products designed to squeeze out that last ounce of sound quality. You'll also find them in separate power supplies for running car stereo and two-way radio units on AC power. These supplies are simple, but they still cause their share of trouble because they handle all the current used by the product, and linear voltage regulation generates heat.

How They Work

Linear supplies feed incoming AC power through a transformer to step it down to a lower voltage, and then rectify and filter it. Most also include linear voltage regulators, which act like automatically varying resistors that keep the output voltage constant as the device being powered uses more or less current.

What Can Go Wrong

Typical problems include shorted or open rectifiers, bad zeners in the voltage regulator stages and open or shorted series-pass transistors. As always, electrolytic capacitors can dry up or develop high equivalent series resistance (ESR) with age, but that problem occurs much less frequently than it does in switching power supplies because the frequency of operation in linear types is so low (50 or 60 hertz), and the voltage is applied to the caps in nice, gentle sine waves, not fast pulses with steep rise and fall times. So, the caps don't have to absorb rapid voltage spikes, and their internal resistance doesn't heat them up much, greatly extending their life spans.

Is It Worth It?

While older products had linear AC *wall wart* adapters, those are pretty much all switchers now, so most likely you're dealing with an internal power supply in a home theater receiver or other higher-end device or a fairly costly high-current external supply. Unless the power transformer has opened (there's no connection across one of its windings), yes, you'll want to fix it.

The Dangers Within

In a linear supply, AC goes straight to the transformer after passing through the power switch (if there is one) and probably a fuse. As long as you stay away from any exposed wiring or connections on the primary side (the side connected to the wall plug), you should be fairly safe. In the case of a low-voltage supply, such as a typical 13.8-volt unit used for running automotive and two-way radio gear in the house, your only likely potential danger is shorting across the output and dumping a lot of current through the short. It's more startling than hazardous when the short is via a screwdriver, but not so nice when all that current goes through a ring on your finger! With a supply providing voltages high enough to run, say, the complementary (push-pull) amplifiers in a stereo receiver, you'd be wise to avoid touching the supply's output connections, as they could be in the 80-volt range. That can definitely bite.

How to Fix One

As I mentioned, electrolytic capacitors are not the prime suspects in a malfunctioning linear supply. This doesn't mean you shouldn't bother to test them, but you're more likely to run into blown rectifiers, regulator problems and shorted or open series-pass transistors. Especially those.

　　If the unit is completely dead, start by checking the fuse on the primary side of the transformer, where it goes to the AC plug. A blown fuse is a sure sign of a short someplace, either in the supply or in the device being powered by it. What? No fuse? Yes, you may run into AC-powered products without fuses! That would seem incredibly dangerous and contrary to established safety regulations, but, in fact, there is a fuse. You just can't see it because it's a fusible link inside the power transformer! To test it, unplug the machine and check the resistance across the primary winding of the transformer. You can do this right from the unplugged AC plug as long as any "hard" power switch (one that interrupts the power from the wall socket) is turned on. If the ohms scale on your DMM shows an open circuit, the transformer's fusible link or the primary winding itself has opened. As discussed in Chapter 7, sometimes you can slice open the transformer's insulation carefully, find the link and reconnect it. Just be sure to find and repair the short before applying power, or the link will blow again as soon as you hit the switch. Don't bypass the link! You'll have no protection against overcurrent, and smoke is likely to ensue.

Unfortunately, in some designs, the winding itself serves as the fuse. If you don't see a fusible link under the insulation, the winding is burned out, and you will need a new transformer unless you're incredibly lucky and the melted wire happens to be on the outer layer. I've never been so fortunate; all the failures of this type I've run into have been deep inside the winding, out of sight. Just to be sure the transformer really is the problem, unplug the AC cord from the wall and try connecting your ohmmeter across the power transformer after the power switch, in case the switch itself is bad. Of course, you can check the switch, too, with your meter.

Assuming the transformer is good, there should be AC voltage at the secondary windings when the unit is plugged in and turned on. That side is isolated from the house wiring, so it's a lot safer to test with power applied. Exactly how the secondary windings are connected to circuit ground varies quite a bit, depending on the design, so using the chassis as a ground reference may not give you meaningful measurements. If you can't get a plausible reading, try measuring the AC voltage directly across the winding using your DMM's AC scale. As long as the DMM is running on batteries, there's no worry about accidentally shorting the supply voltages to circuit ground through the meter via your house ground, so how the winding is connected in the supply won't matter.

Many products use transformers with multiple output windings to generate the different voltages required to operate the circuit. Each winding has its own rectifiers and filter caps, along with any regulators that may be employed.

Be sure to test each section as if it were a separate power supply, because it is. One could be dead while the others work fine. When there are many wires coming from the transformer, it can be hard to know which ones go with which unless they're color-coded. The separate windings are electrically isolated, so you can figure them out with your ohmmeter as long as power is disconnected and filter caps are discharged. There could be some resistance between unrelated windings due to paths farther down in the circuit, but the mated pairs will have low resistance between their wires, typically tens of ohms. Sometimes, there can be multiple wires tapped off the same winding, and those will all show low resistance between them, but most of the time there are only two. If you find three, one of them will be a center tap in the middle of the winding. That one probably

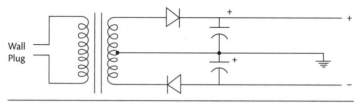

FIGURE 15-1   Bipolar power supply with center-tapped transformer.

goes to a common point in the supply or to circuit ground. Often, that configuration is used to create the bipolar supplies (those with both + and – voltages relative to ground) used in complementary audio amplifiers (Figure 15-1). Center taps used to be common even in unipolar supplies for providing *full-wave rectification* with only two rectifiers, but these days bridge rectifiers are so cheap that there's no reason to spend extra on a center-tapped transformer.

While it is possible for a transformer's secondary winding to be open, it's not common. Most of the time, transformer damage from a short on the secondary side blows the primary winding, its fusible link or, if you're lucky enough that there is one, the fuse between the transformer and the AC line.

After the secondary winding will come the rectifier. It could be just a single diode, a couple of them or a bridge. Older products might have four diodes in the classic ring configuration to form the bridge, but most stuff built today uses an integrated bridge, which is a plastic rectangle or circle with four leads. Two accept the AC input, and two provide the rectified DC output. The internal configuration is the same ring setup used with separate diodes. See Chapter 7 for more about bridge rectifiers.

Following the rectifier stage will be one or more big electrolytics. If any caps might go bad in a linear supply, these are the ones. Unlike with a switching supply, though, faulty caps won't shut things down. Instead, the supply's output will have a lot of AC line-frequency hum, or *ripple*, riding on it.

Voltage regulation in linear supplies doesn't have to be done with a linear regulator (one that acts like an automatic variable resistor), but it nearly always is because it's electrically noiseless compared with a switching regulator—the usual reason for using a linear supply in the first place. Look for zener diodes and series-pass transistors. In a high-current supply, the transistors will be mounted on a heatsink. An open zener or a shorted series-pass transistor will result in too much output voltage. Sometimes there will be ripple, too, even though the filter caps are good. This happens because there is still some hum riding on the top of the waveform, since the filters can't remove it all, but the voltage regulator normally keeps the voltage below where that hum lives, so it never makes it to the output. When the regulation fails, the voltage rises and the hum appears (Figure 15-2).

An open series-pass transistor will result in no voltage output. A shorted zener will result in very low voltage, but it probably won't be completely dead.

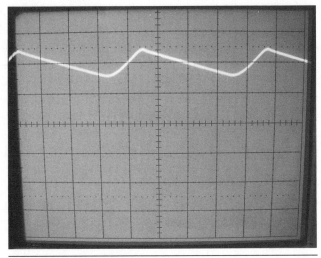

FIGURE 15-2 Ripple on top of DC voltage.

A Crow Walks into a Bar . . .

Some supplies employ a *crowbar* circuit intended to blow the fuse if the output voltage goes too high. Crowbars in linear supplies usually employ a silicon-controlled rectifier (SCR) across the supply's DC output to deliberately short it and pop the fuse. Sometimes resembling transistors, these parts have three leads. Unlike transistors, though, they stay turned on once they've been tripped by applying voltage above a certain level to the control gate terminal. See "Safety Components" in Chapter 7 for more on SCRs.

If the supply keeps blowing fuses, check the voltage on the control gate on the SCR to see if it's getting tripped. Or, just disconnect it and see if the supply runs. If so, it'll have too high an output voltage, so don't try to power anything with the supply while testing the crowbar or you'll probably wreck whatever's connected. A 12-volt supply with a shorted series-pass transistor can have 20 volts or more at the output!

Switching Power Supplies

Because you'll run into issues with switching power supplies in many kinds of products, and today's AC adapters and battery chargers are also little switching supplies, let's cover them first, before looking at the devices in which they take up residence.

How They Work

Switching supplies rectify the incoming AC into DC, smooth it with an electrolytic capacitor and then chop it at high frequency, pulling current through a transformer's primary coil with each pulse. Using a high frequency allows the energy to be replenished on the other side of the transformer much more frequently than with the old linear approach, which was limited to the 60-hertz power line frequency. So, the transformer doesn't have to convert as much power at one time and can be a lot smaller. This approach also keeps the chopper transistor either saturated (turned all the way on) or cut off (turned all the way off) almost all the time, resulting in high efficiency, since it spends almost no time per pulse in its *linear region*, where transistors act like variable resistors, with the attendant heat resistance generates. Efficiency is improved even more by regulating the output voltage with pulse-width modulation (PWM). Instead of making more power than is needed and then throttling it back with a series-pass transistor that operates like a resistor, the pulse width going through the transformer is adjusted to let through only as much energy as is required to keep the output voltage constant as the circuit being powered varies its current demand. No extra power to throw away means no extra heat to dissipate. See Chapter 8 for more on pulse-width-modulated voltage regulation.

What Can Go Wrong

Controlling those high-frequency pulses, with their fast rise and fall times, is tougher and stresses the components a lot more than those in low-speed linear circuits. The fast pulses with high-voltage peaks punch holes in transistors' substrates, and the rapid charging and discharging of filter capacitors wears those out too. Voltage spikes on the AC line are more likely to do damage because semiconductors are present on the AC side of the supply—notably the chopper transistor. Consequently, switchers fail significantly more often than do linear supplies. Nonetheless, very few products still use the old technology—switchers are everywhere.

Is It Worth It?

Unless the transformer is shorted or open, it's generally worth repairing a switcher, especially if it's inside a product you want to get working again. It might be a waste of effort on an AC adapter that could be replaced easily. The transformer rarely goes bad, though I've seen it happen with that once-ubiquitous, now thankfully obsolete cousin of the switching supply, the fluorescent backlight inverter. The high voltage of an inverter's output used to break down the insulation between the transformer's windings quite often. In a normal, AC-powered switcher used to create low-voltage DC output, shorted windings are an unlikely scenario.

If multiple semiconductors have blown from a chain reaction feeding voltage from one stage to the next, fixing the supply may be more trouble than it's worth. Especially if the PWM chip is dead, getting the part can be a hassle.

The Dangers Within

To service a switcher, disconnect the power first! *Never* work on one with power applied unless you have an isolation transformer. Even then, don't work on powered switchers until you gain a fair amount of experience. The dangers are real and significant. Did you take off your wristwatch and all jewelry? It's especially important now. Be sure to wear shoes, too.

Switchers are organized in two sides: the primary side, connected to the AC line and with no ground connection to the rest of the product's circuitry, and the secondary side, isolated from the AC line by the transformer and connected to circuit ground in the rest of the device.

The primary side of a switcher is where most of the trouble is, and also most of the danger. The chopper circuit stresses its transistor harder than any other component in the supply, leading to frequent failures. The "hot" side's direct connection to the house wiring makes service hazardous because any contact between you and ground completes the circuit. The secondary side is at lower voltages, and its isolation from the line means it's a lot safer. Some switchers go as far as having holes in the circuit board between sides for some extra protection against loss of isolation from arcing over during a power line voltage spike.

How to Fix One

Most switchers fail from bad electrolytic capacitors, blown rectifiers or a dead chopper transistor. Check the metal oxide varistor (MOV) surge suppressor at the AC input. If it's burned, the supply has taken a serious voltage spike, perhaps from a nearby lightning strike, and is likely to be badly damaged. Look at the capacitors. Any bulges? Change them. Leakage? Change them. Anything at all unusual about their appearance? Change them! They look fine? Check 'em with your ESR meter anyway.

Checking the rectifiers is easy enough if they're separate diodes. When you have a bridge rectifier, with all four diodes in one package, each diode must be tested as if it were a separate part. Take a look at the bridge rectifier symbol back in Chapter 7, which shows its internal construction. With all power disconnected and the big electrolytic near the bridge discharged, desolder the bridge from the board and use your DMM's diode function to test each diode in it, as described in detail in that chapter. If you find an open, a short or leakage in any of the diodes, replace the bridge. Shorts or leakage will apply reverse-polarity voltage to the high-voltage electrolytic capacitor connected right at the bridge's DC output terminals, so have another look at that as well.

The chopper is the big transistor, probably heatsinked, on the primary side of the transformer. Some choppers are bipolar transistors, but most are power MOSFETs. Small supplies might use a chopper in an IC package, and it can be integrated with the PWM chip. If the fuse is blown, it's a good bet the chopper has shorted out. The transistor can fail open, too, in which case the fuse might still be good. The transistor may have shorted and then opened, and the fuse may or may not have survived the momentary overcurrent. Check the chopper using the out-of-circuit techniques discussed in Chapter 7.

If you have an isolation transformer, you can do some powered tests before pulling parts. Check the voltage across the big cap on the primary side of the supply, near the chopper. Remember that you can't use circuit ground on this side. The negative terminal of the cap will be the reference point to which you'll connect the meter's black lead. You should see at least 300 volts. If it's much less, suspect a bad bridge rectifier. If it's zero, the fuse is probably blown, which could mean a bad bridge, a shorted cap or a bad chopper, or that the crowbar circuit tripped due to a voltage regulation problem. Usually, it's the chopper.

It's best not to try to scope the chopper directly because the voltages are very high, which could damage your scope and be a shock hazard if you touch the wrong spots. The safer approach is to scope the secondary side of the transformer, hooking the low-voltage side's circuit ground to the scope's ground clip. Many switchers have multiple taps on the secondary winding. Any of them will do as long as it's not the one connected directly to circuit ground. If the chopper is running, you'll see pulses at a significantly lower voltage than what's on the other side. They won't be tiny, though. Expect anything from 10 to perhaps 60 volts from baseline to peak. No pulses? The thing ain't running.

If the chopper is good but isn't operating, suspect the PWM chip or the regulation circuitry near the output. Open zener diodes on the secondary side can allow the output voltage to rise too high, activating protection circuitry and shutting down the PWM, or even tripping the crowbar, deliberately blowing the fuse. No pulses, no chopper, no operation.

If the supply is running but not putting out proper power, caps on the secondary side are the primary suspects. Scope them. If you see much of anything but DC on an electrolytic that has one lead going to ground, change it. Either its capacitance has declined, its ESR has risen, or both. If you change the cap but the waveform still looks noisy, look for a leaky rectifier feeding the cap.

Finally, remember that most switchers will shut down if output current demand exceeds their safe limits. Some may blow their fuses for the same reason. A short somewhere else in the device being powered may be pulling too much current and causing the supply to act like it's broken when it's really just protecting itself.

Not all switchers employ a crowbar, but it's common in high-current types. Sometimes, the crowbar is found on the AC line side of the circuit instead of at the output. In that case, a TRIAC is used because it conducts in both directions when energized, so it's effective at shorting AC current. The TRIAC is placed directly across the supply's AC power input, but just *after* the fuse. The crowbar normally doesn't conduct, so it has no effect on the circuit. If the output voltage goes abnormally high, a detector circuit sends voltage to the TRIAC's control gate, turning it on. The short across the AC line blows the fuse and stops the supply, protecting whatever gear it's powering, along with the supply itself. If you run into a switching supply with a blown fuse, but the chopper and bridge are good, it's reasonable to suspect that the crowbar tripped, and there's some problem with the voltage regulation system, such as the bad zener diodes mentioned earlier.

Because the AC line side and the output side of a switcher need to be electrically isolated, the signal to trip the crowbar is transferred between sides via an *opto-isolator*, which is an LED and a photodetector (light-sensitive transistor) in one IC-like package.

Audio Amplifiers and Receivers

Audio amps and receivers are at the center of all home theater setups. The units have to produce significant power to drive speakers, so they include a fair amount of heat-generating circuitry that is prone to failure.

How They Work

Though today's audio amplifiers and receivers employ digital signal processing for surround-sound decoding, delay effects and sophisticated tone controls, power amplification is still an analog process in many of them. The exception is the class D digital amplifier, which we'll discuss shortly.

Before home theaters replaced simple stereo systems, the chain was straightforward: preamp, through tone controls, to drivers and outputs. Not anymore! Input may come from analog jacks, digital coaxial cables or digital optical cables, with varying sample rates and bit depths (number of bits per sample), or through HDMI from a cable box or a streaming device. Several formats of multichannel audio encoding are used, too, and the unit has to be able to handle all of them. Once the desired signal processing has been accomplished, the data stream is converted back to analog and applied to conventional audio circuits. Capacitive coupling, with each stage connected to the previous and next stages via capacitors, is rarely used because it causes subtle distortions in the audio. Most of today's amplifiers are directly or resistively coupled for maximum fidelity.

Because modern digital formats brought audio with more than two channels to the home, the conventional stereo receiver is less common than it used to be; newer units usually have 5.1 channels, meaning there's a center front channel for movie dialog, two fronts, two rears and one to drive a subwoofer. Some have 7.1 channels. What's the difference between a woofer and a subwoofer? One of them can operate under water! No, seriously, a subwoofer is for reproduction of only the lowest frequencies, typically under 100 hertz, while a woofer's range may extend into the hundreds of hertz. With today's small satellite speakers unable to reproduce low frequencies much at all, the subwoofer is often really a woofer, but the term has stuck. One defining characteristic is that there's only one of them, as opposed to the usual separate woofer for each channel, because very low audio frequencies have little directionality, filling the room regardless of the speaker's position. Thus, there's no separation, stereo or otherwise, and the sense of spatiality comes from the higher frequencies being reproduced by the satellites.

For every channel, there is a complete amplifier chain culminating in an output stage. Many receivers use power amplifier modules for their outputs, but higher-end units still go with discrete transistor stages because they're reputed to sound better.

The Pulse of Things: Digital Amplification

Up to now, all-digital amplification has been popular only in automotive applications and Bluetooth speaker systems used with phones and tablets. It's gradually making its way into home theaters as the fidelity improves to the point of being competitive with analog techniques.

In an all-digital amplifier, incoming audio is used to pulse-width modulate a pulse generator operating at a frequency well above the audio range. If the input signal is analog, it'll be applied to a PWM that varies the width in step with the varying voltage of the audio signal. If the signal is digital, the pulse widths can be calculated in the digital domain, without ever converting the signal to analog.

Once the pulses are modulated, they're current-amplified without the issues inherent in analog amplifiers. There's no distortion or noise to contend with, no bias and no thermal drift. The output stages are often power MOSFETs, and they're driven between saturation and cutoff, just like in a switching power supply, so they don't get very hot, and efficiency is quite high.

So, why aren't all amplifiers made this way? Well, playing around with fast, high-current pulses leads to new problems. Essentially, a digital audio amplifier is like a switching power supply on steroids; there's a lot of power being switched on and off very fast.

The hassles come when those fast, powerful pulses are turned back into audio at the speaker end of the amplifier. Naturally, speakers require analog signals to move their cones in and out to mimic the audio waveform. A low-pass filter consisting of inductance and capacitance is used to integrate the pulses into a changing voltage, just as is done in a switching supply. The wider the pulses, the higher the resulting voltage; the narrower, the lower. Here, though, it needs to be done fast and accurately. Filters aren't perfect. They can ring (exhibit *resonance*) and cause other distortions. Plus, all those fast pulses can generate massive radio-frequency (RF) noise, and they also may *alias* against the desired audio frequencies, depending on the pulse frequency chosen, causing an especially jarring distortion.

Only recently have digital amplifier designs solved most of these problems well enough to sound truly great. Eventually, all home audio gear—except, perhaps, for high-end audiophile equipment—will go completely digital from input signal until the speaker wires, just as portable Bluetooth speakers are now.

What Can Go Wrong

The power supply works mighty hard, at least when the volume is turned up high. Moving a lot of air with the subwoofer, necessary for the serious bass frequencies found in movie explosions and such, takes a particularly large amount of power. While modest home systems may have a hundred watts available for this, self-amplified subwoofers with several *thousand* watts have been marketed. Of course, people exercising the full power of those things can't actually *hear* their movies anymore, but that's what subtitles are for, right?

Plenty of receivers still use linear power supplies because switchers can introduce high-frequency noise that's hard to eliminate. To power five channels of 100 watts each takes some serious iron in the transformer, along with high-current diodes, hot-running linear voltage regulators and huge electrolytic capacitors. There's a reason the expensive receivers are so big and weigh so much! All that heat and high current take their toll, especially on the diodes and capacitors. Even with the generally higher reliability of the linear approach, power supply failures in receivers and powered subwoofers are common.

After the power supply, the most trouble-prone parts of a receiver are the output stages. Whether modules or separate transistors, they are where the current is. Plus, in standard analog designs, they operate in their linear regions, neither saturated nor cut off at any time (one hopes!), so they're essentially resistors, dissipating power supply current as heat. Look for large heatsinks, and you'll find the output stages.

With all those jacks, input switches and interconnections between boards, signals can be impeded by bad connections, causing them to crackle or drop out completely. Phono preamp sections are especially vulnerable to this because they handle the very tiny signals generated by magnetic phono cartridges, and it doesn't take much to stop those. Cartridges put out around 5 millivolts peak to peak, compared to the 1-volt standard for line-level audio. Still, even with the higher-level signals, ratty connections in the signal path cause many receiver problems.

Speaker protection circuits sense when there is significant *DC offset*, or variance at the midline of the signal waveform from 0 volts. When offset occurs, there's a fault somewhere in the amplifier, and the protection circuits disconnect the speakers to prevent excessive power supply current from burning out their voice coils. At least that's how they're supposed to work. Now and then the protection circuit malfunctions, going into protection mode when nothing is really wrong. Also, the relays used in these circuits can develop bad contacts, causing a channel to cut out randomly. A receiver of mine had that very problem.

Is It Worth It?

Not much in a receiver's power supply or amplifier chain is especially expensive or hard to get, so most receivers of any real value are worth repairing. Output modules are available from online parts houses, as are power transistors. You aren't going to find the digital signal-processing (DSP) chips, but it's highly unlikely you'll need them anyway. In receivers, the power-handling areas are usually where the mischief lurks. I've never had to change a DSP chip.

The Dangers Within

Most output stages are *complementary* push-pull types fed by positive and negative power supply lines, or they're quasi-complementary, or unipolar, but still push-pull. The supply lines may have 30 to 80 volts or so on them, so they are capable of shocking you. And, should you be unlucky enough to touch both at the same time in a true complementary design, you could come in contact with 160 volts or more.

Receivers with fluorescent display panels have small inverters supplying the panels with the few hundred volts required to light them up. The inverters are typically located on boards just behind the displays. Most of 'em have a small, round transformer to step up the voltage. Heatsinks can get mighty hot, so avoid touching them if the unit has been on for a while, especially at high volume.

How to Fix One

Servicing a receiver is very much a process of elimination. The first thing to consider is whether the problem affects all the channels. If so, head for the power supply. One channel could have a short pulling everything down, but the supply is the first thing to check. If only one channel is affected, that's where you'll find the trouble; the supply has to be okay because it's powering the other channels properly.

Power Supply Problems

With a linear supply, testing is pretty painless, compared with the ordeal of working on a switcher. The transformer comes early in the chain, and what's between it and the AC line is easy to evaluate without power applied. If the unit is dead, check the fuse. As with other products we've examined, a blown fuse almost always means something pulled too much current through it. Look for shorted rectifiers or a short somewhere in the amplifiers being powered, on the other side of the transformer. If there's a connector you can pull to isolate the supply from the rest of the unit, try doing that to see if the overcurrent situation persists. For more info, see the section on linear power supplies earlier in this chapter.

Output Stage Problems

A shorted output stage will pull the supply's output down toward ground, possibly blowing the fuse. Sometimes there's enough resistance between the shorted transistor or module and the supply to limit the current to a value below the fuse's rating, so it survives. The telltale sign of a shorted output stage is a loud hum through a speaker connected to that stage. However, other channels' outputs may also produce the hum because they're fed from the same supply line that's being pulled down by the bad one, and the hum comes from that. They won't hum as loudly, though. Scope the output. If you see DC with something resembling a 60-hertz sawtooth wave riding on it, you've got a short.

Frequently, changing the output transistors or the module takes care of everything. Sometimes, and particularly with discrete designs (no module), the driver transistors may be shorted too. Also, always check the resistors in series with the emitters of the power transistors in discrete stages. The overcurrent can really cook those babies, altering their values or even cracking them. Most techs replace the emitter resistors as a matter of course when they change output transistors.

In direct-coupled designs, shorts in one stage may blow surrounding stages. I've seen a shorted output transistor take out every transistor in its channel, right back to the input jacks! This used to happen frequently in early designs, and it's still possible even now. Ironically, the costliest units are more likely to have extensive damage, thanks to their closely coupled stages with few or no components to isolate them. That kind of circuit sounds the best when it works, but it's a mess when it breaks.

At the inputs to the driver stage, look for diodes or zeners. They're used to set the bias points, keeping the positive and negative halves of the output stage just slightly turned on even when there's no signal or the signal is very small. An open zener or a shorted diode can make the bias go wild, causing the outputs to conduct themselves to death. Changing the transistors without checking the bias just means you'll need more transistors in a few minutes.

The damage can proceed the other way, too, from input to output. Again, designs with DC coupling are the most susceptible because the bias of their output stages can be affected by the DC offset of previous stages. If the input transistor shorts, it can drive DC right to the outputs, upsetting their bias and sending them to semiconductor heaven. Or, considering the heat, perhaps it's the other place.

Although it seems like diagnosing an amplifier should be very straightforward, the interdependence of DC levels from stage to stage can turn a romp into a nightmare. If you find yourself going around in circles, disconnect the output stages from the power supply, and scope their input lines. You should see normal audio with some DC offset (the bias). Most receivers with analog amplifiers use bipolar output transistors, so the bias is a matter of base *current*, not voltage. With no power to the transistors, the offset may be significant, but it shouldn't be close to the voltage of the power supply *rails*.

In a digital amplifier, all you should see are big pulses whose widths vary with the audio signal. If the output FETs are okay and you see the pulses on their output terminals (usually the drain), check the filter inductors and capacitors between them and the speaker.

Small-Signal Problems

Small-signal issues are like those in any other device. You could have a bad cap, a bad transistor and so on. To determine whether the problem is in the small-signal sections, test the various inputs. If *any* of them work properly, the power supply and output stages are doing their jobs. Unless the receiver is all-digital, using a class D amplifier, the various signals—digital, analog, radio—wind up as analog audio at the inputs to the power amplifier stages. If no input works, go back to the beginning of the signal path, using the simplest one possible: an auxiliary analog jack. Feed it a signal from an audio player, select it on the front panel as the active audio source, and scope your way through the low-level stages to see where it stops. Oh, and be sure to turn the "tape monitor" switches off! They interrupt the audio path so a tape deck can be inserted in line with it. Most receivers still offer the tape monitor function, even though tape recorders are dead and gone. If the button is engaged with no deck connected, the audio comes to a dead end, never reaching the power amplifiers. Also, some receivers have preamp output and power amp input jacks on the back, and jumpers (short cables or connecting rods) have to be installed between these, or you'll hear nothing.

With their multiple input paths, receivers have more connections through switches than most devices, offering plenty of opportunities for bad connections. Newer units often have electronic switching, rather than the old rotary or push-button switches used for decades. Noncontact switching is more reliable, but a blown analog switch chip will stop things dead. The most common chip used to route audio signals is the 4066B, and I've seen plenty of 'em go bad. It's pretty rare for just one of the chip's inputs to fail, instead of the whole chip, but it does happen.

Hum isn't always a power supply or output stage problem. If the sound of the hum is thin, with high-frequency content that's a little bit buzzy, and it isn't loud enough to wipe out the audio, there may be a bad ground connection somewhere in the low-level circuitry. As in many products, grounding from the board to the chassis can be mechanical, provided by the pressure between the board's ground lands and the metal tabs into which the board is screwed. Over time, oxidation increases the resistance of those connections, and hum can result.

If the level of the hum varies with the volume control, the source has to be early in the chain, before the control. Try unplugging the audio source from the input jacks. If the hum disappears, you have a bad cable feeding the jacks or a *ground loop* between the audio source and the receiver. It's still there? Pull the board and clean the ground lands and the chassis mounting tabs until they're nice and shiny. If this doesn't solve it, check for bad solder joints that may be causing poor grounding.

Digital Still Cameras

Once upon a time, cameras were mostly mechanical, with just enough electronics to open the shutter correctly and fire the flash. My, how things have changed! Today's cameras are all digital, crammed with large-scale integrated (LSI) chips and support components. Most of the better ones still have mechanical parts, though, including shutters and motorized zoom lenses. After all, if all you want are a fixed lens and an electronic shutter, you can just use your phone. In fact, phone cameras are getting mighty good these days. Camera makers know this, so they aim their product lines toward more sophisticated photography, which requires some mechanisms.

How They Work

Digital cameras are based around an image sensor chip that takes the place of film. Charge-coupled device (CCD) imagers have largely been replaced by CMOS versions, though both are still used. Most pro and semipro cameras have mechanical irises and shutters, emulating film cameras and permitting fine control over shutter speed and exposure.

Pocket cameras, which are being phased out, sport zoom lenses that extend to take pictures and retract when the camera is turned off. Some especially small models had folded optics, with everything contained within the body, but they're getting rare. Fancy digital single-lens reflex (DSLR) cameras use the same permanently protruding zoom assemblies found on old-style film cameras.

The image is read out of the chip and processed for color correction, exposure and other factors and then encoded into a JPEG file. Some expensive cameras can save the raw pixel data without JPEG compression. Many digital cameras also take videos, saving them in various file formats, most of which are some variety of MPEG4. The image or movie data is stored on a flash memory card.

What Can Go Wrong

Most digital camera problems are mechanical. In particular, the retractable zoom lens assembly gets sand or dirt into its gears and becomes stuck. The shutter can stick as well. Other issues include poor battery and memory card contacts, along with oxidized internal connectors. All of these problems are environmental in origin; the cameras get tossed around, dropped into beach sand, splashed on, sat on and otherwise abused.

Is It Worth It?

If the imaging chip goes bad or there's a defective component on the tiny, dense circuit board, consider the camera a loss unless it's expensive enough to be worth sending off for repair. If the LCD has broken from a fall or being sat on, you might find a parts unit or a new LCD online. Replacement isn't hard if you can get the part. If the case is bent badly, however, be sure the rest of the camera works before you shell out for an LCD.

I've had little luck getting sand out of stuck zoom lens assemblies, but it's not out of the question. It costs only time and patience to try. If the zoom is otherwise broken, you'll need a parts camera. The gears in most of these things are plastic, and they crack easily. Bad battery and memory card contacts can be cleaned. Often, that does the trick.

The Dangers Within

There's only one significant danger in a digital camera, but it can bite you like a thirsty vampire. In cameras with a real flash tube (not an LED), the flash capacitor stores a few hundred volts, provided by a tiny inverter. The cap can hold a charge for quite a long time! Also, you don't know while opening the camera where the cap connects to the board, and those contacts might be under your fingers as you take off the case. As I mentioned in Chapter 9, I've gotten zonked by digital cameras more than by any other product. Watch your fingers!

How to Fix One

Typical camera construction has everything mounted to the front. When you take off the back, the LCD and circuit board are easily accessible. Getting to the lens assembly is a bit tougher. You have to remove the board, which involves popping off all of those tiny ribbon cable connectors. In some cameras, the battery contacts are built into the case and soldered to the board. It won't budge until you unsolder them.

The optical assembly is sealed to keep dust off the imaging chip's face, where it would show up as blobs in photos. Once you disconnect the chip's ribbon connector and all the other connectors to the lens assembly, you can unscrew the assembly from the front of the chassis and have at it (Figure 15-3). Construction varies among models and manufacturers, but you'll see a couple of tiny motors. I've run across a few assemblies that had a motor hiding inside the barrel, where you'd never expect it.

One motor extends and retracts the lens. Usually, this is the one that gets stuck. Look for grains of sand or dirt in the gear teeth. The obstruction may be in gears buried deep inside the lens assembly. If you can clean them all out and get the thing back together, you have a shot at restoring proper operation. If the gears are cracked, the camera is not going to work without replacement parts. Your best bet is to replace the entire assembly, in one piece, from a parts unit.

 A can of compressed air can be useful here. Just be sure not to blast it directly on the imaging chip's surface, because the gas can create a static charge that'll cause dust to stick to the chip, and the can's propellant can mar the optical surface, too. When clearing dust from the imager, keep the nozzle at least a few inches from the chip.

FIGURE 15-3 Digital camera lens assembly.

If the camera reports a low battery when you know the battery is charged, either the battery is getting old or there's resistance between the battery and the circuitry. The problem shows up mostly when the flash is charging, because that takes a fair amount of current, so the voltage drop across the unwanted resistance goes up at that moment and the camera's low-voltage sensor gets tripped.

Old lithium batteries exhibit increased internal resistance, so the battery itself could be responsible for the drop. If you're sure the battery is okay, check the contacts on both the battery and the camera. Most of them are gold-clad, but they still get dirty and cause this problem. Clean them with a pencil eraser, and then use an alcohol-soaked swab afterward to remove the rubber residue.

Also, check the spots where the battery contacts connect with the board. Some are soldered, and that's best, but the solder job might need retouching. Some forgo solder and use pressure to contact pads on the board, and these types cause the most trouble. Clean up those contact points, and the low-battery indication should go away.

Beyond bad imager chips, random circuit failures and liquid damage, the only other common problem with cameras is unresponsive buttons, caused by oxidation. Those buttons can be mighty tiny! I don't recommend trying to spray inside them with contact cleaner, mostly because you won't get the stuff inside the button, and it'll wind up all over the rest of the board or, much worse, in the optics. You sure don't want it there! If you can't scrounge a replacement button, you can try taking apart the existing one and scraping its contacts with an X-Acto blade. It takes a steady hand and a magnifying glass, but I've done it without removing the button from the board and had it work. The hardest part was getting the button back together again.

Disc Players and Recorders

Discs are getting pretty obsolete, but some people have significant investments in their collections, so they keep using them. Plus, streaming services often don't offer older favorites, and public libraries still lend out DVDs and Blu-Rays. Most of us just stream now, but discs have not gone away.

Though inexpensive, CD and DVD players are not simple. They are actually little computers combined with the mechanical and optical sections necessary to retrieve data from the disc.

How They Work

To find, follow and decode the disc's microscopic optical tracks requires a focused laser beam, a three-axis servo system and a fair amount of computing power. An awful lot has to go right for the data to be read from the disc and transformed into your favorite movie or music.

First, the disc must be accepted into the player, properly seated on the spindle and clamped down. The machine moves the optical head assembly, or *sled*, to the center of the disc, where playback will begin. Then, the laser turns on, and the lens moves up and down while the photodetectors in the head look for a reflected beam from the disc. If no reflection is detected, the player stops and displays "No disc." Once it sees a reflection, the machine stops the lens when proper focus is found. This is determined by reception of maximum beam strength in the head's center detector with minimum beam strength in the side sensors, indicating optimal spot size. Only if focus is achieved will the player spin the disc and try to read the track. If the lead-in track is found, the sled motor starts moving the head away from the disc's center, and playback commences. As the disc spins, its normal wobbles and eccentricities far exceed the size of its tracks, so the tracking and focus servos keep the lens dancing around in three-dimensional lockstep with their movements. The disc's speed gradually decreases as the head proceeds toward the outer rim because the linear distance for each rotation increases. The object of the technique, called *constant linear velocity*, is to keep the speed of the track constant as it goes past the head so the bits can be crammed in as tightly as possible for maximum disc storage capacity. The spindle speed is controlled by the microprocessor, which monitors the rate at which data is being read from the disc. Only when all these systems work in concert can a disc be stably tracked and played.

What Can Go Wrong

The disc may seat incorrectly, resulting in more wobble than the focus and tracking servos can handle. Many tray-loading players move the tray with a belt that stretches over time, preventing the door from opening or closing fully. They use nylon gears that can fracture, sticking when the mating gear's tooth hits the crack. The leaf switch telling the micro when the door is open or closed can bend or become oxidized, so the door motor keeps running even after the door hits its limits, and the micro never starts the playback sequence.

Portable players with top-loading lids have their own quirks. To prevent you from looking directly into a running laser, they use a small interlock switch to sense when the door is closed. A bad switch makes the micro think the door is open, resulting in no operation. This is a very common failure in these machines.

The sensor indicating when the head is at the starting position can malfunction. Most players use a leaf switch, though some use an optical sensor. The optical variety is pretty reliable, but leaf switches get oxidized or corroded and stop passing current, so the microprocessor never gets the message that the head is positioned. The result is a clacking sound made by the head as it slams over and over into the end of the sled's track.

Unlike LEDs, laser diodes have finite life spans, and they get dimmer as they age. If the laser is too dim, the reflected beams will be hard to detect and decode properly, and the machine will have trouble starting discs. It may also skip on discs that do play, although there are other causes of skipping. And even bright lasers can develop odd internal reflection modes resulting in optical impurity; the beam stays bright but can't be focused to the required spot size.

If the ribbon cable's connections to the head are oxidized, the weak signals from the photodetectors will be erratic, confusing the machine badly and causing symptoms much like those of a failing laser. If the sled motor has a flat spot on its commutator (where the brushes transmit power to the rotating coils) or the slide or gears need lubrication, the head will stick as it scans across the disc, resulting in skipping or freezing.

It's important to note that the sled motor does not move the head in the tiny increments required to advance it along with the spiraling track; such fine mechanical motion would be impossible to achieve in an affordable product, if at all. Instead, the *lens* moves sideways, even as it bobs up and down to maintain focus, until it approaches the point where the beam would miss the sensors. The micro detects when the lens is near its limits and pulses the sled motor, advancing the head. The lens then moves back to the center of its travel, and the process begins anew. Essentially, the lens and the sled play a game of inchworm as they follow the recorded track across the disc. So, intermittent, tiny jerks of the sled motor are completely normal. If you observe the sled assembly of a working player, you'll see the worm gear shaft twitch every few seconds as it pushes the head outward ever so slightly.

If the spindle motor, which spins the disc, is worn out or gummed up, the disc may not come up to proper speed, resulting in a slow data rate the machine can't process into normal audio or video content. And, of course, all of the servo and decoding circuitry has to work properly. Playing an optical disc is a feedback process; the data rate tells the spindle motor how fast to turn, and the reflections from the disc surface tell the tracking and focus servos how to keep a grip on the track. A loss of any of these systems can keep the others from doing their jobs.

Is It Worth It?

If the laser is dead, forget about repairing the machine. The alignment of the laser diode in the head is critical, so you can't pull the laser and replace it; you need a whole new head. When disc players were costly, replacement heads were available from parts houses. These days, the players are so cheap that there's no market for the heads, so their sources have dried up.

By the same token, if mechanical parts like motors or gears are broken, your only hope is a parts machine; you're not going to find replacements. Disc players incorporated into game consoles are the notable exception. Game units still cost enough that heads and even entire sled assemblies are available. Check the internet for parts houses supplying these items.

Just about anything else can be fixed. You're not going to find a source for LSI chips and such, but those are unlikely to be causing the trouble anyway.

The Dangers Within

Never look directly into the laser beam! The wavelength used for reading CDs is in the infrared, but the purity isn't perfect, and there is some visible red as well. The red portion of the output is dim compared to the primary infrared energy, so looking at it gives a false sense of what your eye is receiving. It's like staring at a solar eclipse: Your poor retina is getting blasted but you don't know it. DVDs are played with a visible red laser. It's just as damaging, but at least you can see what's coming at you. In either case, you might burn a permanent hole in your visual field if you look straight down the bore.

It's routine to check for a working laser, but always look from off to the side. There's enough reflection in the lens to let you see if the beam is there. When you see how bright a DVD player's beam is, even from the side, you'll get a good sense of just how damaging a full-on view from either a DVD or a CD player could be.

Under no circumstances should you ever try to view the laser of a disc *recorder* when it's in record mode, even from off to the side! The optical energy output is high enough to pop balloons; imagine what even a momentary reflection might do to your eyes.

How to Fix One

As with all remote-controllable shelf gear, power supply capacitors may be shot if the unit has been plugged in for a few years. Beyond that, the most common problems are failure to accept and seat the disc properly, inability to play at all, and skipping while playing. If the door won't open, look at the display. Is the digital control system working? Do normal numbers or messages appear on the display? If not, head for the power supply and check all the usual things like capacitors and output voltages. If the supply is good, scope the clock crystal at the microprocessor. Players typically have multiple crystals, but you should find one right next to the biggest chip on the board. See if it's running. It should have a sine or square wave on it of at least a few volts, at or very near the frequency marked on the crystal.

Door Problems

Assuming the power and control systems are working, check for leaf switch problems around the door. The exact layout varies from machine to machine, but the door motor, gears and belt are usually located under the platform (the entire mechanism), and the leaf switch will be buried in there someplace, with wires going to it (Figure 15-4).

Reaching the door mechanism requires removing the platform from the frame. Look for largish screws at the corners, seen from above. You'll probably have to remove the front panel as well, although sometimes you can pull the platform toward the back and lift it out without doing so.

Some players use two leaf switches, one for the fully out position and one for fully in, but most use one single-pole, double-throw (SPDT) leaf. The center blade gets pressed against one outer blade when the door is open and against the other when it's closed. Look for three wires. Check that the blades aren't bent and that their contact points are clean. If

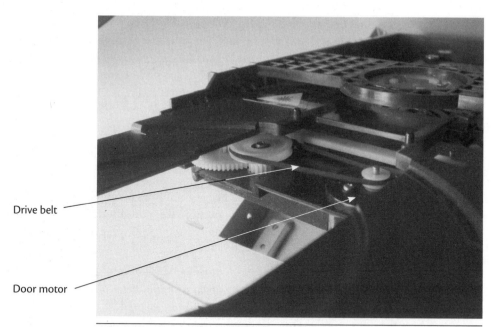

Drive belt

Door motor

FIGURE 15-4   **Door mechanism with belt on underside of platform. The leaf switch is not visible.**

the contacts are black, their lubrication may have dried out and become insulative. Gentle application of fine sandpaper or careful scraping with your X-Acto knife will take off the black coating. Don't go too hard on those contacts! If you grind off their plating, they'll work for a while but will fail again pretty quickly. A few gentle brushes with some no. 600 sandpaper should get the black coating off without doing any damage.

If the door opens and closes but not all the way, either the belt is shot or there's a broken nylon gear. Nylon gears crack easily. Any sort of abuse can break them, and sometimes they just fail with time. If the door is sluggish, the belt is the likely culprit. If it moves at normal speed but stops abruptly at a consistent spot, look for a broken gear. Check between the teeth; a grain of sand can get in there and jam the gears without breaking anything.

Clamp and Spindle Problems

A working door mechanism should pull the disc in, plop it on the spindle and then lower the clamp onto it. Believe it or not, in a shelf-type player, the only thing keeping the disc in contact with the spindle is friction. The clamp contains a magnet that's attracted by the metallic spindle, holding the disc in position and facilitating the friction grip.

Portable players rarely have clamps. Instead, three or four spring-loaded ball bearings or tiny tabs in the spindle grab and hold the upper edge of the disc after it is pressed down firmly. I've never seen one of these be a problem.

In players with sliding doors, improper disc clamping is common. The clamp's upper portion hangs loosely in its holder and should float when the disc is spinning. If the holder is misaligned, the clamp will rub against it, making an obvious noise and dragging the disc speed down, perhaps preventing playback. Don't be surprised if the clamp wobbles at the top a bit without rubbing. That's normal. Rubbing, of course, is not.

Dirt on the spindle can keep the disc from sitting flat, leading to focusing errors and skipping. If the spindle gets greasy, the disc may slip. Make sure it is clean.

When the disc spins, it shouldn't wobble much. You may see some shimmering of the surface, but the edges should sit pretty flat. If it wobbles, try another disc, to be sure the first one isn't warped. Assuming a good disc, any significant eccentricity is caused by either a bent spindle shaft (unlikely in door-type players but more plausible in top-loading portables because of the need to press on it when inserting discs) or misaligned clamping.

Playback Problems

If the disc is properly seated but won't spin, the startup sequence of head positioning, disc detection and focusing has failed. In a door-type player, it should begin as soon as the disc is seated, after the leaf switch signals the micro that the door is closed. If the switch doesn't work, you'll hear the door motor strain as it continues to try to shut the already-closed door. In a portable unit, there's no leaf, but there's another element: an interlock switch. When the lid is closed, a little plastic finger on it protrudes through a hole in the player's body, pressing on the switch. If the finger breaks off or the switch fails, the micro will never know the door has been closed, and nothing will happen. I've seen numerous bad interlock switches in these players.

Is the head all the way in toward the center of the disc? If not, there's a sled problem. Check that the rails are greased at least a little and that the start position limit switch is okay. Sometimes that leaf switch gets bent just a tad and will still work, but not until the head is closer to the spindle than it should be. Focus will be achieved, but the head won't find the lead-in track, and it'll just sit there until the machine gives up and stops.

Before going further, check that the lens is clean and not scratched. Portable players with exposed heads are especially subject to dirty and damaged lenses. The lenses in door-type machines are well protected from scratches, but they can still accumulate dirt. If the machine has been in a smoker's home or was installed in a kitchen, there may be a film on the lens, preventing proper focus. Even a protected lens could be scratched if the user tried one of those nasty cleaning CDs with brushes. *Never* use those things!

Getting to the lens in a portable is easy. Pop open the lid, and there it is. Take a cotton swab and wet it with water. Do *not* use alcohol! The lenses are plastic and will be permanently blurred by it. Blot most of the water from the swab with a tissue so that it's damp but will not drip water into the optical head. Wipe the lens gently, and then wipe it again with a dry swab. Don't put pressure on the lens while doing this. Believe it or not, the cotton fibers of swabs can scratch plastic lenses if you press down while wiping. Use the gentlest of strokes.

To clean the lens in a door-style player, you may have to disassemble the clamp assembly and remove it. Sometimes you can reach under it by bending a swab's head at an angle.

With the head at the starting position, the lens should move up and down as the player searches for proper focus. The lens is mounted on coils of very fine wire called *voice coils*, so named because the arrangement of a coil over a magnet is similar to what's found in a speaker. Current passing through the coils generates a magnetic field that interacts with a permanent magnet in the head, permitting the control system to move the lens toward or away from the spindle and up and down. During playback, the lens assembly floats on this field, bobbing and weaving as necessary to follow the disc's track. Look at the head from the edge of the disc, and you'll see the lens. If it's not moving, there's a problem with the focus circuits or the cable to the head.

If the lens doesn't move after you hit "Play," the startup sequence has not been initiated. Even with a dead laser, the lens should bob up and down a few times until the micro figures out that there's no reflected beam. There could be a digital control problem, or the door's leaf switch might not be signaling the micro that the door has been closed. The door motor may still be straining, but you might not have noticed it. In portables, check that interlock switch!

Laser Problems

Is there a beam? Its side reflection will be red (except in a Blu-Ray player, where it'll be blue). In a CD player, the beam will look dim because most of the energy is infrared. In a DVD player, it's very bright. If it's there, the laser is probably okay. It still might have problems, but at least you know it's not dead.

If the lens moves up and down but there is no beam, you've hit on the trouble! Alas, many disc player failures are due to a bad laser. If there's no light, the laser is probably shot, and you have just become the proud owner of a parts machine. To be sure, you can trace back from the laser diode to see if it's getting voltage. Laser diodes are driven by a few volts DC, and there's usually a tiny trimpot right on the head at the diode that'll help you find the right connections. If the DC is there but the beam is not, it's bye-bye laser and bye-bye player.

You may see two trimpots in DVD players. They use two lasers, an infrared for playing CDs and a visible red for playing DVDs. If the machine will play one but not the other, one of the lasers may have died. Be sure to test the unit with the type of disc it won't play.

Even if there is a beam, it might be too dim for proper operation. It's hard to tell with a CD player because most of the energy isn't visible anyway, but you can get a good idea by comparing the brightness to that of a good player. Remember, look from the side only. In a DVD player, the beam is so bright that often you can see it right through the disc!

If the beam shines and the lens moves but nothing else happens, it may not be finding focus. Look on the board for a test point labeled "FOK" (yeah, I know), "FOC OK" or "FOCUS OK." Scope it. It'll change state (usually from low to high) when focus is achieved. If it doesn't, then something is preventing proper focusing.

If the disc starts spinning, focus lock is good. The rotational speed depends on where on the disc the head is positioned. At the start, the disc should spin a few hundred RPM. If not, there's a problem with the spindle motor. Either it is mechanically gummed up, the motor is bad or the driving circuitry isn't doing its job.

Hair, both pet and human, can wind itself around the spindle motor's shaft, even in the protected environment of a door-style player. Remove the disc and turn the spindle by hand. It should turn easily and smoothly. If not, check for hair. Sometimes the motor goes bad or its lubrication dries out. If the spindle offers significant resistance, there's a mechanical problem of this nature. Taking apart the motor to try and lubricate it is pretty involved, but if you've ever disassembled a little motor in a toy, it's the same process. Be careful not to wreck the brushes at the back. But before you dig into a sticky motor, try dribbling a little silicone lubricant spray down the shaft. I've had pretty good luck with that on all kinds of small DC motors. Depending on the construction, you might be able to get some into the back bearing as well.

The optical head provides two functions: tracking and data extraction. To get data, tracking has to be working. Either or both of these can be affected by oxidation or corrosion of the head's ribbon cable connections. After a dead laser, this is the second most common head-related problem. Laptop drives, for some reason, are especially prone to this issue. Pull the cable at the board end (be sure to check for a latch on the connector!), and look at its metal fingers. If they're gold, wipe them with an alcohol-moistened swab. If they're silver, they probably have solder on them. These are the kind that cause the most trouble. With a magnifier, look for black pitting. Gently scrape it with the tip of an X-Acto knife, and then use a swab to wipe away the tiny metal flakes. Clean with alcohol and reinsert. Be sure to lock the latch if there is one. If there's a connector at the head end of the cable, do the same procedure there.

Be aware that laser diodes are easily destroyed by static charges. Be sure you and your tools have touched ground just before you begin working on the cable. After reinserting the cable, check to see if proper operation has been restored. I've saved countless laptop drives with this procedure. At least they're external these days, so taking them apart isn't too rough. Back when they were built into laptops, getting at the innards was a real pain.

The primary output from the head is a signal called the *eye pattern* (Figure 15-5). This signal is the actual raw data being read from the disc. All players have a test point at the first preamplifier stage, which you can find by following the head's ribbon cable back to the board. The preamp will be a chip of low to medium density, and you'll see test points very near its pins. Often, they're not labeled, but you can find the eye pattern by scoping the points; no other signal will look anything like it.

The eye pattern should be around 1 volt peak to peak. If it's much less than that, either the player is not tracking well or the laser is dim. Some amplitude wobble is normal as the disc spins, especially when the head is near the outer edge, but it shouldn't be more than 15 percent or so. If the amplitude dips a lot, expect skipping or dropping out. Weak lasers can cause this, as can problems with the focus servo. In players that have focus gain trimpots,

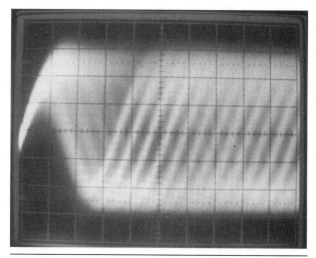

FIGURE 15-5   Laser optical head eye pattern of a CD at a sweep rate of 100 nanoseconds. These are fast signals! Use 50 nanoseconds for DVD eye patterns.

sometimes you can compensate for a less-than-optimal laser by upping the gain a tad. Most newer players don't have servo adjustments, but it's worth looking for a trimpot labeled "F GAIN" or "FOC GAIN," just in case.

If all looks well but the player skips, check for binding in the sled. If it's well greased and clean, with no hair in the gears, there might be a tracking issue. With power off, gently move the lens back and forth with a swab, checking that it moves freely. There could be dirt or hair there, too.

While newer players feature all-digital servos with no adjustments, some older players have tracking servo trimpots. Look for "TR GAIN" or "TRACKING GAIN," and try increasing it a little bit. The more hissing noise the lens makes, the higher the gain. You'll hear a "knee" above which the hiss will suddenly increase a great deal. Be sure to stay below this point or irregular tracking may occur.

If the machine is tracking the disc and the spindle is turning at the proper rate, there should be normal playback. Any other problems will be due to circuit failures that are probably more trouble to find than the machine is worth. Those kinds of problems are rare, though. Most players can be fixed with the procedures we've just examined.

Earbuds and Headsets

Wired earphones of any sort are some of the most easily broken items. If you have a magnifying lens and a steady hand, you can fix most of them.

How They Work

Earphones are just tiny speakers, with the same magnet and voice coil arrangement. Unlike speakers, though, earphones use thin plastic diaphragms instead of paper. Headsets with microphones have the usual condenser mic, with a common ground for both the earphones and the mic. In headsets with only one cable going to one side, the opposite side is fed from a branch of it. When full-sized headsets show no obvious wire going to the opposite side, it's hidden in the headband and goes through the hinges.

USB headsets have an interface board between the analog speaker and mic elements and the all-digital USB plug. The board is either in a little box about halfway down the cord or in the USB plug.

Bluetooth headsets are really low-powered two-way radios. They're tiny, and most have no external wires. There's not much to fix on these things.

What Can Go Wrong

Most headset problems are due to a broken wire in the cable or the headband. The typical spot is right where the cable enters the earpiece, because it gets flexed so much. The second most common breakage point is at the plug, for the same reason.

A broken wire will cause loss of any sound at whichever side it feeds, or intermittent connection as the wire moves around. Sometimes both sides work, but one side buzzes. That's not a connection problem. A foreign substance, usually a hair, has gotten past the foam ear pad or grille and is touching the diaphragm.

On the interface board of a USB headset are a chip and some resistors and capacitors. I've seen a few with bad chips, but most of those headsets have the same problem that simple headsets do: broken wires, usually where they enter the headset.

Active noise-canceling headsets have analog or digital processing boards, internal microphones that listen to the sound in the ear wells, and batteries. Anything from bad parts to cold joints can cause them to stop working, but broken wires still rule. On those using disposable cells, don't forget to check for corrosion in the battery compartment, as always.

Most Bluetooth headset failures are caused by worn-out lithium batteries. Now and then, you might run into a problem with bad pushbuttons. See Chapter 14 for more about those.

Is It Worth It?

For cheap earbuds, probably not. The size scale makes them hard to work on, but if you like them and want to keep them, it can be done. For expensive headsets, it pays to fix them. Given the low power applied to earphones, the chances of a bad speaker element are next to none, so parts aren't a problem. However, the tiny batteries used in most Bluetooth units aren't readily available, so you're pretty much stuck when the cells wear out.

The Dangers Within

The only way you can get hurt on a headset is to blast your ears with high sound levels accidentally while testing the speaker elements. It happens, so keep audio levels low.

How to Fix One

Taking apart a headset can be very easy or pretty messy, depending on the type of headset. Earbuds usually require nothing more than peeling back the rubber ring around the speaker element and, perhaps, pulling up some glue with your X-Acto knife. Be careful not to tear the diaphragm or pull too hard on the wires going to the element.

Full-sized headsets usually have screws under the ear pads. The pads come off by pulling gently around the edges. Once you get to the screws, just take them out, and you're in, except that the wiring is likely to restrict how far you can separate the speaker element from the rest of the headset. Headsets with boom mics and/or USB interfaces can have quite a rat's nest of wires inside.

When one side of a pair of wired earbuds doesn't work, you're going to have to cut the cable and rewire it. Where to cut? That is the question! Operate the headset, and listen as you wiggle the wire at the headset end. If you hear so much as a crackle on the dead side, the break is at the headset. If not, try wiggling the wire at the plug. If neither makes a difference, look for any bulges in the insulation or other damage that might be a clue to where the break lies. Headsets with headbands may have broken wires in the headband or, especially, where the wire goes through the movable joint to the earpiece. You can use your ohmmeter to check for continuity between the plug and where the wires terminate inside the headset.

When the break is at the plug, soldering on a new one is probably not worth the effort. Nor is it likely to stand up to hard use for long. Instead, get a cheap headset from the dollar store and cut off its cable. Wire it in the obvious fashion, using the existing wiring as a guide. Be careful not to mix up the ground (the shield in a shielded cable and the black wire in an unshielded one) and hot wires, or the headset will sound weird and uncomfortable. Both diaphragms have to be moving in the right directions relative to each other for proper sound, and reversing the wires on one will cause it to be out of phase with the other. To verify which is the ground wire on the new cable, check for continuity to the ring on the plug farthest from the tip on three-ring plugs. For the four-wire type used on headsets with mics, ground is the second-to-farthest ring.

When the break is at the headset end, you can reuse the existing cable. Just cut off the broken section and resolder it. I like to cut about ½ inch before the break to be sure nothing is broken inside the insulation that could remain and cause failure of the repair.

Speaking of insulation, sometimes there isn't any! If the wires coming out of the cable don't appear insulated, they are enamel-coated. It is insulation, actually, but you can't strip it off with a tool. Instead, bathe the end of the wire in molten solder by putting a blob on the end of your iron and holding the wire in it. After a few seconds, the enamel will boil off and the wire will tin, ready for soldering.

With USB and active noise-canceling headsets, look for all the usual stuff like bad battery contacts and cold joints. Noise cancelers use condenser mics that can suffer from the oxidized ground problem common to those, so have a look at Chapter 14 for info about them. If a noise-canceling headset plays music okay but exhibits odd behavior regarding the noise-canceling circuitry, check for proper mic operation by scoping the output of the mics while talking into the earpieces. Of course, a broken wire anywhere in the system can cause failure too, but those aren't the wires that get flexed, so they're not likely to be the cause.

Flat-Panel Displays and TVs

This section is about direct-view, flat-panel displays, not DLP or LCD projection TVs. Those are really just projectors and screens in one box, so see the section on projectors if your set is of that type.

Unlike the old CRT (picture-tube) technology, today's flat displays are matrixed. Each *pixel*, or picture element (a single dot), is addressed in *X-Y* fashion, so all of the circuitry required for scanning an electron beam over a screen is gone, and the associated service issues are different as well.

How They Work

In an LCD, a low voltage twists the molecules of a tiny bit of liquid-crystal material in each pixel, changing the orientation of its light polarization. In conjunction with a fixed polarizer at the front of the panel, this change of polarization darkens a pixel, making the pixel block the backlight's illumination.

Remember plasma TVs? Nobody makes 'em anymore, but some have lasted long enough that you might run into one for repair. In a plasma display, a high voltage causes ionization of gas in each pixel, generating ultraviolet light that excites colored phosphors to generate visible light. To produce a picture, first the gas in the pixels has to be ignited by a high voltage, and then the brightness of each pixel is controlled by pulse-width modulation. Essentially, the pixel is flashed on and off faster than our eyes can detect, and the perceived brightness corresponds to the duty cycle (on versus off time) of the pulsing.

Today's displays have millions of pixels. A standard HDTV display of 1,920 × 1,080 resolution contains 2,073,600 pixels, each with three subpixels of red, green and blue. That's 6,220,800 tiny dots, and every one has its own connection. For a UHD or 4K set, it's more like 8 megapixels, or around 24 million dots! How are all those addressed? There aren't millions of wires coming out the back, after all. The process works somewhat as it does with memory chips: Row and column addresses are sent through decoders that fan out to the appropriate connections, eventually reaching each and every pixel through transparent, printed conductors at the edge of the glass. Using a grid formation allows far fewer connections than there are pixels; those 2 million pixels require just 6,840 lines (1,920 dots × 3 colors, plus the 1,080 lines to select the rows). At the edges of the panel, the decoder chips make contact with the glass elements via ribbon cables affixed with pressure and conductive glue.

Virtually all LCDs made today are some flavor of the thin-film transistor (TFT) variety. Instead of addressing the LCD elements directly, the exciting voltage pulses a transparent transistor behind each element. This enhancement lets the pixel store its state after it's been addressed, resulting in much higher contrast than if it got pulsed only momentarily and then left alone until the next frame of video came along, as was the case with the original, low-contrast LCD designs from years ago.

Printing millions of functional transistors over the area of a large screen requires very high-precision manufacturing. In the early years, TFT LCDs suffered from bad pixels; most had a few, and it was considered normal. Today's displays rarely ever show any stuck pixels unless there's something wrong with them. Nearly all new LCD panels are 100 percent functional. It's pretty amazing, really.

Most modern TVs can accept both analog and digital input, with jacks for everything from old-fashioned composite and component analog video to HDMI. Over-the-air reception comes via the sets' *ATSC* digital TV tuners, many of which can also tune unencrypted digital cable channels.

What Can Go Wrong

Loss of a single connection out of thousands at an LCD panel's edge results in an entire row or column that won't darken, leaving a bright line across or down the screen. On a plasma screen, you get a dark line that won't brighten. Individual pixels can also fail, resulting in one dot of color that never moves on an LCD or a dark spot on a plasma panel. Large areas of the screen can turn to colored lines that may be static or can flash and move around. Bars many pixels wide might appear white on an LCD or black on a plasma TV. Plasma sets can burn their phosphors when bright images don't move for an extended period, reducing the brightness of the affected pixels and leaving a ghost of the offending image. That happened most often with panels used for static display of airport schedules and such, but it also occurred at home due to extended video game play and TV network "bugs," or logos, at the bottom of the screen. While plasma is mostly gone, the same problem afflicts OLED TVs, in which each pixel is a set of three or more organic-based LEDs. OLED sets are expensive, too, making burn-in a sad event.

The fluorescent backlights in older LCDs not lit with LEDs were driven by those pesky backlight inverters we've discussed. They were just like the inverters in laptops, only bigger. Because the illumination provided always has to be as bright as the brightest picture could get, the inverters ran pretty hard and hot, and were prone to failing. Sets with LED backlighting have no inverters. Those should last longer, but LEDs can fail, and their constant-current LED driver circuits can go bad as well.

The power supplies providing the low-voltage, high-current power to run all this stuff also generate some significant heat, leading to shortened component life and failure. The low-level circuit boards that process the input signals can malfunction, resulting in a screen that lights up but won't produce a picture. In some sets, the screen stays blue no matter what you feed in.

Is It Worth It?

If the panel develops a bad row or column, there's a chance of repair if the failure is due to bad contact at the edge of the glass. You're more likely to get it working in an LCD set than in a plasma TV. If a row or column driver chip at the edge of the panel has gone bad, you're out of luck.

If the glass is cracked, you can't repair it. Plenty of today's TVs get hit by a kid's toy or the family dog, rendering the sets useless. The cost of a new panel is usually more than the price of replacing the TV. Just get a new one.

Electronic problems like bad power supplies and voltage regulators, blown LED driver circuits, burned-out LEDs, dead crystals, flaky ribbon connections and failing capacitors can be navigated successfully. Luckily, these account for most of what you'll see. If the trouble is on a board but it isn't the usual bulging capacitors, replacing the entire board might be the easiest, most practical repair because working boards, harvested from sets with broken screens, are available online at very low cost.

If you have to take apart an LCD screen to get at a bad LED, you'll need plenty of room to lay out the various layers, especially if you're repairing a big set. Be sure to keep your work environment clean, too, so you don't wind up with dirt obscuring the image after reassembly.

Plasma for home use is an obsolete technology. It took a lot of power, and the picture quality didn't match what you get from a modern LCD. It's not worth putting a lot of money into a plasma set, but there are aficionados willing to do so because the picture looks more like the CRT images they got used to seeing, so they love these sets. In another few years, I expect all plasma sets to be discarded or dead. Considering the awful interference they caused to my ham radios from half a block away, good riddance!

The Dangers Within

Plenty! In fluorescent-lit LCDs, watch out for the inverter and its output cables. They may have more than 1 kilovolt on them. Plasmas are also full of high voltage, and it's fed to the panel's pixels, not just to a few backlight lamps.

LCD screens use no high voltages, but the liquid-crystal material in an LCD is toxic and should not be handled. You'd come in contact with it only if the glass were broken.

How to Fix One

Plasma

In plasma sets, look for bulging electrolytic power supply caps and bad connections. A total failure might indicate a blown chopper or other typical switching power supply issue.

The *Y*-sustain and *Z*-sustain (sometimes called *X-sustain*) boards present the image information to the panel's row and column drivers. Between the *Y*-sustain board and the plasma panel is the *Y*-buffer board that stores the pixel information and feeds it to the display. The *Y*-sustain board is a frequent source of trouble. Check it for bad caps and blown transistors after disconnecting it from the set.

Because high voltages have to be switched rapidly, the circuitry driving the panel works hard and gets hot, leading to failures. The power supply generating the high voltage for the panel is another source of trouble.

The screen itself can go bad, and there's nothing you can do to fix it. Symptoms include missing columns or rows, dark bars the width of many pixels, areas of bad color and red dots. Big black bars, however, can also be caused by bad contact at the ribbon cable connectors. In fact, it's fairly common. Before you conclude that the screen has failed, kill the power, unplug the set and reseat all the ribbon cables going to the screen. Sometimes the connection to the glass at the panel's edge makes poor contact, and you might be able to clamp it down, restoring contact.

A lot of power bounces around in plasma TVs. Keep an eye out for burned components. Even ceramic capacitors, normally among the most reliable components, can short in the power-handling sections of these sets. Also, look for transistors wired in parallel and mounted on heatsinks. Anything in parallel suggests that a lot of power is being manipulated, and there'll be plenty of heat. Shorted parts, both FETs and bipolar transistors, are common. Luckily, shorts are easy to test for without unsoldering the components. Before you apply your ohmmeter, just be sure all power is disconnected and there's no voltage on the transistors from charged electrolytic capacitors.

Be extra careful when trying to fix a plasma TV! There are lethal voltages in there. Unless the problem is something simple like bad electrolytics, you're better off replacing entire circuit boards than trying to troubleshoot down to the component level if it involves snooping around with the power on. Don't go looking for bad panel contact by pressing with your fingers on the glass at the edges while the thing is running, or you could get shocked badly. Use a cotton swab to press on the contact areas.

LCD

In LCD sets, bad ribbon contacts cause similar problems, but the bars will be white, not black, because the disconnected LCD pixels stop blocking light. You also might see a region of the screen covered in colored lines. LCD panels are driven by the timing control (T-Con) board. If reseating ribbons doesn't fix the lines, consider replacing that board. Screen damage can cause the same symptom, but the bad area is likely to be irregularly shaped.

If the set runs but the colors are washed out or show ghosts, look on the T-Con board for test points with voltage markings. Usually, several different voltages are required. If the voltage reading on any of those points is off by any significant amount, try hooking up a power supply and feeding the correct voltage to the test point. If that clears up the picture, there's a bad voltage regulator either on the board or at the power supply.

If there's no picture, check that the crystals on the T-Con board are running. Scope each side of a suspect crystal and see if there's a waveform. If you have a stable RF signal generator, you can inject replacement signals to verify that you've found the trouble. It takes a mighty accurate generator to do this, though. One with digital frequency control probably is good enough, but the old analog dial type isn't.

Individual stuck pixels are internal to the panel. It may seem impossible to do anything about them, but you might have luck by massaging or pressing on an LCD's stuck pixels with a finger or a pencil eraser. Now and then, this will restore a bad connection inside the pixel, and it'll come back to life. Tapping with the back of your fingernail also can work. I've fixed a few that way, but they usually flaked out again in a few days, and I had to tap on them every now and then to keep the sparkly little annoyances at bay.

Pressing too hard can cause further damage. If you've ever seen LCDs with functional but darkened patches with irregular borders, it's because something pressed against them hard enough to deform the innards a little bit. Once this happens, there's no undoing it.

Believe it or not, what is displayed on the LCD can affect the connections inside, especially when they are weak. Flashing a pixel on and off repeatedly can restore its function. Turning it on continuously also might do the trick. There are pixel repair apps you can download and try. Some are free. They can be run on computers, phones and tablets, so you will need to use a cable or a wireless setup to get the images onto your TV. Do they work? Well, um, uh, sometimes. They're worth a try, anyway. I've done better with rubbing and tapping.

Many panel problems are caused by failures on the long, skinny, flexible boards at the edges of the panel itself. Especially if one half of the screen works and the other doesn't, head straight for those boards. Reseat ribbon connectors, and keep an eye out for cracks in components or the board. Many of these can be fixed. You may find test points with voltage markings here, just like on the T-Con board. Check 'em, and sub in the correct voltage if one is missing or way off.

Unlike in plasma sets, the voltages going to the glass are low, and it's safe to press gently on the ribbons to check for bad contact. Be sure to press only on the contact points where they meet the glass and the circuit board. If pushing on a spot restores a dead row or column, you've found the fault. The ribbons are glued to both the glass and the board. One practical fix is to use a shim or a clamp to apply pressure to the bad spot (Figure 15-6). Anything nonconductive that you can use to clamp the bad connection or put enough pressure on it to get it to make contact should do the job. Depending on the conductive adhesive used, you might rejoin the separated spot by heating it. People have used heat guns (essentially hair dryers with smaller nozzles) and soldering irons insulated by a piece of paper. A hot-air soldering station would be a better bet. Either way, be careful not to overheat the ribbon, or

Conductive glue underneath

FIGURE 15-6   Ribbon meets board.

you may melt it. The risk is especially high with a soldering iron or a hot-air station, so keep them on low. Get the spot to around 160°F, and press on the ribbon with a pencil eraser as you remove the heat source, until the glue cools down.

LCD TVs also have the usual power supply and cap problems, but the most frequent cause of failure in older, non-LED versions is the backlight inverter. Nobody makes sets with fluorescent backlights and inverters anymore, but they were pretty robust, so there are lots of them still out there, and you might run across one worth the trouble to fix.

If you're not sure what an inverter looks like, see Figure 10-5. LCDs of smaller size have at least two lamps driven by multiple inverter circuits, often combined onto one board with output transformers at the ends. Big screens can have 6 to 10 lamps, with more inverters to drive them.

If you turn on the set and it lights up for a second before the screen goes dark, one of the inverters has died. You get that one moment of light because the other sections are running their own lamps until the micro senses the loss of the blown one and shuts them off a second later.

The poor output transistors in an inverter work like dogs for countless hours, and eventually, one of them shorts, taking the fuse on the inverter board with it. In most designs, each side will have its own fuse soldered to the board near the connector from the power supply or the microprocessor board. This is very helpful because a blown fuse tells you which side of the board has quit. Try replacing the fuse first, even if just with a temporary arrangement employing two soldered wires, clip leads and a physically larger fuse of the same rating as the original. If you can't determine the original's rating, 3 amps is a reasonable value to try. Now and then, you may find that the fuse has fatigued and failed, but the rest of the circuitry is fine. If the lights come on and stay on, a new fuse is all you need. Be sure both lamps are working. The screen should be at normal brightness and evenly lit. If one end is significantly brighter than the other, one lamp is still out, and it's quite possible that the transistors on the blown side of the inverter became open from the momentary surge current when they shorted. So, they won't blow a new fuse, but they won't light the lamp either.

Take a peek at the inverter's transformer. If you see a burned spot anywhere, the transformer's insulation is damaged and the coil has arced over, either from one winding to the next or from a winding to the core. The increased current draw from arcing usually pops an output transistor. Changing the transistor does you no good because the transformer will just kill the new one. If you see no burns, the transformer still could have internal shorts or arcing, but it's less likely.

It's common for the secondary winding (the one connected to the lamp) to go open. You can easily check for shorts and opens by measuring the winding's resistance and comparing it with what you see on the other inverters. They should all read about the same, with a typical value of around 1 kΩ. A bad one will measure noticeably less than the others if it has a shorted winding and infinite resistance if it's open.

If the transformer looks okay, you can change the inverter's transistors if you can find some. Often, they're oddball output components that aren't easy to locate. Sometimes you can substitute similar transistors, but they have to be a pretty close match, especially in their gain characteristics. Not enough gain will cause the transistors to run more in their linear region than fully saturated, and they'll get very hot and fail in a hurry. Too much gain can cause them to "ring," with the tops and bottoms of what should be a flat square wave having sine wave–like variations. This also puts them in their linear region and overheats them. If you do sub the transistors, scope their collectors or drains, and compare what you see to the waveforms on the good side of the inverter. If they look a lot different, those transistors are not a suitable match. As long as the waveforms look like they're turning on and off all the way and the inverter seems to work, let it run for a few minutes, and then turn it off and touch the transistors. They might be warm but shouldn't be too hot to touch. Compare them to the good side.

Modern, LED-lit LCDs use multiple LEDs driven by a constant-current power source that keeps the current through them from varying by adjusting the applied voltage as they heat up. An LED or its current source can fail, leaving a darkened area or killing the backlighting altogether. Some big sets lit from behind have several hundred LEDs, with multiple series strings of them. Each string has a connector going to the constant-current supply. The supply watches the current drawn by the string and will shut down if any of them isn't pulling current. Edge-lit screens use a few dozen LEDs at most but use the same constant-current driving method.

When the screen won't light up at all, don't assume that the power supply is dead! Pull the connector for each string, one at a time, and connect the LEDs to your bench supply, with your DMM in series with one of the two leads and set to read current. If your supply offers current limiting, set it to 50 milliamps.

The wires to each string might not be color-coded with the usual red and black for + and −. Luckily, LEDs won't be harmed by reverse voltage because they're diodes and will simply block it. So, if the following test yields no results, swap the connections and try again.

Start at 0 volts, and slowly turn up the voltage, watching the current as you do. If the LED string starts to light at a voltage below 60 volts or so, it's good. Don't let the current draw exceed 50 milliamps. If you're using a non current-limited supply and the current starts creeping up to that level, reduce the voltage to protect the LEDs.

If your power supply doesn't provide high enough voltage to do this test, put a couple of fresh 9-volt alkaline batteries in series with one of the leads, + to −, so that the total available voltage will be in the range of 60 volts or so when the supply is turned up near its maximum voltage.

If you find a bad string, you'll have to test each LED in it to find the dead one. The hard part is figuring out which LED belongs to which string, because they may not be lined up in a row or column. Visual tracing is your best bet. To test an LED, disconnect the string from the TV's power supply. LEDs are diodes, so you can test them with your DMM's ohmmeter function, but they may not show as good on its diode test because the voltage drop on bright LEDs is much higher than on a normal silicon diode. Depending on the meter's test voltage, it might not see the LED at all, falsely indicating an open circuit when the LED is actually good. Or, hook your bench supply, with no added batteries to up the voltage, across each LED, starting at 0 volts and turning up the voltage slowly while watching the current to keep it below 50 milliamps. If the LED takes no current and doesn't light, and you're sure you have the polarity correct, there you go!

When you find the bad LED, try shorting across it and testing the TV, after plugging the connector back in where it belongs. Most of the time, the TV's power supply won't notice that small a difference in the current draw, since there are other LEDs in series with the now-shorted one, and the TV will come back to life. You'll probably see a darkened spot on the screen, though. If you're willing to live with a slight brightness loss, you can just leave the short in place or replace the LED with one from a corner of the screen, where the loss will be less noticeable, and short across where that one was.

You can find replacement LEDs at parts houses like DigiKey or on eBay, or you can scrounge one from a discarded set. Most people just toss 'em when they die, so check the free section of craigslist.org. Keep in mind that disposing of the big screen after you raid it for parts might not be easy. Many communities have occasional electronics recycling events, and you may have to wait for one of those.

Hard Drives and SSDs

Hard drives and solid-state drives (SSDs) are used to store just about everything too big for thumb drives these days. We all depend on them, and they endure thousands of hours of use. Most people consider the drives irreparable—they either work or they don't—but that's not always true.

How They Work

Hard drives use a rotating platter coated with ferric material onto which is recorded digital data in the form of tiny regions of magnetism. The head flies on a cushion of air just a few wavelengths of light from the surface, suspended on an arm that flits at high speed across the disc, controlled by the drive's microprocessor as it reads and writes sectors of data. A hard-drive head never touches the recording medium. The disc speed is held constant by a servo that monitors motor speed and locks it to a crystal reference. SSDs use the same sort of flash memory as thumb drives do, except it's faster and built to withstand more writes before wearing out.

What Can Go Wrong

A mechanical hard drive's motor or its drive circuits can fail, in which case the disc won't turn no matter what. The signals from the heads are very tiny, and a poor connection between the circuit board and the heads can cause them to drop out or get too weak to read. The drive will recalibrate with a clacking noise, desperately trying to find the data. It may also write incorrectly, severely corrupting itself.

SSDs wear out eventually, just like thumb drives. The drive may still read what's on it but be unable to write new data. More often, though, it just dies with no warning. One day it's there, the next it's gone and the computer doesn't even know a drive is installed.

Is It Worth It?

Well, sure, if there's important data on the drive, you're going to want to recover it. But is there really anything you can do to repair a hard drive? With an SSD, no. I've tried scoping the circuit boards of a few dead ones and haven't found any way to get 'em going again.

On a mechanical drive, opening the case is out of the question; once you expose the platter and allow room air and its dust particles in, the drive is wrecked. Believe it or not, there is one thing you can try, and I've saved numerous drives with the procedure. It's fast and easy, and it doesn't involve breaking the seal.

The Dangers Within

No danger here. Voltages are low, and all the moving parts are sealed.

How to Fix One

If the drive is recalibrating often or returning errors, take off its circuit board and look at the connection points interfacing the board with the body of the unit. You'll see two sets— one for the motor and one for the heads. Rarely are the motor connections problematic.

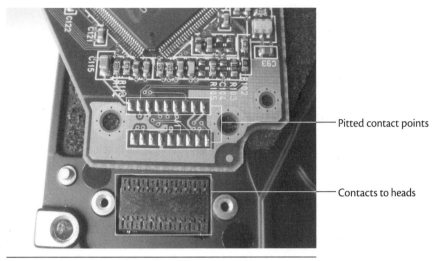

FIGURE 15-7   Hard drive head contacts.

The signals from the heads, though, are so small that it doesn't take much resistance to lose them (Figure 15-7).

Some brands of drives use gold-clad connectors, with the female side on the board. Pins on the body of the drive fit into tiny sleeves when the board is put in place. This is the most reliable method, but over a period of years, the sleeves can relax, making poor contact with the pins. To restore contact, take off the board, and use the tip of an X-Acto knife or a very small screwdriver along the edges of each sleeve to squeeze it together a little bit. You have to get the tool into the crack between the sleeve and its surrounding plastic, and it can be difficult. Don't worry if you damage the plastic slightly, as long as all the sleeves remain isolated from each other.

Many manufacturers save a few cents by replacing the connectors with sharp pins on the body of the drive that press into solder pads on the board. These are the ones that cause the most trouble. After a few years, the solder gets oxidized, and its resistance goes up, impeding the delicate signals generated by the heads when they read data. If yours has the pads, look at the indentations made by the pins. Are they blackened at the center? That's the sign of this malady. Gently scrape off the oxidation with the tip of an X-Acto knife, and then wipe the pads with a dry swab to remove the metal flakes.

Reassemble the drive, and be prepared to be surprised! The darned thing just might work. If the drive has corrupted itself too much, it may be unrecoverable, even though the electronics are now functional. Sometimes, a drive repair utility can fix that, but sometimes you can restore it only by reformatting, wiping out the data. Other times, nothing works. But, heck, it wasn't doing you any good before anyway, right? By the way, if you do this procedure on an older drive *before* it starts exhibiting the "clack of death," you can extend the drive's life quite a bit. I have some hard drives that are 15 years old and still going strong, thanks to my prudent use of this procedure at around the 5-year mark, before trouble arose.

Laptop Computers

Laptop computers have largely replaced desktop machines in most people's homes and workplaces. They offer numerous advantages in terms of required space, power consumption and heat generation, but they're a lot more prone to breaking.

How They Work

Laptops are functionally just like desktop computers, except that everything is much smaller and runs on less power, and the LCD monitor is built into the unit. Also, laptops incorporate power management systems for safe, efficient charging of lithium batteries, and their AC power supplies are not internal.

What Can Go Wrong

Laptops are enormously complex, with most of the circuitry on the motherboard. That board has many layers and lots of LSI chips, along with a zillion tiny support components crammed together.

As discussed in Chapter 2, some of those huge chips are connected to the board with a *ball-grid array* (BGA), which is a bunch of tiny, ball-shaped contacts soldered to pads underneath the chips. If you see a chip with no leads, it has a BGA (Figure 15-8; also see Figures 2-28 and 2-29). BGAs provide hundreds of contacts in a small space, so they're used for microprocessors, video graphics chips and other very high-density devices. Some of these items run pretty hot, unfortunately, and can degrade their solder joints with thermal flexure of the board over time, resulting in intermittent connections and a machine that keeps crashing.

How can you repair those things when you can't even see the contacts? BGA rework stations have come down a lot in price, making them available for hobby use. Even with a rework station, reballing (resoldering) BGAs is an exacting task. It's probably not worth buying a station just to fix one machine, but you won't be successful without one. Don't try fixing a BGA problem with a soldering iron! I foolishly tried resoldering one barely visible pin at the edge of a video graphics chip's BGA once, long before there was any affordable alternative, and succeeded only in destroying the chip and the motherboard.

Many motherboards sport a surprising number of surface-mount electrolytics, with the usual problems those cause. When changing them, be extra careful with the heat because you're dealing with a multilayer board you could easily wreck. Review Chapter 12 for info on handling situations like this.

FIGURE 15-8   **LSI chip with BGA contacts underneath.**

Battery charging is controlled with MOSFET power transistors. An open one will result in no charge reaching the battery. All modern laptops with external batteries use *smart batteries* featuring their own microprocessors that tell the machine the state of charge, how many cycles have been used over the life of the pack, the battery's model number and so on. Also included is the BMS (battery management system) for safe charging, like those described in Chapter 7. If that micro gets scrambled, the BMS goes bad or the contacts between the battery pack and the laptop malfunction, the battery may not charge or even be recognized. Of course, worn-out cells will cause the same problems. Most laptops won't even recognize that a battery is installed if the pack has completely discharged, regardless of whether the cells are still good. The BMS prevents it because trying to charge a completely dead lithium cell can cause it to swell, resulting in a fire hazard. Laptops with internal, non-removable batteries typically put the BMS on the main circuit board.

The power supply input jack is a frequent source of laptop malfunction. Pulling on the plug, tripping over the cord and even just normal insertion and removal can crack the jack's solder joints, resulting in failure to charge the battery or no AC operation at all. If you have to wiggle the plug or push it to one side to make it work, the joints are cracked.

At one time, a dead backlight inverter was one of the most common laptop problems of all. Today's machines use LED backlights, so they have no inverters, but a few older laptops out there still have fluorescent lamps. The lamp itself can get weak, and dropping the machine may break the bulb with no outward physical sign of damage. If the laptop has been dropped and suddenly won't light up, but you can see the image by shining a bright light on the screen, the long, thin lamp tube is probably shattered inside the bottom of the LCD.

The screen is connected to the body of the machine via cables running through the hinges. After the lid has been opened and closed hundreds of times over a few years, a cable can break. In those old models with an inverter, it was located in the screen assembly to avoid having the high-voltage output wires going through the hinges. The power and control lines feeding it did go through them, and those were often the ones that broke. In a modern laptop with a dead backlight, the problem could still be in the cable, or an LED might have gone bad.

A bad RAM (memory) module can cause random crashes. Sometimes the module isn't actually defective, but its contacts have gotten oxidized, resulting in weak signal transfer to the motherboard. Faulty RAM chips soldered on the motherboard will cause crashes too.

The keyboard can develop bad keys or entire bad rows. It can also drive the machine crazy with a stuck key that sends the same signal indefinitely.

Machines with removable batteries often include an internal rechargeable backup cell soldered to the board or plugged in on a connector. This keeps the clock and some other data intact when you take off the battery. Like any battery, that cell goes bad eventually, resulting in CMOS errors when you try to boot the machine. When the main battery is internal, there's no need for the backup cell because some voltage is always present.

Is It Worth It?

A bad motherboard is usually not worth your time, but there are exceptions. If you see signs of dying electrolytics, you can change those and probably restore normal operation. If the battery won't charge but everything else works, and you're certain the battery is good, you may be able to find the open MOSFET and replace it. They are big enough to remove and resolder. If there's some obscure logic failure, it's toast. You're not gonna find it.

Power supply jack problems are easy fixes, once you get to the darned thing! You'll spend far more time on disassembly than on the repair. In some laptops, removing the back gets you right where you need to be. Others require total disassembly to get anywhere near those precious solder joints.

Be glad that backlight inverters are in the past! Repairing one is tough, mostly because the tiny output transistors they use are hard to find, as are the soldered-in fuses. Replacing the inverter, however, is pretty easy on most models. Usually, all you need to do is open up the screen assembly, and there it is! Unplug the connectors, replace the board and you're done—that is, unless the problem is a broken screen cable. When inverters were in common use, it was easy to find replacements. Now, it's pretty much impossible without a parts unit from which to scrounge one. A laptop old enough to have an inverter probably isn't worth fixing anyway unless there's some data on it you really need to recover, in which case it's easier to remove the hard drive and put it in a drive case than to deal with the inverter.

Ah, the screen cables. Very often the real reason the LCD won't light up is because the cable has a broken wire inside or a fractured conductor, in the case of a ribbon cable.

The problem can also be caused by a failure at the video graphics chip's BGA. So, it can be tough to tell whether the cable or the motherboard is the true culprit. If moving the screen through its range of angles makes it turn on and off, one of the cable's wires or printed conductors is broken, and the ends are touching each other just enough to make contact when the cable is at certain positions.

Bad RAM can be easy or really tough, depending on whether it's a module or soldered to the motherboard. If it's soldered on, you're looking at hot-air soldering in a very dense, small area. Luckily, lots of machines use removable RAM modules, and all you have to do is swap in a new one.

Hard drives and SSDs are easily changed in some machines and hard to reach in others. The drives fail often enough that many manufacturers make them readily accessible, but some disregard that reality and bury the darned things so deep that it takes an hour or two to get to them.

Dead backup cells aren't too tough to change. You'll have to order one from a computer store or a parts supplier. Replacement is easy as pie if it's on a connector and a little harder if it's soldered to the board.

The Dangers Within

Most of a laptop operates at low voltages, so it's pretty safe. Backlight inverters put out a high voltage, so stay away from their output areas. LED-lit screens are pretty safe because LEDs operate on just a few volts. Switching converters on the motherboard can generate some other voltages that it's best to avoid, but they shouldn't be high enough to injure you. You could possibly get a shock if you touched the wrong point while the machine was running.

How to Fix One

Let's look at a few common laptop problems and how to approach their repair.

It Won't Start Up

Unlike with most electronics, where a dead device is the easiest to fix, a completely dead laptop may be irreparable, at least without replacing the entire board. Lots of problems can cause the computer to fail to complete its bootup sequence, but something should happen when you press the power button.

Are you certain the machine is getting power? A failed AC adapter or a shorted or broken DC cord can prevent power from reaching the unit, making it appear dead. I've seen some laptops that wouldn't start up without a battery installed, too, even when run on AC power! The battery didn't have to be good, but it had to be there. I've even seen a few that wouldn't start because the internal backup cell had gone dead. Disconnecting it made the system come to life.

Check the machine's power jack for cracked solder joints. That's a very common problem. Also, be sure to examine the board for signs of liquid intrusion. People spill drinks on laptops often, and if the liquid hits the motherboard, it'll usually kill the poor thing. Sometimes, washing the board with distilled water and letting it dry out will restore operation, but most of the time the machine is ruined, never to start up again. The old "put it in a bag of rice for a week to dry it out" trick is highly controversial. Some people swear by it, and others claim it's nonsense. I've never tried it.

If the machine does start but won't boot, check that the RAM modules are inserted correctly and that their contacts are not dirty. Checking the RAM is part of the boot sequence, and the thing won't get very far if the RAM test fails.

Sometimes, failure to boot is caused by a data corruption in the power management system, and holding down a certain key combination while pressing the power button can clear the bad data and restore operation. Check the internet for specifics related to the laptop model you're servicing.

It Won't Boot

Hanging on bootup can also be caused by a malfunction of anything on the system bus, such as a USB port or the hard drive. Try disconnecting the drive and restarting. You won't be able to boot the operating system without a hard drive, but you should get a normal startup sequence up to the point where the operating system is supposed to load.

If nothing works, the board may have failed, and you're not likely to find the problem. The density and sheer number of components on a laptop's motherboard are staggering. Beyond reworking a broken BGA connection, pretty much nobody fixes these things down to the component level, not even their manufacturers. Before you give up, though, look around for bulging capacitors, burn marks or other obvious signs of failure. If it's something simple, like a bad cap, you might be able to change it. The cost and hassle of reballing a BGA may be worth it if the machine would cost a lot to replace, whether you invested in the equipment to do it at home or you have to send the board to a board repair house. It might be cheaper just to find a replacement board on eBay.

It Crashes

If the machine works but crashes randomly, pull the RAM modules and use a dry swab to clean their contacts on both sides. Even if they're gold-clad, which most are, the swab may show a surprising amount of grayish dirt when you're done. Pop the modules back in and test.

If this doesn't solve the problem, use diagnostic software to check for a bad RAM module. If one comes up as bad, replace it. That's about all you can do. RAM is pretty reliable, but it does fail now and then. Some motherboards have RAM soldered on, with modules used only for expanding the memory above the stock configuration. If the motherboard RAM is bad, you can replace the chip if you're good at hot-air soldering, but the size scale makes it difficult. Some of those chips are so small that it's impossible, though. The big trick is avoiding desoldering tiny components next to the chip.

If the machine crashes so often that you can't even run the software, check for bulging or leaking capacitors on the motherboard. Some laptops crash after they warm up. Usually, this means a bad chip or an intermittent solder joint, probably in a BGA somewhere under the microprocessor or the video graphics chip. I've seen hard drive interface chips do that too; the machine works fine until the chip warms up enough to malfunction, and then it won't read the drive or it corrupts the data on it.

It Won't Charge

If the machine runs but the battery won't charge, check first to see if the battery is any good. The cells' chemistry wears out eventually, and if the machine senses a bad battery, it won't try to charge it. Also, the internal charge control circuitry can fail. As I mentioned earlier, sometimes a battery stored so long that all of its charge has leaked away may not be recognized because its internal microprocessor won't signal its existence to the laptop without at least a little power to run. Some systems offer a battery resetting utility that can hunt for a dead battery and try some charge to see if it's there and working. Check online for a battery recovery app you can download.

Short of cracking the battery open and applying a little charge directly to the cells, there's nothing you can do. And frankly, I don't recommend doing this. It's rather difficult to accomplish without wrecking the battery, and the lithium-ion cells inside are pretty dangerous if a screwdriver pierces them or causes a short while you're breaking open the plastic casing. Any breach of their seals can create a fire hazard because lithium reacts violently with water, including water vapor in the air. Unfortunately, applying power to the pack's outside terminals won't work; the BMS inside allows connection to the cells only when the laptop gives it the go-ahead.

Even if you take the risk to open the pack and apply low current directly to the cells to get enough charge on them for the BMS to recognize their presence, there's no telling if the battery will still work. The micros in some lithium packs have software in RAM that disappears once the charge leaks down to zero. Without the code, the computer won't recognize the pack and charge it. Or, if it does, the machine may not turn on with that battery powering it. There's no fix for this situation.

If you're certain the battery is good, check to be sure the power supply is working! Many times the adapter fails, and it's just assumed that the computer is the culprit when in fact it's not getting any power. If the adapter has an LED, it should be lit. These LEDs are driven by the adapter's output. So, if the light comes on at normal brightness, the supply is working. If it doesn't light up or is very dim, unplug the adapter from the computer and see if it comes back to normal. If so, there's a short in the computer dragging down the voltage, or the adapter's capacitors are shot, so it can't provide enough current to run the laptop. If the LED doesn't light at all, even without the computer connected, either the supply is dead or its output cable has a short. I've seen that happen a few times, and repairing the cable restored normal operation, thanks to protection circuitry in the supply that prevented its destruction or a blown fuse from overcurrent.

Once you're sure the supply works, check for a jack issue on the laptop and in the supply's plug. Does the jack move when you wiggle the plug? Does indication of charging come and go when you do that? If you're watching for software indication of charging, remember that it may take 10 or 20 seconds for the operating system to recognize the change.

To ascertain whether the trouble is in the plug or the jack, hold the plug steady with one hand while moving the wire with the other. If the connection cuts in and out, the plug's the problem. If not, it's the jack. You may have to do this a few times to be sure, because it's easy to move the plug slightly in the jack while trying to hold it still.

When you're sure the power supply, jack and battery are okay, it's time to consider a motherboard problem. Just because the computer runs doesn't mean everything on the board is okay. Power management on laptops is complicated, involving firmware (software encoded onto the machine's chips), system software and the power management unit, a specialized microprocessor used only for controlling the flow of power to the various parts of the machine. A problem with any one of these could prevent charging. If there's a reset procedure for the power management unit, try that.

If nothing works, open the machine and look in the area of the battery connector. Because significant current of up to several amps gets passed in charging, the power transistors controlling it are usually located near the connector to avoid having to waste space with wide circuit board traces that can handle the juice. The transistors could be on either side of the board. They might be bigger than most of the components around them, but these days some rather small MOSFETs are used. They don't look like transistors, though. Instead, they are eight-pin flat SMD packages that look like chips. The giveaway is that many of the pins are connected together. Typically, all four on one side are connected, and that's the drain terminal. Three on the other are the source, and one on that side is the gate. You'll find these types of power-switching MOSFETs on BMS boards, too. Use a shield or other obvious ground point, such as a chassis screw connected to a wide trace on the board, for circuit ground. With the AC adapter connected, try scoping the transistors' terminals while you insert and remove the battery. If you find one with constant voltage on one terminal and a signal that changes state on the other shortly after you pop the battery in, but nothing shows on the third terminal, the transistor is probably open. If it has a number on it, you have a chance at finding a replacement from one of the big parts houses.

Display Problems

Unlike a desktop computer monitor's HDMI, DVI or VGA connection, the interface from a laptop's LCD to the motherboard is specific to the particular make and model. Signals may be carried on a bundle of wires or a printed circuit ribbon cable. Most newer models use ribbons. Those rarely break from normal flexure, although some get brittle after a number of years. Who keeps a laptop that long anyway? Wire bundles do break, resulting in all kinds of display symptoms.

With the laptop running, gently move the screen back and forth through its entire range of angles. If you see even a flicker of backlight, the problem is almost certainly in the cable.

Testing it isn't hard. Unplug it at both ends, and use the ohms scale on your DMM to check for continuity of each wire or printed trace. With wire cables, to get a connection to those tiny holes in the connectors, connect a clip lead to one lead of a small component from your stash and push its end into the connector, with the other end of the clip lead going to the DMM. Let the second lead of the component hang loose. Ribbon cable ends can be touched with your DMM's probes directly.

If the cable is good, an LED may have failed. Unlike in a big-screen LCD TV, the strings of LEDs have only a few of them in each string, due to running on a lower voltage than is available in a TV. It should be pretty easy to check each LED and find the dead one.

If the screen lights up but the video isn't normal, plug an external monitor into the machine and see if video works properly on that. If not, the motherboard has a serious problem at the graphics chip. If it looks okay externally, either the screen cable or the screen itself is causing the trouble.

A single bad line on the display cannot be the fault of the cable because no one wire is specific to such a small area of the screen. A bad line or two is caused by the row and column drivers inside the LCD or their connections to the glass. Unless you get lucky and find a spot along the panel's edge where clamping or reheating restores contact with the glass, you can't fix this, but finding a reasonably priced replacement screen online is easy enough for many laptop models. Changing it is entirely a mechanical job; no soldering will be required. Take apart the screen bezel, get the new LCD in the frame, plug in its connector and you're done. On most models, you don't even have to open the main body of the computer.

When video is severely distorted, with large areas of the screen a total mess, suspect the cable or the graphics chip. A bad screen can cause this too, but it's less likely unless you see obvious cracks from a fall. I once worked on a laptop with video that started shaking back and forth after it had been on for about half an hour. Eventually, the image would tear, looking a lot like an analog TV with its horizontal hold misadjusted. I proved the fault was with the graphics chip by spraying component cooler on its heatsink. As soon as the chip cooled even a little bit, video returned to normal for a few minutes, until it got hot again.

Some LCDs have thin circuit boards on the back, with ribbon cables connecting them to the row and column driver chips along the screen's edges. If they use sockets, check the fingers on those cables to be sure they're not oxidized. Clean them and test the screen again. Beware the thin, printed conductors wrapping around the edges of the screen! Those are the connections to the drivers, and they're especially fragile because of their density. Pressing on one is likely to ruin the LCD. Keep that in mind when installing a new panel, too.

Drive Problems

As described in this chapter's section on hard drives, there is an issue with some brands of drives, especially after they've been in use for a few years. The contacts from the board to the head assembly inside the metal casing become intermittent, causing read failures and sometimes severe data corruption. For some reason, laptop drives seem especially susceptible to this problem. If yours is recalibrating a lot, having trouble reading, and returning errors, it's worth taking it out and trying the procedure described in that section.

The 5-volt DC power to the drive needs to be quite steady and clean for the drive to function properly. Drives pull up to 500 milliamps, which is not insignificant. If the motherboard's electrolytic capacitors are starting to get weak, the voltage may dip or develop spikes when the drive turns on, making it corrupt itself or causing malfunctions in other areas of the board. As the drive spins up from a dead stop, current draw can be a full amp for a second or so. See if malfunctions occur at the moment spin-up begins. If so, suspect voltage regulation or bad caps on the motherboard.

External DVD recorder drives take a lot of current when recording. Although designed to operate from a USB port, some significantly exceed what a standard port can provide. The ports are supposed to shut down if too much current is pulled from them, but I've seen some that continued to work but got damaged. Once that happened, they could never support such a drive again, but they might still offer enough current to run a thumb drive.

Other Problems

In most laptops, the Wi-Fi antenna cable goes through the hinges, with the antenna in the screen assembly. Every time you open and close the screen, the cable gets flexed. A broken coaxial cable will cause severely reduced wireless range. So will poor contact at the tiny coaxial connector (one with a center conductor surrounded by a shield) with which the antenna system connects to the Wi-Fi card. To check the cable, unplug it *gently* from the Wi-Fi card by rotating it a little to break any oxidation and pulling it up slowly. Don't yank on the connector or you may tear the socket off the board and ruin the entire assembly.

Many machines have two antennas, with a small duplexing board in the screen assembly that combines their signals, so you can't test for continuity from the card end of the cable right to one of the antennas. Be sure to check from the card connector to the duplexer. If you follow the cables from the antennas, which are little flat things next to the screen, they'll meet at the duplexer. The third cable, going down to the hinges, is the one you want. Use your DMM's resistance scale to test from its solder contact on the duplexer back to the other end inside the main body of the laptop. Be sure to check both the shield and the center conductor, as either or both could be broken. Usually, it's the center conductor that breaks.

If the cable is good, suspect the connector. Those little guys have two fingers in the center hole that are supposed to grip the pin on the board's socket (Figure 15-9). Over time, the springiness of the fingers relaxes, and the connection gets poor. While you have the connector off, use the tip of an X-Acto knife to push those fingers closer to each other. Quite often, this is all it takes to restore proper Wi-Fi performance. Don't overdo it; the fingers still have to fit around the sides of the pin.

Laptop keyboards are mostly mechanical, with a conductive rubber button under each key. The connections are arranged in a grid, but the irregular layout can lead to some pretty obscure patterns. You can't assume that it's all rows and columns.

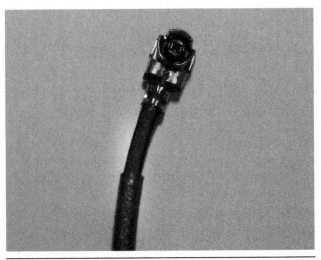

FIGURE 15-9   Micro coaxial Wi-Fi antenna connector.

When one key stops working, there's a contact problem underneath it. Time and oxidation can do it, but, more frequently, someone has spilled liquid into the keyboard. If you're really intrepid, you might be able to take apart the keyboard and clean it out, but it's a lot of trouble, and keyboards for most machines can be had pretty inexpensively. Replacing one is just a matter of popping it off, disconnecting the ribbon cable and installing the new one.

If a whole bunch of keys die and there hasn't been a liquid spill, check the ribbon cable connections at the motherboard because one entire line may be out. If not, there could be a motherboard problem with the keyboard decoder. Or, the break might be in the keyboard.

To check it, remove the keyboard and do a continuity test on every combination of lines coming from the keyboard, keeping in mind that good contacts may show as many as a few dozen ohms, as is normal for conductive rubber switches. There aren't that many lines, perhaps six or eight, so the test isn't that rough. If a bad key produces no continuity on any combination, the keyboard is the problem.

If the machine goes nuts, acting like someone is typing the same key over and over, that just might be the case! That "someone" was the person who spilled coffee or soda into the keyboard, shorting one or more of the contacts with conductive goo. Disconnect the keyboard and fire up the machine. If the stuck key goes away, you know somebody got sloppy with the drinks. This happens quite often, and the poor computer will act like it's mondo loco when all it really needs is some peace and quiet from its keyboard connector. The easiest fix is a new keyboard. I've tried washing a few in warm water and then letting them dry for days. Sometimes it worked, but more often the problem returned pretty quickly because there was still some conductive gunk in there.

Remote Controls

Factory-supplied infrared remote controls are all built pretty much the same. They use a matrix array of conductive rubber switches, a simple micro to read them and an infrared LED to flash the code generated by the chip. The micro is clocked with a ceramic resonator at a low frequency, usually in the hundreds of kilohertz.

How They Work

When you press a button, the micro senses the intersection of a particular pair of lines and cross-references it to whatever code is to be generated. The codes are hardwired into the chip's memory and are specific to the product to which the remote belongs. The code is sent as a series of rapid flashes of the LED. That's about it.

What Can Go Wrong

Beyond corroded battery contacts, the most common problem with remotes, by far, is erosion of the conductive coating on the business end of the rubber buttons (Figure 15-10). A button stops working while others continue to function. This can also be caused by grime on the conductive surface or, more often, on the carbon coating it contacts on the circuit board (Figure 15-11).

FIGURE 15-10 Conductive end of rubber buttons.

FIGURE 15-11   Carbon contacts on remote control board.

Other problems include bad solder joints where the battery springs meet the board, cracked joints at the LED, bad vias on the board and a dead ceramic resonator. Those are pretty reliable, but they can fracture inside and stop working when the remote is dropped onto a hard floor.

If liquid has been spilled into the remote, its residue can prevent button contact. Or, if the liquid or its residue is conductive, it can make one or more buttons act as if they're being pressed continuously, locking out all other commands.

Is It Worth It?

Because remotes are specific to each product, it's almost always worth trying to save one unless you have a cheap, easy source of a replacement. For many common devices, you can find exact matches online at reasonable prices, or a universal remote will work. For others, original remotes are shockingly expensive or just plain unobtainable, and no universal remotes can emulate them.

The Dangers Within

None.

How to Fix One

First, be sure the remote really isn't working! That's clear when one button doesn't function, but it's not so definite when a product doesn't respond to any commands. The problem could be at the receiving end.

There's an easy way to tell. Almost all digital cameras, including those in phones and tablets, can see infrared light. Make sure the remote has good batteries, and then point it at an active camera while looking at the camera's display. If the LED is flashing without your pressing any buttons, liquid almost certainly has been spilled into the remote, shorting out some of the contacts. The fix usually requires nothing more than a good internal cleaning. If you see nothing from the LED, press a button and see if the light starts flashing. If it does, assume the remote is working. I've almost never seen one flash incorrect codes, except on universal remotes that were incorrectly configured. The sole exception was a remote for an HDMI switch that had a bad via between a couple of the buttons. The missing connection made the micro read the button presses as the wrong ones, sending the wrong codes.

If you see no flashing, make sure your camera can see infrared by testing it with a known good remote. If the camera sees it but not the one you're working on, the remote is indeed not functioning. The number one cause is bad contact with the batteries. Because remotes use so little power, people leave the batteries in until they rot. Check for clean contacts before trying to take apart the remote. If the contacts are crudded up, scrape them clean with a screwdriver or an X-Acto knife to see if this gets the remote going.

Unless you're lucky enough to be working on one of the rare remotes with screws holding it together, probably your biggest difficulty will be opening the case. Almost all remotes are snapped together, and very tightly. Some have a screw or two but still won't pop open because they're snapped together as well.

Remote controls are not made to be taken apart. They're considered nonrepairable items by their manufacturers. The plastic is stiff enough that pressing along the sides won't pop the snaps. At least, I've never managed to get that to work. Sometimes, putting a screwdriver in the seam around the edge of the remote and then twisting it will pop a snap, and you can work your way around. Most of the time, you'll have to go with the same approach used on ultrasonically welded AC adapters: carefully tapping in the screwdriver to make a slot in the plastic and then twisting. Either way, expect a few broken snaps and some plastic damage along the seam.

Once you get the two halves to start separating, keep the button side face down while you proceed with the opening to avoid having any loose buttons or switches fall out and disappear. Many remotes have all the buttons in one sheet, but some use little slide switches or separate buttons, and they're easy to lose. In fact, it's not a bad idea to take a digital photo of the front of the remote before you start opening it, in case some buttons fall out and you're not sure which goes where.

On most remotes, the battery springs are soldered to the board and slipped through slots in the back of the case. I've seen a few that made contact only with pressure between

Solder
not
bonded
to wire

FIGURE 15-12   Cold solder joint on battery spring contact.

the springs and pads on the board, and those can develop oxidation and fail. Sometimes, even the soldered ones stop working, usually because the soldering wasn't so great in the first place (Figure 15-12).

If there's no infrared light from the LED, and you're sure the battery contacts are working and power is getting to the board, check the solder joints at the LED itself. Lots of remotes have the LED sticking out the front without a cover over it, and it can get wiggled or pressed on, breaking its solder joints. If those look okay, scope across the LED and see if the micro is pulsing it. If so, the LED has gone bad. If not—which is much more likely—either the ceramic resonator is dead or the buttons aren't making contact. The micros in remotes are highly reliable, so it's safe to assume the problem is not the chip itself. Sometimes there's a transistor between the micro and the LED for current amplification, and that could be open. And, of course, any part of the circuit could have a cold solder joint.

Using the negative battery terminal as circuit ground, try scoping the ceramic resonator's leads while pressing a button. If you're sure the button is making contact, you should see some sort of waveform. If not, either the micro or the resonator is shot. Most likely, it's the resonator (Figure 15-13). You probably can replace it with one from another remote made by the same company, because the frequency is likely to be the same. Resonator frequencies are not standardized across different manufacturers, so it's less likely you'll find a working replacement from a different maker. It can't hurt to try one of the wrong frequency, though. Even if the remote doesn't put out correct codes, at least you might determine whether the resonator really is the problem by seeing if the flashing of the LED resumes. Some resonators are marked for frequency, but don't count on it. If you can find one of the same frequency, it should work.

If some buttons work and others don't, almost certainly the trouble is with the buttons. Dirt and grime can interfere with contact, but the usual problem is that the conductive coating on the buttons has worn off. The buttons that get used the most, like those for

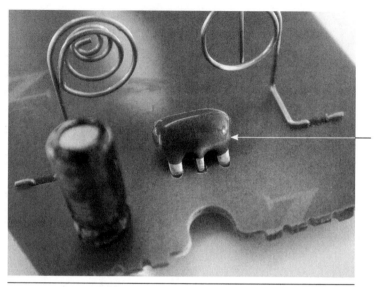

FIGURE 15-13   The ceramic resonator can be on either side of the board. Look for three leads.

volume and TV channel, are the ones that stop working. They can be fixed, but it's not always as easy as it might appear. Here are some approaches you can try.

Replacements

If you have some old remotes that have similarly shaped and sized buttons, you can cut them from their sheet and put them in place of the bad ones. It might look a little funky, but they'll work. You can also cut and move buttons from the same remote you're trying to fix if there are some you never use. Perhaps the remote can control multiple devices and has buttons for a disc player or a TV that you don't own. If you swap those with the bad ones by cutting them from the sheet, the remote will look and work fine, except for the unused buttons. I've saved a few this way.

Graphite

Rubbing graphite from a pencil tip onto the contact surface of the bad buttons will get them going, but I haven't seen this approach last very long. It might do in a pinch, but expect those buttons to quit making contact again in short order.

Conductive Paint

There are conductive paint kits made especially for restoring remotes. The reviews are mixed. The paint flakes off after a while because the buttons are flexible, causing the conductive ends to bend a little when you press them. The paint isn't flexible, so it cracks and falls off.

Stick-on Contacts

You can buy button repair kits for cordless phones that can be adapted for fixing remote controls. They work well and last a long time because they stick on the board, not on the flexible rubber surface. You just peel and stick the conductive dots over the carbon-coated contact surfaces on the board. The button presses on the dot. Because the dot completes the electrical contact, the button's conductivity no longer matters.

These stick-on contacts are not made for remote controls, though, because remotes have such a wide variety of shapes and sizes of buttons and contacts. To use the stick-ons, which are for considerably larger buttons, you'll need to cut them to size. Also, they have to make contact only when pressed; only the adhesive on the edges, which is not conductive, can touch the board's contacts when the button is not engaged. Shaping the stick-ons to work just right takes a small pair of scissors, a magnifying glass, time and patience. Nonetheless, I've had better results with this method than with any other.

Smartphones and Tablets

Probably the most ubiquitous electronic gadget in the world today, the smartphone is incredibly dense and sophisticated. It's a full-fledged computer that can do just about anything a laptop can, all in your pocket. Oh yeah, you can also make phone calls on it!

How They Work

Smartphones are based around a touch-screen interface. Inside are a CPU, flash storage, some RAM, cameras, a position sensor, a GPS receiver and three two-way radios: cellular, Wi-Fi and Bluetooth. It all runs on a flat lithium battery, managed by a BMS. There are at least two speakers, one for the earpiece and one for speakerphone operation, and one or two microphones, which are condenser or MEMS mics like the ones discussed in Chapter 14. There is also a vibrating module, which is really a tiny motor with an offset weight, and several antennas. All of the peripheral items are likely to be connected to the main board by ribbon cables with pop-off connectors, except for the antennas, which have tiny coaxial plugs and jacks. Those just pop out too. Sometimes pushbuttons or their ribbons are soldered to the board, and they may be held to the case with double-sided tape. You may also find buttons, mics and jacks on their own little boards, with metal fingers connecting everything when the case is snapped shut. Also, several peripherals may share one cable and connector, so they all have to be removed together.

What Can Go Wrong

Most phone troubles involve broken screens, bad connections at the headset and USB jacks, and failing batteries. Damage from liquid getting into the USB or headset jack is pretty common as well. Liquid can also enter around the edges of pushbuttons or an external SIM or SD card tray. Sometimes, the speaker used for the speakerphone burns out from being played at high volume for long periods. Far and away, broken screens and dying batteries are the most frequent problems.

Is It Worth It?

Smartphones were never meant to be repaired! They're put together with lots of tape and glue, and getting them apart can be a chore. Yet, there is a thriving industry of people fixing phones, and many of these repairers wouldn't know a transistor from an inductor or a scope from a DMM. It's a different kind of repair process, but, yes, you can do it.

The most common repair is battery replacement. For a long time, lots of phones had removable batteries behind a pop-off rear cover, and all you had to do was insert a new cell and you were done. The trend, though, has been toward internal, non-replaceable batteries. They can be changed, but it's a job.

If the screen broke, you can replace it with an aftermarket part for not a lot of money. Bad connections at the jacks also can be repaired by replacement with aftermarket jack assemblies, and you might even be able to resolder the existing jacks and save them if the plastic holding their contacts in place isn't cracked. Internal lithium batteries are fairly inexpensive. Some require soldering, while others just plug into tiny connectors. If the radio sections fail or the computer circuitry becomes flaky from a bad BGA, it's not likely you'll effect a repair without a BGA rework station and a lot of experience. Liquid damage is often fatal, but not always. You're more apt to save a phone on which drinking water was spilled than one that got hit with wine, soda or milk. The most damaging liquid of all is salt water. If the phone went into the ocean, don't waste your time.

The Dangers Within

The only hazards are shards of glass or oozing liquid-crystal material from a broken screen, and the usual issues with lithium batteries. For the most part, working on phones is safe as long as you don't short or puncture that battery. Also, be careful not to shove a metal tool in to pry up the back of the case because it could short the battery contacts and cause a quick fire.

How to Fix One

Repairing phones rarely involves soldering, simply because everything is so tiny and densely packed that board-level repair is almost impossible. Not even the manufacturers do it! Instead, you replace modules such as cameras, mics, screens and jacks by unplugging their connectors and popping in a new assembly. The one exception is that sometimes you can resolder a jack. But don't count on it.

The hardest part of fixing a phone is getting inside. See Chapter 9 for generalized disassembly details, but also look online for specifics of your model phone. The screen is glued to the case, and getting it off without destroying it requires finesse. If the LCD is broken, you may not care what happens to it, since you'll be replacing it anyway. Many times, however, the LCD is fine, and the outer glass, which contains the digitizer for touch sensing, is shattered from being dropped. In screens with the glass bonded to the LCD, you have to replace them both together, but some are separate pieces that can be replaced individually. Digitizer glass is a lot cheaper than the actual LCD. You don't want to wreck the LCD while trying to replace the glass! The same goes for getting inside to work on other elements like the jacks.

In devices that open from the back, reaching anything in front of the board means taking it off. Removing the board involves popping off all of the peripheral connectors and perhaps prying up the items, which may be held on with two-sided tape. Look for online tutorials that will guide you through the process step by step. There are lots of YouTube videos for specific models.

To replace the battery, first run it down as much as you can by keeping the phone on until it quits. That way, if you do cause a short, there's a lot less energy available for mischief. Once inside, you might have to remove an internal plastic cover, depending on the model of phone. Most of the ones I've seen were held on with several tiny screws.

The battery will have a pop-off connector that mates with the circuit board. It's best not to remove it just yet because it could hit something and short, heating the battery rapidly. If you do have to pull the connector, put some tape around the battery side so it can't short.

Once you get to the cell, you may find it glued to the phone. Do *not* try prying off the cell! The manufacturers use strong adhesive for this, and the battery will bend before it lets go, causing a catastrophic short inside. Someone I know caught his phone on fire by forcing a screwdriver under the battery to do exactly that.

To remove the tenacious glue, try passing dental floss under one end of the cell and pulling it up from both sides to break the glue bond. Put in the new battery, using double-sided tape to hold it in place. Plug in the connector and replace whatever cover you had to remove. Reassemble the phone, and it should work fine.

To replace the screen, you'll have to pry off the old one, probably destroying it in the process. That's fine if you're replacing the entire thing, but if you want to replace only the digitizer while preserving the LCD, you'll need to be extra careful and lift the broken digitizer off with a plastic shim after heating the screen's edges, working your way around slowly, or use an LCD separator, as described in Chapter 2. Any flexing of the screen will wreck the LCD. Figure 15-14 shows what can happen if you bend the screen even a little bit. If you do break the LCD, be careful not to contact the liquid-crystal material, should any ooze out.

If liquid got into the unit, you may see corrosion on the board. This is a bad situation, but it's worth disconnecting the battery and all the connectors, and washing the board with distilled water. Then, let it dry thoroughly. You can use a hair dryer, but don't get it too

FIGURE 15-14 Ruined LCD on an iPod with the same construction as a phone.

close. Just blow a gentle, warm breeze on the board and keep it there for long enough to dry things out. If there was corrosion, check the connectors, and wipe their contact areas with a dry cotton swab. Your chances of success are not high, but sometimes you can get the thing working.

Once you replace the screen or whatever module has broken, or cleaned out the connectors, hook it all back together in reverse order of the disassembly, and see what happens. Don't glue a new screen back on before checking to be sure everything works. The last thing you want is to have to take a brand-new screen off again. This is one time you'll want to pay especially careful attention to ATE. Test every module before regluing the screen and closing up the phone.

Tablet Differences

Tablets are built just like smartphones, only bigger. The issues and repair techniques are the same. When removing the screen, there is more leverage from edge to edge, due to the larger size, so you have to be even more careful and go more slowly to avoid flexing the screen and wrecking it. An LCD separator is your best bet here.

As in a phone, the battery may be affixed to the back of the unit with double-sided tape or glue. Use the dental floss trick, as described earlier.

You have a better chance of resoldering jacks because of the larger size and better accessibility. But even at the scale of a tablet, things are pretty small.

Video Projectors

LCD and digital light processor (DLP) video projectors are quite popular, and each type has its characteristic failures. Let's look at projectors and how to work on them.

How They Work

Projectors use a bright lamp or three LEDs, a *microdisplay* device that forms the image, and a series of lenses to magnify the results. While LEDs are now bright enough to light up a credible home theater projector, hot arc lamps are still found in projectors intended for business presentations, classrooms, auditoriums and other large gatherings, especially those that might take place in a room with significant ambient light. Plenty of home theater projectors still use them as well for their higher brightness. Eventually, LEDs will be so bright that arc lamps will be an expensive thing of the past.

In hot-lamp LCD projectors, there are three small LCD panels, one for each primary color of red, green and blue. Light from a very high-intensity arc lamp is filtered to remove ultraviolet (UV) energy and then split into three beams, with color filters for each color. Each beam illuminates its own panel. The resulting three images are recombined with a prism and focused on the screen with the projection lens. When LEDs are the light source, each panel is lit individually, and there's no need for a UV filter.

DLP projectors have no LCD panels. Instead, they use a special chip called a *digital light processor*, invented and manufactured by Texas Instruments Corporation. On the surface of the DLP chip is a matrix array of movable microscopic mirrors, each separately addressable. Today's high-definition chips have millions of mirrors. Feeding power to a mirror makes it flex, deflecting the light at an angle and reducing the amount reflected straight toward the lens. The result is a projected video image of high contrast.

That'd be all there was to it, except for one small detail: color. DLP chips are expensive, and they require a fair amount of circuitry to drive them. Some very pricey, professional-level projectors have three DLPs, combining their outputs optically like LCD units do. The home units you're likely to service, though, have only one chip and accomplish color projection by rapidly flashing the three color images in sequence. Lamp-based units place a high-speed rotating color wheel of red, green and blue segments between the lamp and the DLP chip. Your visual system, which can't keep up with anything coming at it that quickly, combines them into one full-color image. DLPs can move their mirrors much faster than is required for normal video rates, so it's possible for them to flash two or even three complete sets of tricolor images in the time span of one frame. When the specs say the unit has a 2X or a 3X color wheel, each frame of video is being flashed at that rate, compared to a normal video frame, with three flashes of color each time. So, a 3X-rate projector flashes nine images in the time it would take a CRT to scan one frame.

In LED-lit DLP projectors, the same rapid frame-sequential color system is used, but there's no color wheel. Instead, red, green and blue LEDs are flashed in sequence. Unlike lamps, LEDs can be turned on and off so fast that rapid frames are no problem. The only moving parts in such a projector are the cooling fans, if you don't count the slight movement of the micromirrors in the imaging chip.

The fast frame rate helps to diminish the *rainbow effect*, an annoying consequence of the single-chip, frame-sequential color projection method, that occurs when the viewer's eyes move. Especially in darker images with bright points of light, like night scenes with streetlamps, the visual trail left by the bright spot can break up into its component colors as the eyes change position, because each color frame strikes the retina in a slightly different spot, so they don't blend together. Some people find the effect very distracting, so manufacturers keep speeding up the frame rates to minimize the time between projection of the different colors, keeping them closer together in the moving eyes of the viewer.

What Can Go Wrong

There's lots to malfunction here. The most troublesome elements in a hot-lamp projector are the very expensive lamp and the circuitry powering it. The brightness required is so extreme that only a high-pressure mercury vapor arc lamp will do the job. Operating an arc lamp is not as simple as just applying power. First, it has to be "struck," or started, by applying a fairly high voltage until conduction across the arc is achieved. Then, once current starts flowing, the mercury inside vaporizes and makes the lamp conduct much more readily, with lower resistance. The voltage must then be reduced to typically less than 100 volts. The circuitry driving the lamp is called the *ballast*, though it's much more complicated than the simple ballast that starts an old-fashioned fluorescent lamp.

Wait, there's more. Arc lamps exhibit some pretty odd behaviors. As they age, they tend to develop bad spots on their electrodes, increasing the resistance of the most direct path across the arc. Because the vapor conducts, other paths arise, and the arc can jump around, causing flickering of the light. To combat this annoying malady, the ballast may adjust the operating voltage or add pulses to keep the lamp at its best. Even with all this effort, lamps go bad, they go dim, they fail to strike, and now and then they explode violently.

The second most failure-prone part differs between LCDs and DLPs. In LCDs, the *polarizers*, sheets of special light-polarizing plastic film in front of each panel, get burned by the residual UV output of the lamp, even after its light has passed through the UV filter. In particular, the blue polarizer tends to burn, resulting in a yellowed image or splotches of yellow.

In hot-lamp DLPs, the color wheel, whizzing around so fast, often experiences bearing failure or catastrophic disintegration. Because of lamp heat, color wheels are made of glass, not plastic, and most are assembled with nothing more than glue! In time, the glue degrades, also from lamp heat and UV radiation, and the delicate red, blue and green segments fly off, smashing against the inside of the projector's case and shattering into a million pieces.

DLPs also use a light guidance arrangement quite different from that in LCDs. LCD panels are considerably larger than DLP chips, and the lamp's output must be spread over enough area to illuminate them fully. That's easier than the DLP scheme, in which the light has to be formed into a small beam to match the much smaller size of the DLP chip. To do so, DLPs use a mirror tunnel, also called a *light tunnel*, made of four mirrors arranged to form a rectangular channel. Like the color wheels, the mirrors may be glued together, and the glue can fail, collapsing the tunnel.

Cooling the light source is a critical function in all projectors—even LED-lit ones—so they all have fans blowing air through the lamp housings or across heatsink fins. Most projectors have multiple fans. LCD units usually have one just to cool the panels because, as blocking elements, they absorb a lot of heat from the lamp. DLPs, which reflect light instead of absorbing it, don't overheat their imaging chips, but they may use fans to cool the rest of the optical chain, along with the lamp fan. Some projectors have power supply fans as well. It's common for projectors to include four to six fans.

A failed fan, especially if it's the lamp fan, will cause the projector to overheat and shut down. It takes a few minutes for the thermal sensor to heat up enough to cause shutdown, so the projector will run for a short time before it turns off.

Most hot-lamp projectors have dust filters on the lamp housings, and they get clogged with room dust to the point that airflow is severely restricted, triggering an overheat shutdown. LED-based units blow the air across a heatsink, and its fins can get clogged as well, but you can't clean them thoroughly without taking the projector apart.

And, of course, projectors suffer from the usual power supply issues, especially bad capacitors. The units are remote-controlled, so at least part of the power supply runs all the time, even in standby. After a few years, a cap or two is shot, and the projector stops turning on.

Is It Worth It?

An expired lamp might seem like an obviously worthwhile repair, but the lamps cost so much—from around $100 to more than $400—that you must consider whether the rest of the projector will survive long enough to use up a new lamp. Those hot lamps put tremendous stress on the other optical components, and many home theater projectors are designed to last about as long as one lamp. It's no fun to spend two thirds the cost of a new projector for a replacement lamp, only to have a polarizer or a color wheel go bad 100 hours later. Expensive, pro-level projectors are built to last through several lamp replacements, making putting in a new bulb more viable.

The bright LEDs used in lampless projectors are considered permanent parts of the units and are not easily replaceable like a lamp. They're typically rated for 20,000 to 30,000 hours of operation, so you shouldn't need to change one. Naturally, some LEDs will fail prematurely, and you'll be looking for a replacement. High-power LEDs are not cheap, but they're not as costly as lamps. They're also not nearly as easy to find because they are not intended to be user-replaceable.

Burned LCD polarizers are pretty much a dead end unless you can scare up a parts unit. It's almost always the blue polarizer that goes, so a unit old enough to be cut up for parts probably has the same bad polarizer and will be of no use. Manufacturers don't sell the polarizers separately; they want you to buy the entire light engine (optical system) or replace the projector. Believe me, you do not want to spend what a new light engine would cost.

Color wheel costs vary greatly among manufacturers. The wheels for some of the old rear-projection DLP TVs could be had for $50, while those for some front projectors cost an eye-popping $500. There's no basic difference between the parts; it's all a matter of marketing and volume. Parts for those old TVs, including lamps, were generally lower than those for front projectors. Light tunnels can often be repaired for nothing with some high-temperature epoxy and a steady hand.

The Dangers Within

The lamp is not your friend! If you look directly into it while it's running, even momentarily, I hope you like dogs, because you're probably going to need one. The brightness is higher than anything the human eye can withstand. There's a fair amount of UV, too, which is very damaging, even after most of it is absorbed by the lamp housing's UV filter. Even today's LED light sources are bright enough to damage your eyes. Seriously, don't ever look directly into a running projector!

Arc lamps get hot enough to burn you badly, too. After it's been operating, let it cool down quite a while before going near it. When it's hot, the glass is more fragile as well. The actual lamp envelope is only about the size of two pencil erasers, but it's under tremendous pressure. I've seen one explode, and it's not pretty. They go off like a shot, and a fine mist of glass particles tinged with mercury gets ejected from the projector's fan vent. You wouldn't want your eyes or lungs in the vicinity.

The voltages used to strike and drive the lamp are hazardous. Figure around 1 kilovolt for striking and 80 to 100 volts during normal operation. Don't go poking around with your scope in the lamp supply (ballast), especially while the lamp is striking.

A DLP's rotating color wheel could cut you if you contact its outer edge while it's spinning, but it's more likely you'd destroy it.

How to Fix One

Figure 15-15 shows the optical path, or light engine, in a typical lamp-based DLP projector. Most problems are found there or very nearby. The three most common failures are no operation at all, no lamp strike, and overheating with subsequent shutdown. If the unit won't turn on at all, suspect the usual power supply issues. Projectors spend virtually their entire existences plugged in, waiting for a remote-control signal, so bulging power supply capacitors are pretty much a foregone conclusion eventually.

One difficulty in servicing projectors is that restarting a hot lamp damages it badly. So, once you turn the unit on, you don't want to turn it back off, take a few measurements or check a few parts and then fire it right back up again. Always let the lamp cool before

Lamp compartment   Light tunnel   Fan   DLP chip   Light condenser

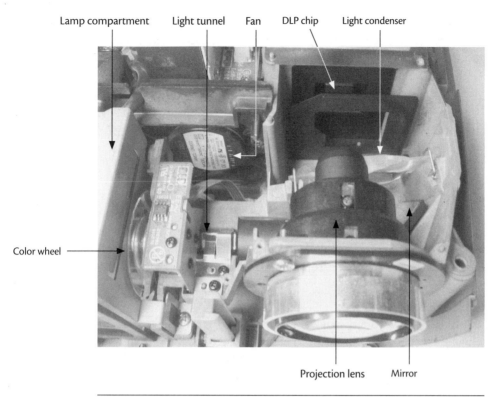

Color wheel

Projection lens   Mirror

FIGURE 15-15 DLP optical path.

restriking it. That can take a half-hour or so. There's no issue with LEDs, though; you can turn them on and off all you want.

Lamp Problems

If the projector powers up but blinks a warning light on the control panel, the lamp or its ballast may be bad. A sensing circuit checks for current draw through the lamp; that's how the thing knows the lamp has struck and it's time to reduce the striking voltage to its normal running level. If the lamp won't strike, the warning light is as far as it'll go.

You can't check the lamp for continuity with a meter. At room temperature, the bulb is an open circuit until it has 1 kilovolt or so applied across it. Assess the lamp's condition visually. Being certain it is cool, remove the lamp. When working with a high-pressure bulb, it's a good idea to wear goggles, just in case you bump it or drop a tool on it, because it could explode in your face. The lamps in most projectors are enclosed in a housing with a UV filter in front, but you could still get showered with glass blown out the sides of the housing through the dust filters.

It's important *never* to touch the bare lamp or the UV filter, because skin oils will make them crack when they get up to operating temperature. A cracked UV filter may leak

UV radiation, and a cracked bulb . . . well, you know what will happen. Kaboom! If you should happen to touch the wrong spot, clean it with alcohol, and check the glass carefully afterward for any skin oil or other residue.

To evaluate the lamp, you need to get to it. The housings come apart. Do so carefully! Once you can see the bare bulb, which is a glass stalk, look at the pea-sized envelope at the back of it head on, slightly from the side. You're looking down the focal point of a parabolic reflector, so it's hard to see the arc gap, but if you look a bit off-axis, you can see it. It'll look greatly magnified by the reflector, and this is helpful. If the envelope's glass looks clear and clean, the lamp doesn't have lots of hours on it and is probably good. If it looks charred, it's an old lamp and may be shot. Examine it carefully, and you can also see the actual electrodes and their condition.

Almost all projectors have a time counter in their menus telling you how many hours are on the lamp. Without a working lamp, of course, you can't see it! Better LED-based units have time counters too, but some don't bother including them because the LEDs are supposed to last the life of the projector, and users can't change them anyway.

Ballast Problems

The ballast is really a pretty fancy power supply of its own. It has to supply the high lamp-striking voltage and then the lower operating voltage. Operation can require a few amps, for around 150 to 250 watts of lamp power. Output transistors supply it, and they can pop. Also, a fair amount of heat is generated in the output stages of some ballasts, so check the circuit board for burn marks and degraded solder joints.

Don't try scoping the ballast's output stages. It's far safer and easier to disconnect power and pull and check the output transistors. There may be an onboard fuse, too. Be sure to discharge any large electrolytics before desoldering anything.

Look for *opto-isolators* in the path between the ballast board and the main board. The output of one of them will change state when the lamp strikes successfully, relaying the information that it's time to lower the voltage. If you can't find this signal, the lamp is not being struck, or it's bad.

LEDs are operated through a constant-current regulator. As they heat up, the LEDs' resistance decreases, causing a rise in the current through them. The constant-current supply adjusts the voltage to keep the current from changing, because too little current will make the LEDs dim and too much can lead to *thermal runaway*—more heat means more current, and more current means more heat, so both keep rising until the LED is destroyed. If one color goes out, trace back to see whether that LED's current supply is working.

While newer LED projectors use single-chip LEDs, one for each color, a lot of them have employed LEDs that are actually many small chips on one substrate to achieve the required brightness. Over time, some of them burn out, resulting in greatly reduced brightness but not complete loss of the color. If you remove such an LED from the projector, it'll be obvious that there are many small chips in a grid arrangement. With a single-chip LED, greatly reduced brightness without complete failure is highly unlikely.

Overheating Problems

Overheat shutdowns don't happen instantly; it takes a little while for the heat to build up. If the projector runs for 5 or 10 minutes and then quits, it's probably overheating. A lamp very near the end of its life can run excessively hot and cause this condition, but most of the time it's due to lack of airflow over the bulb, or to clogged fins on an LED assembly. LEDs don't get nearly as hot as arc lamps, but they still generate significant heat that has to be carried away. If their heatsink fins are clogged, they are easily cleaned with a blast of compressed air. Check for blocked dust filters. Some projectors, especially LCDs, have them at the air intakes for both the lamp and the LCD assembly. Blockage of either can trip the shutdown. DLPs usually have filters right on the lamp housing. Most LED types have no filters, but it's not out of the question. Even some hot-lamp units are designed to be dust-resistant and use no filters. This type rarely clogs up, but check the air channels at the lamp housing just to be sure nothing has gotten in and blocked those.

There's a thermal sensor over the lamp. It's usually connected in series with the AC line, so don't touch its contacts! If the fan doesn't do its job or the filters are clogged, the lamp overheats, the sensor trips and the unit shuts down. Never defeat this sensor, even just for a repair test. Even if you don't cause a fire, you'll probably destroy the lamp, and it could get hot enough to burst. LED-based projectors have thermal sensors on the LED assemblies to control the fan speed.

A lot of projectors don't wait for an overheat condition to shut themselves down if a fan isn't turning; the fan rotation itself is sensed, and if it stops, the projector switches off power to the lamp and comes to a halt, blinking an error message on the power light. The fans themselves can fail, but sometimes the problem is the circuitry powering them, especially in units with variable-speed fans. Most projectors have normal and economy settings for brightness, with increased lamp or LED life at the lower setting. In economy mode, they don't get as hot, so they don't need as much cooling. To reduce noise, they slow the fans down, using power transistors to control the voltage. Over time, a transistor may fail, and the fan will go dead. All of the fans should always be turning when the projector is on. If one of them isn't spinning, disconnect power, and try turning the blade by hand. It should spin freely. If not, it's gummed up with dirt or its lubrication has dried out. That can be fixed. Peel back the label, remove the clip holding the fan blade shaft in place, pull the blade assembly straight out, and clean and lube the bearings as needed. I've saved lots of them that way using silicone liquid or spray. Light machine oil will also work, but it won't last as long.

These fans are brushless, with a little circuit inside to convert the DC power they run on to a rotating AC field that makes them spin. To see if the circuit is good, pull the fan's connector and hook it up to your bench power supply, with the voltage set at whatever the fan label states. Be sure to connect red to positive and black to negative. To get a connection to the fan plug's terminals, insert a capacitor's leads into the plug's holes. Use a 0.1 µF or smaller value, *not* an electrolytic! The small capacitance will act like an open circuit to the applied DC, and you can connect your power supply to its leads to power the fan. It should spin, nice and fast.

If the fan is good, the projector's driving circuit is out. Trace the fan's wires back to the board, and look for a small power transistor. Typically, the transistor is in series between the fan's negative terminal and ground, with supply voltage going directly to the fan's positive terminal. You can verify this configuration by looking for one terminal of the transistor going to ground. Most likely the transistor is open (blown), so no current is getting from the fan's negative lead to ground. After making sure the circuit is indeed wired this way, short the fan's negative lead directly to ground, and the fan should run fast. A new transistor ought to fix it right up. All the units I've seen have used an NPN bipolar transistor for fan control, and any old NPN, such as a 2N3904 or a 2N2222, will work. See, there really is a reason to keep those parts on hand!

The transistor's collector goes to the fan, the emitter to ground, and the base to the third pad on the board, which feeds it the control voltage to turn the fan on and regulate its speed.

Light Tunnel Problems

If the light tunnel collapses, the projector will run but you'll see a darkened area along an edge or even large portions of the image blacked out. Gluing it back together is a chore, but the price is right! Carefully disassemble the optical path, and remove the tunnel's mirrors. The shape of the tunnel corresponds to the aspect ratio of the imaging chip. You should be able to deduce how it went together from the position in which you find the pieces and from the glue remnants. Clean off the old adhesive, and glue the tunnel back together with epoxy. Don't use instant glue—it will not survive the lamp's heat. Use good epoxy that takes a while to dry, not the 5-minute quick-set type. If you can find high-temperature epoxy, that's your best bet. Let it dry for a night or two, and then reassemble.

Color Wheel Problems

A lamp-based DLP's color wheel is driven by a small motor, with a position sensor to tell the video circuitry when to flash the correct image for whatever color is in front of the lamp. Most of them use a Hall-effect sensor that picks up a magnetic field from a magnet embedded in the assembly's rotor, generating a pulse for each revolution of the wheel. Without this signal, the unit's microprocessor assumes that the wheel isn't turning properly, or at all, and stops operation, shutting down the lamp.

Treat the color wheel gently. It's delicate, and breaking it means the end of the projector unless you get really lucky and a petal comes off intact, in which case you can glue it back on. When the unit starts up, see if the wheel is turning. It should spin very fast. If it seems sluggish or isn't moving at all, unplug the projector and gently turn the wheel by hand. If it doesn't turn freely, the motor's lube may have dried out. Often, dried lube will cause a horrible screeching sound from the wheel, too. The normal remedy is replacement, but the cost can be ridiculous.

I've saved a few screeching color wheels for nothing, though, and the fix has lasted a long time. Get some silicone lubricant spray. Don't use anything else, like oil or other kinds of sprays. Remove the wheel, and spray into the motor bearings, saturating the motor as best you can. You'll probably be spraying under the color disc itself into the top of the motor. Let it soak in for a few minutes, and then dump out what's left into a tissue. Very gently clean excess lubricant off both sides of the color disc with an alcohol-soaked swab. Don't put pressure on that disc or you may snap off the glass petals, but be sure to get all the lubricant off the petals so they won't crack from the lamp's heat. Give the motor a few careful spins, and then reassemble the unit. Sometimes it works.

Video Processing Problems

The circuitry driving the LCD panels or the DLP chip is dense and complex. It's all digital and not very serviceable. If you're losing a color on an LCD or the image is flickering or freezing on a DLP and the problem is not the light source, check for bad cable connections to the boards driving those imaging devices. Also, look for leaking surface-mount electrolytic capacitors. The heat of a projector can really shorten their life. If there's no video at all, look for onboard voltage regulators, and scope to see if they're putting out voltage. Beyond that, you'd need a service manual and some serious high-speed digital test gear to make much sense out of the signal processing. These kinds of problems aren't common, though. Most projector failures can be traced to the power supply, the ballast, the lamp or LEDs and the optical components.

Blobs in the Image

These are caused by dust on the LCD panels or the DLP chip. It's a more common issue with hot-lamp projectors than with LEDs because air has to be blown directly on the lamp, so dust has an easier time getting into the optical path. LED optics can be sealed, with the air blown on external metal heatsink fins. A little carefully applied compressed air will take care of the blobs. You may have to open a housing or remove the lens to gain access. Be careful not to spray the air too close to the imager, in case some of the cold propellant winds up on the optical surface, where it could mar the transparency. The actual mirrored surface of a DLP chip is behind a sealed window, so there's no chance of harming the mirrors. LCD panels and their polarizers are more easily damaged.

DLP Dots

If your DLP is starting to show random white dots, the DLP chip itself is failing. The only remedy is a new chip. They're not cheap, but you can find replacements online. For an older lamp-type set, it's probably not worth the cost because the color wheel is likely to need replacement soon anyway. On an LED-lit unit with plenty of life left, it might pay to replace the DLP.

LCD Lines

Stationary lines on the screen of an LCD projector are caused by a bad column or row driver chip or a bad connection to the glass of the panel. You'll have to replace the panel.

As with DLP replacement, it might not be worth it in an older projector. You can't buy those panels separately, and an entire light engine would cost more than replacing the whole machine, so your only hope is a parts unit. If you're lucky, you might find one online and scrounge its panel. The positional alignment of the panels relative to each other is pretty critical, so it makes more sense to swap out the entire optical assembly instead of just one panel.

Yellow LCD

If your LCD's image has a yellow cast or yellowed areas, the lamp has cooked the blue polarizer. If you can find a parts unit that doesn't have the same problem, changing the entire assembly is your easiest route to a successful repair.

Have At It!

I hope you've enjoyed our extended romp through consumer electronics and its repair. Armed with the techniques presented in this book, you should be ready to dig in and have some fun! Just always remember to put safety first, be sure to ATE, and keep in mind that expertise at repair develops over time, as with any skill. If you wreck a product while trying to fix it, toss its remains on the parts pile and chalk up the loss to a valuable lesson learned. We all lose a few along the way. The more you practice, the better you'll get.

Ready to rock? Scope on? Soldering iron warmed up? Ignition sequence start . . . T minus 3, 2, 1, . . . blastoff! Stay safe, and have a grand adventure!

Glossary

Here are some of the common terms you'll encounter when working on electronic devices. There are plenty more, but you'll see these often.

active elements   The semiconductors at the heart of a *circuit stage* that perform amplification, switching or other signal-processing functions defining the stage's operation. Transistors and integrated circuit chips (ICs) are active elements. Before solid-state electronics, the active element was the *vacuum tube*.

ADC   Analog-to-digital converter. A device that converts analog signals to a digitally coded representation.

alligator clip   A toothed clip-on metal terminal on each end of a wire, used for the purpose of making temporary connections. The complete assembly is called a clip lead.

alternating current (AC)   The polarity periodically reverses direction. Reversals may occur at any rate above zero cycles per second, or *hertz*, up to billions per second.

amperes (amps)   The volume of flow of electric charges per unit of time, or *current*. This corresponds to how much electricity is flowing, independent of how hard, and is represented as *A* or sometimes *I*. In small-signal circuitry, it is usually specified in *mA*, or milliamps (thousandths of an amp).

amplitude   The voltage strength of a signal. Mostly an analog term, rarely used for digital *pulse trains*. For audio, amplitude represents loudness. For video, brightness.

analog   A method of representing information by varying a voltage over time in a continuous pattern resembling the information, distinctly different from the either/or, on/off states of *digital* representation.

ATSC   American Television Systems (or Standards) Committee. The group that determined the American standard for over-the-air digital television broadcasting. Also, the name of the standard itself.

audio   An analog signal representing sound by varying a voltage in the same pattern as the original mechanical vibrations of the sound.

audio frequency (AF)   A signal in the frequency range of 20 Hz to 20 kHz, the approximate *spectrum* of human hearing.

audio taper   The logarithmic resistance curve of a potentiometer used in audio applications to provide a seemingly constant rate of change in rotation of the knob versus perceived loudness. Also called *log taper*.

ballast   The circuit used for starting a lamp. Ballasts range from very simple, as with old-fashioned fluorescent lamp tubes, to complex, as with the high-pressure arc lamps used in video projectors.

barrel connector   See *coaxial plug*.

baseband   A signal not being carried by another. In particular, analog audio or video that is not modulated on a radio-frequency (RF) carrier or that has been recovered from one by *demodulation*.

BGA   Ball-grid array. The grid of tiny ball-shaped solder pads under high-density chips like CPUs and graphics processors. BGAs often fail due to thermal expansion and contraction of the circuit board, especially with chips that run hot during normal operation.

bias   The DC current or voltage supplied to the input of an amplifying element to keep it in its conducting state during the required portion of the signal waveform.

bipolar   A circuit with both positive and negative power supply polarities relative to ground. Also, the internal construction of an NPN or PNP transistor.

bit   In digital data, a binary digit, either a 1 or a 0. The two states are represented by the presence of voltage for a 1 and the lack of it for a 0.

blown   A component that no longer passes current. It has become an open circuit, usually because too much current passed through it and destroyed it.

BNC connector   A type of coaxial connector used for RF (radio-frequency) cables, such as oscilloscope probes.

bridge rectifier   Four diodes in a diamond-like configuration, with two terminals for alternating-current (AC) input and two for direct-current (DC) output. See *full-wave rectification*.

bus   A large connection point used to distribute power or provide ground to multiple areas of a circuit. Also, the shared connections used to distribute data to multiple points in digital devices.

bypass capacitor   A capacitor connected to shunt a signal's AC component to ground without having an effect on the signal's DC component, if there is one.

byte   A group of digital *bits* representing a unit of information such as a letter, number or audio sample.

capacitance   Specified in farads, capacitance is a measure of how much charge can be stored on the two plates of a *capacitor* for a given applied voltage. The more charge stored, the more current will pass back out of the capacitor when it is discharged into a circuit.

Numerous factors affect capacitance, including surface area of the capacitor's plates, their distance apart and what separates them. Most capacitors are specified in microfarads (μF, or millionths of a farad) or picofarads (pF, or trillionths of a farad). In European gear, capacitors specified in nanofarads (nF, or billionths of a farad) are sometimes used.

carrier   A signal on which another, lower-frequency signal is imposed. In radio, the high-frequency signal emanating from the transmitter, on which is imposed amplitude or frequency modulation of audio information. See *modulation*.

CCD   Charge-coupled device. A type of imaging chip in which each pixel's brightness value is represented by a voltage that gets transferred to an adjacent cell, bucket-brigade style, as the voltages are read out from the chip in sequence.

CCFT   Cold-cathode fluorescent tube. The type of lamp used in liquid-crystal display (LCD) panels before they were lit by LEDs. CCFTs have no filaments. They are driven with high-frequency, high-voltage pulses applied across the length of the tube and do not require a *ballast*.

chopper   The transistor used in a switching power supply to convert incoming power into high-frequency pulses that will be fed to the conversion transformer.

class A   An audio amplifier that processes the entire waveform without splitting its positive and negative halves. A class A amplifier is not *complementary*.

class AB   See *complementary*.

class D   An amplifier employing pulse width modulation instead of linear amplification. See *PWM*.

clipping   The state of nonlinearity in an amplifier that occurs when an input signal attempts to drive its output to voltages greater than those provided by the power supply. The amplifier's output signal can't swing as far as the input signal demands, so it stops at its limits and clips off the tops and bottoms of the waveform, causing severe distortion. See *linear*.

CMOS   Complementary metal-oxide semiconductor. This type of integrated circuit (IC) construction is widely used because of its wide voltage tolerance and low-power operation. Once limited to small-signal circuits, CMOS devices now handle serious voltage and current, and are found in metal-oxide–semiconductor field-effect transistor

(MOSFET) power transistors and output modules. Some CMOS devices are easily damaged by static electricity because it can punch holes through the oxide layer, causing an internal short circuit.

coaxial plug   A common plug used to connect DC from the output of an AC adapter to a product. The plug has a center sleeve, or sometimes a pin, surrounded by a shell that provides the other connection. This protects the center terminal from shorting out if the plug touches metal.

complementary   Opposite in phase. Also, an amplifier design with separate transistors operating on opposite halves of the signal waveform. A true complementary design uses PNP and NPN (opposite-polarity) transistors. One using the same polarity transistors for both waveform halves is called *quasi-complementary*. Both schemes are used frequently in audio power amplifiers, with the two transistors shown one above the other on the schematic diagram. Complementary amplifiers are called *class AB*.

constant-current source   A source of power that adjusts its voltage automatically to keep the current through the circuit being powered constant as its resistance changes. Constant current is used to run high-powered light-emitting diodes (LEDs) and for many other purposes.

conventional current   An incorrect but convenient way of imagining current flowing from positive to negative. It's especially useful with a negative ground configuration, as found in most modern circuitry. The term is so named because early electrical experimenters thought current flowed opposite to the way it actually does and established positive-to-negative flow as a standard convention. Oops!

coupling   The passing of a signal from one circuit stage to another. Typically, the function of a capacitor when it transfers the AC component of a signal from one stage to the next, blocking the DC component.

CPU   Central processing unit. In today's gear, CPUs are microprocessors.

crowbar   A component, usually a silicon-controlled rectifier (SCR), used in power supplies to short the incoming AC line deliberately and blow the fuse to stop improper operation that might harm the supply or the circuit being powered. Used in high-current linear supplies and quite common in switchers. The term also may refer to the entire protection circuit, including the stages that sense improper conditions and trip the SCR.

CRT   Cathode-ray tube. A picture tube, as used in older TVs, computer monitors and oscilloscopes.

current   The flow of electric charges. Though it is semantically incorrect to describe the "flow of current" because that is the "flow of the flow of charges," most of us in electronics think of current as the "quantity of electricity," and describing current as flowing is common and considered acceptable by most practitioners. See *amperes*.

curse words   Utterances sometimes emitted by technicians immediately following the appearance of *magic smoke* or the accidental destruction of valuable or hard-to-obtain components.

cutoff   The state of a transistor when it is fully turned off, allowing no current to pass through it.

cutoff frequency   The signal frequency at which a transistor's gain has decreased to 1, or unity gain.

DAC   Digital-to-analog converter, sometimes pronounced "dack." A device that converts digital codes to analog signals.

damper diode   A diode connected across an inductor to short out reverse-polarity current generated when the inductor's magnetic field collapses, which could damage other components. Damper diodes are connected cathode to positive, so they have no effect on the current applied to the inductor. They are most often used with relay coils and motors. Some relays include the diode inside.

Darlington pair   An arrangement of transistors with the output of one directly feeding the base of the next to multiply their gains. With NPN transistors, the collectors are tied together, and the emitter of the first transistor is connected to the base of the second. Some Darlington pairs are built into a single-transistor package, providing gain unattainable with just one normal transistor. Darlington pairs are often found in audio power amplifiers.

DC-DC converter   A small *inverter* used to step up a DC voltage to a somewhat higher one. Typically used to generate 15 to 30 volts in devices operating from 3 to 6 volts.

delay line   A circuit that delays a signal intentionally, using a length of cable or a ceramic device. Common in analog TV and video tape recorder color processing, the effect can be replicated digitally with a FIFO (first in, first out) memory buffer.

delayed sweep   A special function on some oscilloscopes that permits you to zoom in on a small feature of a waveform and see it in detail.

demodulation   The extraction of information from a signal. Also called *detection*. See *modulation*.

dielectric layer   The insulator between the two conductive sides of a capacitor.

digital   A method of representing information by encoding it into a pattern of voltages having only two states, on or off, representing the binary numbers 1 and 0.

digitize   To convert an analog signal to a digital representation. See *ADC*.

dikes   Diagonal cutters.

direct current (DC)   The current moves only in one direction. Current moves from negative (an excess of electron charge) to positive (a dearth of it). Most modern gear uses negative ground, so the current is really moving up from the circuit ground point, through the components and then into the positive power supply connection. It's convenient to think of ground as the sinkhole into which everything is flowing, though, and this is perfectly acceptable regardless of polarity. Conceptually, the actual direction of electron charge flow is irrelevant. In physical terms, however, the direction of flow is quite important, as you'll disastrously discover should you try to run a product from a power supply with the polarity reversed.

discrete   Made of individual parts, as opposed to ICs. An audio output stage with separate power transistors, instead of an integrated module, is a discrete output stage.

distortion   The unfaithful reproduction of a signal by an amplifier, in which the output does not accurately mimic the input. See *clipping* for one example.

DLP   Digital Light Processor. The trade name of a Texas Instruments video projection technology employing a chip containing microscopic mirrors that can be flexed with applied voltage, aiming reflected light toward or away from a lens. Practical DLPs may have millions of mirrors.

DSP   Digital signal processing. A method of performing real-time calculations on digital signals in order to remove noise, enhance speech or video clarity, decode surround sound, etc. The acronym is applied to both the process and the specialized chips performing it.

DUT   Device under test. The circuit or component on which you're taking measurements.

duty cycle   The percentage of time that one cycle of a signal's waveform spends in its "on" state. This is usually used to specify the percentage of time a square wave spends turned on. A wave with equal on and off times has a 50 percent duty cycle.

DVI   Digital video interface, the all-digital video connection commonly used to connect computers to monitors. It carries the same digital video format as a high-definition multimedia interface (HDMI) but on a different connector and without audio. See *HDMI*.

dynamic range   The range of signal strength, optical intensity or sound pressure a transducer or signal-processing circuit can accept before it is either too weak to process or too strong to handle without overload. Dynamic range is usually specified in decibels (dB).

ECO   Engineering change order. Oops, we made a boo-boo and need to change the circuit design, so here's what to change when units come in for service. ECOs are issued by manufacturers to authorized service centers.

elastomeric strip   A flexible plastic strip with conductive vertical channels used to connect a numeric LCD to the circuit board driving it.

electret   The permanently charged diaphragm inside a condenser microphone.

electrolytic   A type of capacitor employing a liquid electrolyte, or separating layer.

emitter follower   A current amplifier, or *buffer*, in which the output is taken from the transistor's emitter. Emitter followers offer only current gain, not voltage gain, and they do not invert the signal. The output is a replica of the input, except that it has higher current-driving capability, so it can feed a greater load.

envelope   The overall contour of a waveform, viewed over many cycles.

ESR   Equivalent series resistance. The internal resistance of a capacitor that limits how fast it can be charged and discharged. Some ESR is normal, but electrolytic capacitors often develop increased ESR as they age, eventually causing circuit malfunctions.

eye pattern   The distinctive envelope of the output signal from an optical disc player's laser head. See *envelope*.

fall time   The time it takes for a waveform to drop from 90 to 10 percent of its maximum value.

feedback   A signal sent from a circuit's output back to its input. The path the signal takes may involve intermediary components and is sometimes called a *feedback loop*.

filter capacitor   A large bypass capacitor used to smooth the output voltage of a power supply.

FPGA   Field-programmable gate array. A chip whose internal configuration can be changed via programming. FPGAs are used in *SDR* receivers so they can be reconfigured for receiving different types of signals without changes in hardware.

frequency   Specified in hertz (Hz), how many times per second an event occurs. See *alternating current*.

frequency synthesizer   A method of generating the many frequencies used for tuning a radio receiver from one fixed-frequency quartz crystal.

full-wave rectification   Changing AC to DC using both halves of the waveform, routing the reversing polarities to the appropriate output connections so that only one polarity results. A *bridge rectifier* is commonly used, but full-wave rectification can be accomplished with only two diodes when they are connected to a center-tapped transformer, with the center tap as ground and one diode on each end.

fusible link   A piece of wire used as a fuse without being enclosed in a package. The wire size is chosen to melt at a specific current level, as with a normal fuse.

gain   The ratio of the strengths of the output and input signals of an amplifying circuit. *Voltage gain* indicates that the circuit creates a replica of the signal, but with a bigger voltage swing from peak to peak. *Current gain* means that the circuit's output signal may have the same voltage swing as its input but is able to supply more current, as would be required, for instance, when driving a speaker.

Also, the ratio of the collector-to-emitter current to the base-to-emitter current in a transistor. For example, if 2 mA of base current will cause 100 mA of collector current, the transistor has a gain of 50. A transistor's gain value is engineered into its manufacture, and it decreases with increasing signal frequency. When the ratio decreases to 1:1, the transistor is said to be at unity gain and is no longer amplifying the signal.

ganged   Mechanically linked, but not electrically, as in switches and potentiometers.

giga   Billions.

GPSDO   GPS-disciplined oscillator. A GPS receiver that outputs a reference frequency locked to the highly precise and accurate timing signals used for navigation. Most GPSDOs output 10-MHz and 1-Hz signals.

ground loop   A condition in which the ground terminals of two connected pieces of equipment aren't at the same voltage level, so current passes between them, causing AC line noise to corrupt signals. Most ground loop problems occur with analog audio and video gear, resulting in audio hum or bars moving up the screen of a video display. In digital systems, ground loops may result in corruption severe enough to cause dropouts or complete loss of data transfer.

half-wave rectification   The changing of AC to DC by blocking the undesired polarity with a diode.

Hall-effect sensor   A semiconductor *transducer* whose passage of current is affected by nearby magnetic fields. These sensors are commonly used to detect rotation and position in servo-controlled motor systems, as in projector color wheel assemblies. Unlike inductive sensors, Hall-effect devices will respond to the steady field from a permanent magnet not moving relative to the sensor.

harmonically related   Two waves that are related in frequency by a whole-number ratio such as 2:1, 3:1 and so on. They are considered related because their cycles will have starting and stopping points that correspond in a regular, repeating pattern.

The term comes from musical harmony because tones whose frequencies are related as multiples sound harmonious, or pleasant. Electronics and acoustics, from which music arises, embody pretty much the same concepts where waveforms and frequencies are concerned.

harmonics   Also called *overtones*. Parts of a wave whose energies are at frequencies that are a whole-number multiple of the fundamental frequency of the complete wave. A square wave, for example, contains energy at all the odd harmonics (third, fifth, seventh, and so on) but none at the even harmonics.

HDMI   High-definition multimedia interface, the most commonly used HDTV connection. This all-digital interface carries both audio and video and is copyright-protected by a handshake, an exchange of an encryption key between the connected devices when the connection is established that ensures the video and audio data are going to a device, typically a display, certified to receive them without the possibility of interception or copying.

heatsink   The metal structure that carries heat from an attached component, radiating some of the heat into the air and keeping the component cooler than it would be on its own. Heatsinks usually have fins to maximize surface area for greater cooling. Sometimes the metal chassis of the product can serve as a heatsink, with power-handling components bolted to it.

heatsink grease   Also called heatsink compound, a special grease used to facilitate heat transfer from a component to its heatsink.

helical scan   The method of video and data recording on magnetic tape in which recording heads mounted on a cylinder are rotated against the tape to produce diagonal tracks. The method is so named because the tape is wrapped around the cylinder in a helical configuration. Also called slant-track recording.

hot   Having voltage, but particularly something connected directly to the AC line. The hot side of a switching power supply includes all of the circuitry between the AC input and the transformer.

impedance   Usually specified in ohms, like *resistance*, impedance is a complex quantity describing the opposition to AC currents. Along with resistance, it includes the effects of capacitance and inductance, known as reactance. Think of it as resistance, but for AC. Unlike pure resistance, the impedance of a circuit varies with the frequency of the applied signal. Capacitive reactance drops as signal frequency rises. Inductive reactance does just the opposite.

inductance   Specified in henries, inductance describes the voltage that builds up in a coil of wire when a changing current passes through it. The effect arises from the magnetic field the current creates and how that field impinges on the coil, opposing its passage of current. It is determined by physical factors inherent in the coil, including the number of turns of wire, whether it is wrapped around an iron core and other conditions. In small-signal circuits, inductance is specified in millihenries (mH, or thousandths of a henry) or microhenries (μH, or millionths of a henry).

input stage   The low-level circuitry used to amplify small signals from a *transducer*.

integrated   Having many components formed on the same *substrate*, or base layer, as in an IC or module.

intermediate frequency (IF)   The mixing product of an incoming signal and a *local oscillator* in a *superheterodyne* receiver. See *superheterodyne*.

inversion   A signal stage flips the signal upside down, so the output voltage drops as the input voltage rises, and vice versa. Amplifiers providing *voltage gain* often perform inversion as well and are known as *inverting amplifiers*. This term is unrelated to use of the word *inverter* to describe a step-up voltage converter of the sort used to drive fluorescent lamps in LCD screens.

isolation transformer   A transformer with a 1:1 turns ratio and no electrical connection between its input and output. Used to enhance safety by isolating items being serviced from the AC line.

JPEG   Joint Photographic Experts Group. The committee that created the standard for image compression used for most digital images. Compression (data reduction) is achieved by examining redundancy in the image, such as a blue sky, and encoding it such that it requires less data to describe than would specifying each pixel.

kilo   Thousands.

land   The copper area on a circuit board where components are connected, distinct from the *traces* connecting such areas. Some techs call all conductive elements on circuit boards traces.

latching or latch-up   The destructive condition that may occur when an IC's signal input voltage exceeds the power supply voltage, causing reverse current through the *substrate*.

LCD   Liquid-crystal display. An electronic light valve in which the molecules of a special liquid with crystalline properties can be twisted to alter the polarization of light passing through them, causing darkening when a fixed polarizer placed in line no longer matches the polarization of the altered area. Practical LCDs may have millions of individual light valve cells.

lead dress   The placement and securing of wires and cables in a device.

leaf switch   A set of flexible metal contacts pressed on by a mechanism and used to sense its position, found in disc players and other devices with moving parts.

light engine   The entire optical assembly of a video projector or projection TV, from where the light enters from the lamp, up to the lens. In LED-based projectors, the LEDs are part of the light engine, but the lamps in arc lamp–based units are not because they are replaceable separately.

line voltage   The voltage of the AC line as it comes from the wall socket.

linear   A graph of a linear circuit's input on one axis and its output on the other creates a straight line, indicating that the circuit is faithfully reproducing the signal without distortion. Especially in high-fidelity amplifiers, linearity is an important design goal. Some circuits, such as radio mixers, are deliberately nonlinear so that two signals will influence each other, creating new signals. Also, a general term meaning analog as opposed to digital.

linear power supply   A power supply that operates at the AC line frequency and/or regulates voltage without the use of pulse-width modulation, in contrast to a switching power supply. See *switching* or *switch-mode power supply in Common Circuits.*

linear region   The state of a transistor or other amplifying element when it is neither *saturated* (fully turned on) nor cut off. In its linear region, the amplifier can respond to signal variations and produce corresponding changes in its output.

load   How much *resistance* (DC) or *impedance* (AC) a device presents to the circuit driving (powering) it. The lower the resistance or impedance, the greater the load because more current will be needed to drive the device. Most speakers, for instance, offer an 8-ohm load to the amplifiers driving them.

local oscillator   The oscillator in a *superheterodyne* receiver that is mixed with incoming signals to produce the *intermediate frequency* (IF). See *superheterodyne.*

log taper   See *audio taper.*

LSI   Large-scale integration. A large-scale integrated circuit chip may have thousands or millions of elements.

magic smoke   The essence that actually runs electronics. Let the magic smoke out, and the circuit won't work anymore. All pro techs know this because they've made it happen, usually to something expensive or made of "unobtainium." See *curse words.*

matrixed   Addressable in an *X-Y* grid, as in an LCD or a keypad.

mega   Millions.

micro   Millionths.

microcontroller   A microprocessor built into a device and dedicated to running a specific program to operate it. Sometimes called an *embedded controller.*

microdisplay   The small image-forming device at the heart of a video projector.

milli   Thousandths.

modulation   The altering of a signal to impose information on it. In amplitude modulation, or AM, a *carrier's* strength changes with audio information. In frequency modulation, or FM, a carrier's frequency is shifted higher and lower to convey the information. Many other forms of modulation are used, especially in digital systems. See *demodulation.*

MOV   Metal oxide varistor. A ceramic device whose resistance is high when low voltages are applied, and low when high voltages are applied. MOVs are used at the AC inputs of power supplies to suppress potentially damaging voltage spikes.

MP3   MPEG layer 3, the audio data compression scheme originally devised for the soundtracks of MPEG-encoded movies and now ubiquitous for music storage and internet transmission.

MPEG   Motion Picture Experts Group. The committee that created the standard used to compress moving video images so that they require reduced data. There are various flavors, such as MPEG2 and MPEG4, that evolved to permit greater compression or better image quality. The DVD and digital broadcast standard is MPEG2. Most mobile devices play some variety of MPEG4.

multiplexed   A circuit in which information to or from separate areas is carried on common lines through time sequencing, or scanning. Computer keyboards and remote control keypads are multiplexed, with a microprocessor scanning a small set of lines in an *X-Y* grid, looking for intersecting connections indicating which button has been pressed. LCDs are also scanned in such a grid. The technique permits a small number of lines to serve a much larger number of switches or screen pixels. Also, a *subcarrier* system used in analog FM radio to encode stereo onto a single carrier.

mu-metal   A special metallic alloy particularly effective at blocking electromagnetic fields, used in shields over sensitive components and input stages.

nano   Billionths.

NC   No connection. Mostly used to denote IC pins that either have no connection to the inside of the chip or are not connected to anything in the rest of the circuit. Also, *normally closed*, as a switch or relay contact.

negative feedback   A signal fed from a circuit's output back to its input, but upside down, or 180 degrees out of phase, for controlling distortion or motion, or any other corrective action.

NO   *Normally open*, as a switch or relay.

nonlinearity   When an analog circuit's output doesn't accurately mimic its input. This may be deliberate or a deficiency, depending on the circuit's intended function. See *linear* and *clipping*.

normally closed   A switch or relay contact that is closed (connected) when the switch is in the "off" position or the relay is not energized. See *NC*.

normally open   A switch or relay contact that is open (not connected) when the switch is in the "off" position or the relay is not energized. See *NO*.

NTSC   National Television System (or Standards) Committee. The group that created the now-obsolete analog color television broadcasting standard. Also, the name of the standard itself.

Occam's razor   The principle, by William of Occam in around the twelfth century, suggesting that the simplest solution to a problem is most likely to be correct.

Ohm's law   Georg Simon Ohm's crucial contribution to the electrical art, stating that voltage (E) equals current (I) times resistance (R). Knowing any two of the variables makes it easy to solve for the third one. Thus $I = E/R$ and $R = E/I$. Using the more common terms of volts, amps and resistance, $V = A \times R$, $A = V/R$ and $R = V/A$.

open circuit or open   A disabled circuit path preventing passage of current required for proper circuit operation, as through an open component. A switch in the "off" position is also said to be open because no current passes through it.

opto-isolator   An LED and a phototransistor in one package, used to pass information without an electrical connection between the two sides.

output stage   The high-level circuitry used to deliver sufficient power to a final transducer or device such as a speaker, motor, transmitting antenna or backlight lamp.

overtone   See *harmonics*.

parallel   Components are connected together across the power source so that current goes to each one, independent of the others. In a parallel circuit, the voltage is the same to each element, and the current that passes through each one is proportional to its resistance.

passive elements   The non-amplifying components supporting the operation of a circuit stage's *active elements*. Resistors, capacitors and inductors are passive elements.

PC board or PCB   Printed circuit board.

phase   The relative position in time of two signals, expressed in degrees as a fraction of one cycle's 360-degree total. When the signals are half a cycle apart, they are 180 degrees out of phase.

photodetector   A light-sensitive component, usually a phototransistor, that conducts when light strikes it.

pick and place machine   The robotic system used during manufacture to place surface-mount components on circuit boards, after which all the parts are soldered at once. See *SMD*.

pico   Trillionths.

pictorial   A drawing of a device's internal structure in physical terms, including appearance and layout of its parts.

piezoelectricity   The property of some crystalline substances, including quartz and various ceramics, that they flex when electricity is applied and generate electricity when flexed. The piezoelectric effect is exploited in quartz crystals and ceramic resonators, both widely used in digital and analog devices. Older technologies relying on the effect included crystal microphones and earphones, and crystal and ceramic phonograph pickups.

pin basing   The order of leads on a component, typically a transistor or IC.

PIV   Peak inverse voltage. The maximum voltage a diode or rectifier can withstand in the direction in which the part does not conduct.

pixel   A single picture element, or dot, especially in a *matrixed* display like an LCD, or in a digital camera's image sensor.

polarity   The direction of current, as indicated by + (positive) and − (negative).

polarized   A component that requires a specific arrangement of positive and negative voltage applied to its terminals. For example, most electrolytic capacitors are polarized and will be destroyed if voltage is connected with the polarity backward.

power transistor   A large transistor capable of passing significant amounts of current and dissipating more than a few watts as heat. Power transistors are used as output elements in amplifiers, as motor drivers and anywhere else their large power-handling capabilities are required.

pull-down resistor   A resistor used to tie a signal line to ground while keeping enough resistance between the line and ground that the line can be pulled up to a higher voltage (typically the power supply voltage) by another circuit element. See *pull-up resistor*.

pull-in current   The current required to activate a relay.

pull-up resistor   A resistor used to tie a signal line to the power supply voltage while limiting current so that the line can be pulled down toward ground, typically by a digital chip. Pull-ups are common in computers and other digital systems with parallel data lines and are often implemented with *SIPs* or other resistor packs.

pulse train   A sequence of pulses representing information. Commonly used to refer to digital signals.

push-pull   A common form of analog audio amplifier in which the positive and negative halves of the signal waveform are amplified separately and recombined at the output. See *complementary*.

PWM   Pulse-width modulation. The technique of varying the *duty cycle*, or on/off ratio, of a series of pulses to convey information or control the flow of power. Switching power supplies use PWM to regulate their output voltages, and PWM is also used to vary the speed of motors in some applications.

quiescent current   The current drawn by an amplifier stage when no signal is present. Quiescent current is set by the active element's *bias*.

radio frequency (RF)   Generally from around 100 kHz to many gigahertz (GHz, or billions of cycles per second).

rail   The power supply line feeding a circuit. Though some circuits with inductors can generate voltages exceeding those of the rails powering them, most circuits cannot, and the rail's voltage defines the circuit's limit.

reballing   Resoldering the ball-shaped contacts under BGA (ball-grid array) chips. Reballing requires specialized equipment. See *BGA*.

rectifier   A diode big enough to handle the current in a power supply, used to convert AC to DC.

rectify   To convert AC to DC, either by blocking one half of the waveform or directing opposite halves to the appropriate + and − terminals so that neither output terminal changes polarity as the incoming current alternates.

resistance   Specified in ohms, resistance opposes the current moving through any conductor or circuit, limiting the total amount. It is essentially atomic friction, and the power lost to it is converted to heat. It is represented as R and by the Greek letter omega (Ω). Resistance has no polarity, just as a crimp in a hose has the same effect regardless of the direction of water flow.

resonance   The state in which a component or combination of components builds up a voltage or current when an AC signal of a specific frequency is applied, because the transit time of the energy through the circuit is such that energy peaks reflecting back and forth through the components coincide with the incoming peaks of the applied signals, reinforcing them. The electrical effect is similar to the mechanical effect that occurs in a musical instrument's string, and resonant circuits are said to be *tuned* to their resonant frequency.

rework station   A desoldering and resoldering system used for surface-mount component replacement.

ribbon cable   A type of internal interconnection made with multiple conductors side by side to form a ribbon. Three basic types include wires separated by plastic, flat copper conductors with similar plastic separator and thin-film printed conductors on a plastic substrate.

ring tester   A test instrument used to diagnose high-voltage transformers, also called flyback transformers, in CRT TVs. Ring testing can be used on other types of transformers as well.

ripple   Unwanted variations in the output of a DC power supply due to insufficient filtering of the voltage conversion frequency (50 or 60 hertz in a linear supply and much faster in a switcher).

rise time   The time it takes for a waveform to rise from 10 to 90 percent of its maximum value.

RoHS   Restriction of Hazardous Substances. The European standard specifying lead-free soldering and other environmentally safer construction characteristics. Devices built to this standard will display "RoHS" on a label or stamped into the case.

sampling   The process of taking rapid measurements of analog voltages to convert them to digital data representation. See *digital*.

saturation   The state of a transistor when it is fully turned on, allowing maximum current to pass through it with minimum resistance.

sawtooth wave   An asymmetrical waveform whose shape resembles the teeth on a saw, typically used in the beam-sweeping circuitry of CRT TV sets and oscilloscopes.

schematic   A circuit diagram showing the symbols and interconnections of components without regard to their size, appearance or physical layout in the device.

Schottky diode   A diode formed from a semiconductor-to-metal junction instead of between two semiconductors. It has a lower forward voltage drop than a standard diode and can operate faster as well.

SCR   Silicon-controlled rectifier. SCRs have three leads and can be turned on with a signal, much like transistors, but stay on once they are tripped until power is removed or changes polarity.

SDR   Software-defined radio. SDR is a digital technique in which incoming radio signals are *digitized* and processed entirely in the digital domain. Wi-Fi and Bluetooth radio chips are SDRs, and the technique is rapidly supplanting the traditional analog *superheterodyne* receiver scheme in most applications for reception of both digital and analog signals.

selectivity   The ability of a receiver to separate adjacent stations.

semiconductor   A component, usually made from silicon, such as a transistor, diode or IC chip. Some exotic semiconductors are made from other materials like gallium arsenide or germanium, which behave similarly to silicon in their ability to control the flow of electrons in response to a control signal.

series   Components are connected end to end, so current must pass through one to get to the other. A fuse, for instance, is connected in series with the circuitry it's protecting so that all current must pass through the fuse to reach the rest of the components.

In a series circuit, the current is the same through each element, but the voltage reaching each element is reduced in proportion to the other elements' resistance. The lowering of voltage is referred to as an element's *voltage drop*.

series-pass transistor   A transistor used as a variable-resistance element in a linear voltage regulator to hold the output voltage to a specific value, with all of the power supply's current passing through the transistor on its way to the rest of the circuitry.

shield   A metal cover placed over sensitive circuitry to protect it from stray electrical noise. Shields are often soldered in place and must be removed for access.

short circuit or short   A low-resistance passage of current where it shouldn't exist, as from a voltage point directly to circuit ground via a *shorted* component.

shunt   A current path in parallel with another path. A resistor across another component, diverting some current around it, is said to be shunting the component or shunting the current.

signal   A changing voltage representing information.

sine wave   A waveform with a smooth, repetitive shape derived from the mathematical *sine* function. Sine waves have no *harmonics*, or energy at multiples of their frequency, so they are the purest type of signal possible.

sink   An acceptor of current that permits it to pass toward ground. See *source*.

SIP resistor pack   Single inline package. SIPs are resistor packs (multiple resistors in one package) with a single line of connecting leads. See *pull-up resistor*.

sled   The linear track assembly on which a laser optical head rides as it scans across a disc. The sled includes the track rods and the motor and gears moving the head.

SMD or SMT   Surface-mount device or surface-mount technology. These are the tiny, leadless components that have replaced conventional through-hole parts with leads.

SMPS   Switch-mode power supply. Another name for a switching power supply.

SNR or S/N   Signal-to-noise ratio. The ratio of desired signal to residual noise in a circuit, usually expressed in decibels (dB).

solder paste   Liquid, low-temperature solder used to solder surface-mount components with a hot-air rework station.

source   A supplier of current. See *sink*.

spectrum   A range of frequencies.

spudger   A plastic prying tool for opening small products sealed with snaps or glue.

square wave   A waveform with flat tops and bottoms that switches quickly between the two states, with fast rise and fall times. Square waves can never truly be square because it takes some time for them to switch states, so the transition between the top and bottom of the waveform is slightly tilted. See *rise time* and *fall time*.

stage   The distinct section of a circuit that performs one function of signal processing. Circuit stages are organized around one or more *active elements*, with surrounding *passive elements* supporting their operation.

subcarrier   A *carrier* signal imposed on another, higher-frequency carrier. The technique is used to piggyback hidden signals onto others, as in FM stereo and analog color video signals.

substrate   The base on which an IC, transistor or potentiometer is formed. In most modern solid-state components, the substrate is a slice of silicon.

superheterodyne   Also called *superhet*. In this classic analog receiver design, dominant until recently being supplanted by *SDR*, incoming radio signals are mixed with a *local oscillator*, resulting in an *intermediate frequency* (IF) that can be amplified using tuned circuits without interfering with the incoming signal. The IF is then further processed and demodulated to recover the information contained in the radio signal. Superhet receivers can be used for both analog and digital data reception. See *SDR* and *demodulation*.

sweep rate   How fast an oscilloscope's beam or LCD spot proceeds from left to right.

switch mode   The operation of a transistor as a switch—that is, saturated or cut off—with fast rise and fall times between the two states. See *saturation* and *cutoff*.

sync   The pulses in an analog video signal that align the position of a TV's display point (the electron beam's position in a CRT TV or the row and column addresses in a matrixed display like an LCD or plasma TV) with that of the originating signal source to keep picture elements correctly placed.

thermal runaway   A destructive, self-reinforcing cycle in which the heat generated by semiconductors makes their resistance go down, so they use more current, which makes them even hotter. The current increases until the device overheats and is destroyed. Various techniques are used to avoid it, including constant-current power supply regulation and *thermistors* glued to power transistors.

thermistor   A resistor whose resistance varies with temperature. Thermistors are used to compensate for changes in circuit behavior that occur with temperature variation, and especially to avoid *thermal runaway* in power-handling semiconductors such as those in audio power amplifiers.

through-hole   A component with leads that poke through holes in the circuit board, soldered to the opposite side.

time constant   The period of time it takes to charge or discharge a combination of components, such as a resistor and capacitor or an inductor and capacitor.

tough dog   A unit with a repair problem that's difficult to diagnose.

trace   The conductive copper lines on a circuit board providing connection between *lands*, or component connection points. Also, to follow a signal through a circuit with a measuring instrument, especially an oscilloscope, or to follow a connection visually across a circuit board or a schematic diagram. Also, the line or waveform on an oscilloscope's screen.

transducer   A device that takes in one kind of information or energy and converts it to another. Microphones, tape heads, laser optical heads, speakers and phono cartridges are all transducers.

tuning   A receiver's process of separating a desired signal from other signals and noise.

unity gain   See *gain*.

unobtainium   An ultra-rare substance from which things that can't be found are made. Unobtainium is related to "extravagantium," from which absurdly overpriced replacement parts are made.

varactor   A voltage-variable capacitor whose symbol looks like a combination of a diode and a capacitor. Although sometimes called *varactor diodes*, varactors do not exhibit normal diode function.

variac   A device that goes between a wall plug and an item being serviced, for the purpose of providing variable-voltage AC power. Commonly used for repairing vacuum-tube gear, with little to no application today. Variacs have been mistaken for isolation transformers, with fatal results. See *isolation transformer*.

VGA   Video graphics array, the high-resolution analog connection used with computer monitors before DVI and HDMI obsoleted it.

via   The interconnect between layers in a circuit board.

video   An analog signal representing a moving image by varying a voltage to represent the brightness and color values of scanned spots in the image. Video signals are complex and include synchronizing and color reference signals to align the receiver's circuits as necessary to interpret the signal properly and display the image.

voice coil   A coil suspended over a magnetic field, typically from a permanent magnet. When current is applied to the coil, the resulting magnetic field either repels or attracts the permanent field, depending on the applied voltage polarity. This moves the coil toward or away from the permanent magnet.

   The name comes from the arrangement's initial and ongoing use in speakers, but voice coils are also used to move the lenses in optical disc players and the heads in hard drives, as well as in other motion control applications.

volt-amps (VA)   This is somewhat like *watts*, but for AC power. It means "volts times amps." Because of certain effects that occur with AC, the maximum voltage peak and maximum current in each cycle may not coincide, and the amount of work that can be done will be less than when they do. The power requirements shown on the backs of AC-powered devices are often specified in volt-amps. For repair work, it's safe to think of volt-amps and watts as equivalent because actual watts used will never exceed volt-amps, and may be less.

volts   The electric "pressure," or electromotive force. This corresponds to how hard the electrons push and is represented as V or sometimes E. Voltage propels current through resistance, so the higher the voltage, the more current will pass through a given resistance. (See *Ohm's law.*) Vcc and Vdd are references to voltages applied to the pins of a semiconductor and mean the same thing as V except that they are always of positive polarity. Negative voltage is sometimes noted as Vss.

   Like all other motion of mass in the universe, voltage is entirely relative; there is no such thing as "9 volts." A given voltage is meaningful only relative to some other point. Polarity is relative too; a voltage can be positive with respect to one point while being negative with respect to another! In circuits that have both positive and negative voltages relative to circuit ground, that ground point will function as positive for the negative voltage point and negative for the positive voltage point.

wall wart   An AC adapter that plugs into the wall and steps down the voltage to operate a product with no internal power supply. Most wall warts provide DC, but a few contain only transformers, outputting low-voltage AC that is processed further inside the item being powered.

watts   Also called *power*, a measure of how much work can be done by electricity. Power is equal to volts × amps. So, 100 watts could be 20 volts at 5 amps or 10 volts at 10 amps; the amount of work either of those quantities could do would be the same. 746 watts equal one *horsepower*, but if you cause 746 watts of electrical power to flow through an actual horse, you wind up with a dead horse and zero horsepower.

Watt's law   This states that power in watts (P) equals current (I) times voltage (E). As with Ohm's law, knowing any two of the variables makes it easy to solve for the third one. Thus $I = P/E$ and $E = P/I$. Or, in the terms of watts, amps and volts, $W = A \times V$, $A = W/V$ and $V = W/A$.

zener diode   A diode that is designed to break down nondestructively and conduct in its reverse direction when a preset voltage is reached. Below that voltage, a zener diode behaves like a normal diode, conducting in its forward direction and blocking current in its reverse direction.

zener voltage   The voltage at which a *zener diode* begins to conduct in its reverse direction.

Common Circuits

Many devices are constructed from the same basic types of building blocks. Here are some widely used circuits you may find:

amplifier   One or more stages that increase a signal's voltage swing, current or both by shaping power supply current into a larger replica of the signal.

buffer   An intermediary circuit isolating two others. In analog devices, an amplifier stage providing current gain and isolating the more sensitive, earlier amplifier or oscillator stage from the ones following it. Buffers typically offer no voltage gain. In digital processing, a temporary storage area holding blocks of data, perhaps in line for conversion to analog signals.

clock oscillator   The square wave oscillator used to step, or clock, a digital circuit through its series of operations. Many microprocessors used in small products include onboard clock oscillators, requiring only a crystal or ceramic resonator to set the frequency.

frequency synthesizer   A complex circuit used to generate a range of frequencies from the single frequency of a quartz crystal. Older frequency synthesizers used *voltage-controlled oscillators* and *phase-locked loops* to lock an analog oscillator's signals to the frequency determined by a digital controller. Many newer systems use *direct digital synthesis*, in which the required analog waveforms are generated by the digital system, via a digital-to-analog converter, in much the same way a CD player reconstructs the analog audio waveform from digital samples.

front end   The sensitive first stages of a radio receiver that accept energy from the antenna and amplify it in preparation for its delivery to the *mixer* in a superheterodyne design or the DSP in an SDR. Many front ends include tuned circuits to help reject off-frequency signals, but those in frequency-synthesized receivers may not.

intermediate frequency (IF)   The fixed-frequency, tuned amplifier stages in a radio receiver providing most of the selectivity (ability to separate stations) and sensitivity. Incoming signals from the antenna are converted to the IF by mixing them with a local oscillator in the *mixer* stage, a deliberately nonlinear circuit that causes the two signals to interfere, generating a new signal at their difference frequency that still carries the information of the original signal.

The mixing process is also called *heterodyning*, and radios that convert signals to an IF, which virtually all modern, non-*SDR* sets do, are called *superheterodyne* receivers because the output of the heterodyne process is still higher up the frequency spectrum than the information carried by the signal. Also, the frequency at which the tuned IF amplifiers operate.

inverter   A step-up switching power supply typically used to generate the high voltages required for fluorescent LCD backlights, camera flash tubes and other applications needing voltages higher than those provided by the product's primary power supply or batteries.

linear voltage regulator   A step-down regulator (one whose output is at lower voltage than its input) using a transistor element in its *linear region* as a variable resistor that dissipates excess power as heat.

mixer   A circuit that accepts two AC signal inputs and outputs the sum and difference frequencies resulting from their interacting, or mixing, with each other in a nonlinear manner.

oscillator   A circuit that produces a steady signal of constant frequency, or variable frequency in some cases, through the use of constructive, or reinforcing ("positive"), feedback.

phase-locked loop (PLL)   A multistage circuit used to slave the frequency of an oscillator to that of a reference signal or another oscillator. By inserting digital dividers between the oscillator and the rest of the loop, the oscillator can be forced to lock to a multiple of the reference frequency, a crucial technique in frequency synthesis. PLLs are used to tune receivers, to recover color information from analog videotape and data from the laser head's signals in CD and DVD players, and in many other applications.

power supply   Any circuit or power source that provides the power the rest of the device requires. In AC-powered products, it typically converts the incoming line voltage to a steady, lower DC voltage. In battery-operated products, the batteries themselves may be considered the power supply. If intermediate circuitry steps the battery voltage up or down, that circuitry will also be considered part of the device's power supply.

preamplifier or preamp   A low-level amplifier used to boost weak signals enough that further stages can amplify them more.

servo   A circuit that slaves mechanical motion to some reference, typically used to lock the rotational speed or phase of a motor to a signal, as with a VCR's rotating video head drum, which must be kept aligned with the recorded tape tracks for playback.

switching or switch-mode power supply   A power supply operating in *switch mode*, in which the incoming power is chopped into high-frequency pulses before being fed to the conversion transformer. This approach allows a small transformer to provide substantial power output, and it is very efficient compared to a linear power supply, which processes power at the very low line frequency of 50 or 60 hertz. Switching power supplies can be used to step voltage up or down. Step-up versions are often referred to as *inverters*.

switching or switch-mode voltage regulator   A voltage regulator operating in *switch mode*, in which the duty cycle of a pulse is varied to regulate the output voltage. This is similar to the operation of a switching power supply, except that the incoming power is already at a DC voltage in the neighborhood of the desired output voltage. Compared to a linear voltage regulator, a switching regulator generates far less heat and is much more efficient. Plus, it can be used to step the voltage up or down.

voltage-controlled amplifier (VCA)   An amplifier whose gain can be controlled by a DC voltage. VCAs are used in the audio sections of digital devices such as phones, tablets and TVs, with a DC voltage derived from the microprocessor's data setting the speaker or headphone volume.

voltage-controlled oscillator (VCO)   An oscillator whose frequency can be controlled by a DC voltage. VCOs are used in radio tuning circuits, with the DC voltage from the digital circuits setting the frequency.

Capacitor Voltage and Tolerance Codes

Some capacitors, especially ceramic types, are marked with codes for working voltage (the maximum continuous DC voltage the capacitor can stand without breaking down) and tolerance (acceptable deviation from the marked value), rather than direct voltage and percentage ratings. Here are tables of the markings you're most likely to see:

Marking	Voltage
0G	4 V
0J	6.3 V
1A	10 V
1C	16 V
1E	25 V
1H	50 V
1J	63 V
2A	100 V
2T	150 V
2D	200 V
2E	250 V
2W	450 V

Marking	Tolerance
D	.5%
F	1%
J	5%
K	10%
M	20%
Z*	+80%, −20%

*Found on electrolytic capacitors

Index

References to figures are in italics.